Mathematics
for
Computer
Studies

Mathematics for Computer Studies

James T. Sedlock

Rhode Island College

Wadsworth Publishing Company

Belmont, California
A Division of Wadsworth, Inc.

This book is dedicated to my many excellent math teachers at
LaSalle High School, Philadelphia; LaSalle University,
Philadelphia; and Lehigh University, Bethlehem, Pa.

Mathematics Editor: Jim Harrison
Production Editor: Patricia Brewer
Managing Designer: Andrew H. Ogus
Print Buyer: Barbara Britton
Cover and Interior Designer: Albert Burkhardt
Copy Editor: Paul Monsour
Technical Illustrator: Shirley Bortoli

ISBN 0-534-04326-7

Library of Congress Cataloging in Publication Data

Sedlock, James T.
    Mathematics for computer studies.

    Includes index.
    1. Electronic data processing—Mathematics.   I. Title.
QA76.9.M35S43 1985   519.4   84–13135
ISBN 0-534-04326-7

Contents

Preface

Mathematics for Computer Studies has been written for two groups of students. First and foremost, it is intended for students with only a high school mathematics background who wish to take college-level computer science courses. Indeed, I envision the book as the text for a first college math course for computer science majors, preceding the standard upper-level course in discrete structures. Second, the book can be used by students in a finite math course with a contemporary, discrete flavor. In any case, the book's only prerequisite is beginning algebra.

I have chosen the topics of this book to help students develop the mathematical skills and thought processes encountered in elementary computer science. With these aims in mind, I have emphasized the clear statement of definitions and theorems, supported by a wealth of worked-out examples; there is no attempt to provide mathematical rigor or proofs. Generally, each even-numbered exercise is of the same type and degree of difficulty as the preceding odd-numbered exercise. Answers to the odd-numbered exercises are in the back of the text.

The unifying thread of this book is the notion of algorithm. To this end, I have included numerous structured flowcharts to describe certain algorithms, although I have not attempted to teach flowchart construction or algorithm development. The reader is urged to work through as many of these flowcharts as possible. The experience will prove its worth in later computer science courses as well as convey the power and pervasiveness of the concept of algorithm.

Mathematics for Computer Studies can support a range of course outlines, depending on the background of the students and the constraints of time. A course to prepare students for further computer science courses should cover:

Chapter 1: All sections
Chapter 2: All sections
Chapter 3: Sections 3.1, 3.2, 3.3, 3.5
Chapter 4: Sections 4.1 through 4.5 inclusive
Chapter 5: All sections

Chapters 6 and 7 contain diverse supplementary topics that will be of interest to particular groups. There is no overall continuity to Chapter 6, and certain portions may be covered to the exclusion of others. (See the introduction to this chapter for more specific details about the organization.) Chapter 7 introduces topics normally studied thoroughly in courses on discrete structures.

A course in contemporary finite mathematics should cover:

Chapter 1: Sections 1.1, 1.2, 1.5
Chapter 2: Sections 2.1 through 2.6 inclusive
Chapter 3: All sections
Chapter 4: Sections 4.1, 4.2, 4.3, 4.6, 4.7
Chapter 5: Sections 5.1, 5.3
Chapter 6: Sections 6.3, 6.4, 6.5

Such a syllabus places the emphasis on the variety of math structures (different-base arithmetic, modular arithmetic, Boolean algebra, logic, probability, statistics, and linear systems) that relate to computer usage.

Many people have assisted me during the course of this project. To begin, I would like to thank several members of the excellent Wadsworth staff: Regional Manager Richard Giggey, for his encouragement to undertake the work; Managing Editor Richard Jones, for his skillful handling of the writing and review phases; Math Editor Jim Harrison, for his orchestration of the many stages of the project; and Production Editor Patricia Brewer, for her meticulous and perceptive production work. In addition, I am grateful to Susan Langdon for the precision and efficiency with which she typed the manuscript.

I am also indebted to the following reviewers, each of whom provided constructive criticism and insight: John T. Annulis, University of Arkansas; Dick J. Clark, Portland Community College; Stella Daugherty, East Carolina University; Louie C. Huffman, Midwestern State University; Wendell A. Johnson, University of Akron; Paul E. Kenison, Saint Anselm College; Elmo Moore, Humboldt State University; Michael Olan, Saint Mary's College; and Edward C. Polhamus, Danville Community College.

Within my own department at Rhode Island College, I want to sincerely thank George A. Anderson and Robert J. Salhany for their expertise and generous assistance in the preparation of the sections on probability (Chapter 3) and statistics (Chapter 6). Any merits of these sections are directly attributable to them. In addition, I appreciate the consideration afforded me by Helen E. Salzberg, Chair of the Department of Mathematics and Computer Science at Rhode Island College, in arranging my course and departmental assignments during the most intensive periods of writing and revision. Also, I wish to acknowledge the contributions of Richard A. Howland of Dickinson College in helping to formulate the table of contents and general thrust of the work at the outset.

Finally, the book would neither have been attempted nor completed without the constant support, accommodation, and understanding of my wife Barbara and our daughters Anne, Kerry, Clare, and Maura. ("How's the book coming, Dad?") To each of them, many thanks!

James T. Sedlock
August 1984

Mathematics for Computer Studies

CHAPTER 1

Introduction

The following checking account problem, with the accompanying flow diagram in Figure 1 and program in Figure 2, is a typical example of computer usage in the business world.[†]

Problem: Write a program to process the checks and deposit slips for a single checking account at the close of each month. The date, amount, and type of each transaction should be printed out, along with information that summarizes the monthly transactions.

Now, you are not expected to spend time poring over the details of this example. Rather, the point is that before one writes the computer program (this one in Figure 2 is written in BASIC), it is often necessary to discover and then carefully define the step-by-step process (or algorithm) to be used in the solution. One popular device for representing such processes is the *flowchart*, or *flow diagram* (as in Figure 1). In this text, flowcharts are used primarily to illuminate the algorithmic nature of certain processes or techniques. Developing algorithms is not a trivial activity, and often it is a focus of beginning courses on programming. Therefore, we hope that the reader will be able to understand and work through the flowcharts as they are presented, but we shall not assume skill in constructing them from scratch.

Sections 1.1 and 1.2 will introduce the reader to flowcharts and their components. The remaining three sections of this chapter present information about base 10 numbers and arithmetic that will be helpful in constructing and verifying algorithms and understanding their output.

[†] E. B. Koffman and F. L. Friedman, *Problem Solving and Structured Programming in BASIC* (Reading, Mass.: Addison-Wesley, 1979), pp. 154, 157, 159. Reprinted with permission.

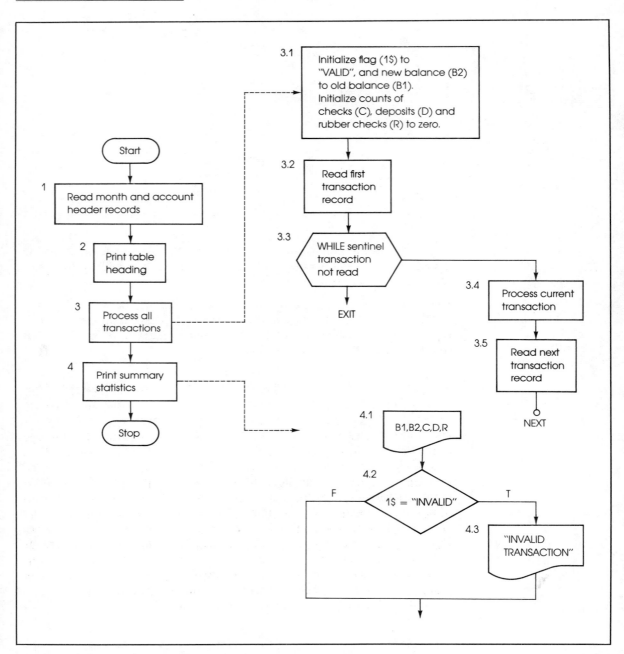

Figure 1 Level one and two flow diagrams for checking account problem

```
110 REM CHECKING ACCOUNT PROBLEM
120     PRINT "CHECKING ACCOUNT PROBLEM"
130 REM
140 REM INITIALIZE PROGRAM PARAMETERS FOR SENTINEL VALUE AND PENALTY
150 REM
160     LET V$ = "Z"
170     LET P = 5
180 REM
185 REM READ MONTH AND ACCOUNT HEADER RECORD
190     READ M$
200     READ A1, N$, B1
210 REM
220 REM PRINT TABLE HEADER
230     PRINT N$, "ACCOUNT NO."; A1
240     PRINT "TRANSACTION RECORD FOR"; M$
250     PRINT
260     PRINT "DATE", "CHECK", "DEPOSIT", "PENALTY"
270 REM
280 REM PROCESS ALL ACCOUNTS AND TRANSACTIONS
290     GOSUB 3010
300 REM
310 REM PRINT SUMMARY STATISTICS
320     PRINT
330     PRINT "STARTING BALANCE"; B1, "FINAL BALANCE"; B2
340     PRINT "NO. OF CHECKS PAID"; C, "NO. OF DEPOSITS"; D
350     PRINT "NO. OF OVERDRAWN CHECKS"; R
360     IF I$ = "INVALID" THEN 370 ELSE 390 [IF I$ < > "INVALID" THEN 390]
370 REM THEN
380         PRINT "INVALID TRANSACTION PRESENT"
390 REM IFEND
400 REM
410     STOP
420 REM
430     DATA "SEPTEMBER"
440     DATA 11385, "GREG LUZINSKI", 85.67
450     DATA 11385, "C", 79.15, "9/9"
460     DATA 11380, "D", 200.00, "9/10"
470     DATA 11385, "D", 3.57, "9/11"
480     DATA 11385, "X", 123.45, "9/11"
490     DATA 11385, "C", 125.67, "9/12"
500     DATA 24077, "Z", 0, "0"
```

Figure 2 Main program for problem

1.1

Algorithms and Flowcharts

From our earliest days of studying mathematics in elementary school, each of us was exposed to the notion of *algorithm*, although the word may not have been used. Loosely speaking, an algorithm is just a carefully stated set

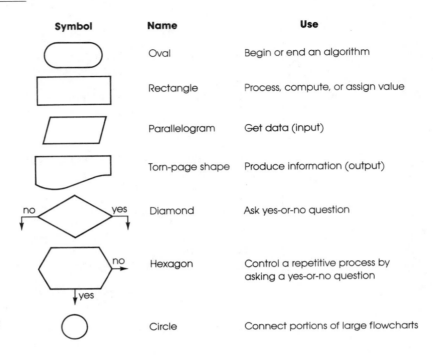

Figure 3 Flowchart symbols

Symbol	Name	Use
	Oval	Begin or end an algorithm
	Rectangle	Process, compute, or assign value
	Parallelogram	Get data (input)
	Torn-page shape	Produce information (output)
	Diamond	Ask yes-or-no question
	Hexagon	Control a repetitive process by asking a yes-or-no question
	Circle	Connect portions of large flowcharts

of instructions designed to accomplish a particular task. Thus, we learned algorithms to add, subtract, multiply, and divide two integers or decimal numbers.

In computer science the notion of algorithm is central, for an algorithm is involved in the development of every computer program.

Definition

> An *algorithm* is a procedure for solving a specific problem that is expressed as a finite number of rules; this procedure must be complete, unambiguous, and guaranteed to terminate in a finite number of applications of these rules.[†]

One way of displaying an algorithm is by means of a *flowchart*, which is a collection of certain geometric figures connected by arrows. While there is no universal agreement on the meaning and use of all flowchart symbols, Figure 3 lists several that are commonly used.

The diamond and hexagon *always* have *two* arrows emanating, although the positions of the arrows may vary. If certain flowcharts in this

---

[†] J. Shortt and T. G. Wilson, *Problem Solving and the Computer: A Structured Concept with PL/1 (PL/C)*, 2nd ed. (Reading, Mass.: Addison-Wesley, 1979), p. 7.

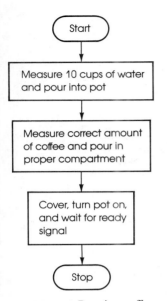

Figure 4 Brewing coffee

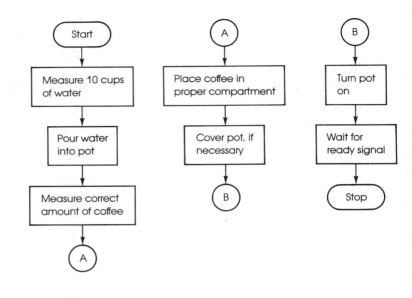

Figure 5 Refined flowchart for brewing coffee

text do not have these two arrows labeled yes and no, then the values are assumed to be those in Figure 3.

Arrows connect the symbols to indicate the sequence for executing the various instructions within the flowchart.

Example 1.1.1   **Brewing Coffee**

Figure 4 is a flowchart showing how to make ten cups of coffee in an electric coffee pot.

This flowchart can be further refined as indicated in Figure 5.   ■

Example 1.1.2   **Winding a Watch**

The flowchart in Figure 6 describes how to wind a spring-powered watch. The problem with the flowchart is obvious: How far down the flowchart goes depends on the given watch and the particular person winding it. This flowchart will be greatly improved by redirecting the first no arrow back to the first rectangle, thus creating a *loop*, and omitting the rest of the flowchart. This is indicated in Figure 7.

Now our flowchart is *closed*—that is, it has no dangling "and so on" portion. As watches and winders vary, *only* the number of times the no arrow is traversed changes!   ■

Although the last flowchart is correct, in this text we prefer to begin and control each repetitive process, or loop, by using the hexagon symbol. This is consistent with the principles of **structured programming**, a technique

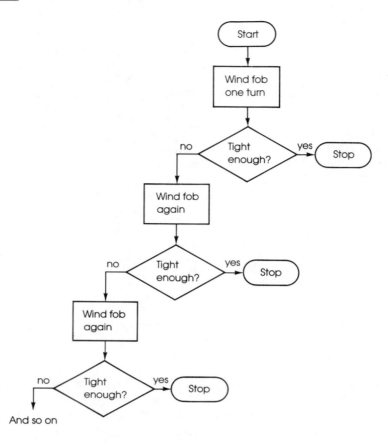

Figure 6 Winding a watch   And so on

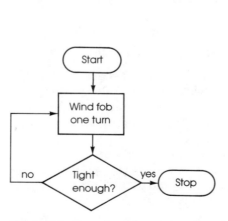

Figure 7 Improved version of Figure 6

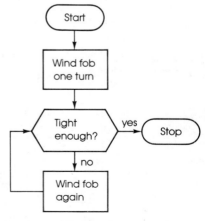

Figure 8 Alternate form of Figure 7

intended to improve the organization, readability, and reliability of programs.[†] Using the hexagon, the flowchart is shown in Figure 8.

Thus, unlike the solutions to many math problems, a correct flowchart is *not unique*. If one has one correct flowchart, then there are many correct variations, one of which may be more efficient in some sense than the others!

While the above flowcharts are filled with words and phrases, it is often preferable to use a single letter or combination of letters and other symbols, called a **variable**, to represent certain objects. When this is done, we shall use the familiar equal sign to describe the meaning attached to the variable. For example,

NAME = name of a state

COUNT = number of people in the room

H5 = height of fifth person

SUM = sum of first two numbers

In computer science, a variable can be considered as the **address** of a particular location in the computer's core memory.

Address

| Contents |

To alter the **contents** of a memory cell, we shall use the **assignment symbol** $\leftarrow$. For example,

NAME $\leftarrow$ Bob

COUNT $\leftarrow$ 7

H5 $\leftarrow$ 5.75

SUM $\leftarrow$ 5 + 8

will result in memory cells looking like the following:

NAME   COUNT   H5   SUM

| Bob | | 7 | | 5.75 | | 13 |

Notice that the operation or process, if any, on the right side of the arrow is performed first, and its value assigned to the variable on the left side.

It is important to understand the distinction between the use of the assignment symbol, $\leftarrow$, and the familiar equal sign, $=$. This is particularly so since many programming languages use $=$ rather than $\leftarrow$ to represent the assignment value. (Pascal uses $:=$ as the assignment symbol.) When we look

---

[†] According to J. P. Tremblay and R. B. Bunt, structured programming is "nothing more than an approach to program implementation in which rigor and structure displace 'seat-of-the-pants' programming" (*An Introduction to Computer Science: An Algorithmic Approach* [New York: McGraw-Hill, 1979], p. 329).

at an assignment statement like

$$x \leftarrow 2x + 6$$

we *should not* think of the variable x on each side as having the same value. Rather, working from the right side to the left, we take the initial value for x, multiply by 2, add 6, and assign the resulting value to the variable x on the left side. The net result is that *variable x has changed in value.* However, when we look at

$$x = 2x + 6$$

as an algebraic equation rather than an assignment statement, then we *do* think of the variable x on each side as having the same value, and we use the rules of algebra to determine that value:

$$x = 2x + 6$$
$$x - 2x = 6$$
$$-x = 6$$
$$\boxed{x = -6}$$

Example 1.1.3   **Compound Interest**

The flowchart in Figure 9 will determine the value of a bank deposit after one year, given the amount of the initial deposit, the bank's rate of interest, and the number of times per year that interest is computed.

DEP = Value of deposit

RATE = Bank's interest rate

T = Number of times per year interest is computed   ■

In this example the variable COUNT, for obvious reason, is called a *counter*. The reader is urged to work through this flowchart for some sample values of DEP, RATE, and T. For example, if one inputs DEP = 1000, RATE = .05, and T = 4, then the adjacent table summarizes variable values for DEP, INT, and COUNT at point Ⓐ. (This table is called a *trace* of the flowchart at point Ⓐ.) The flowchart results in an output of the number 1050.94, which is the deposit value after one year.

COUNT	INT	DEP
1	12.50	1012.50
2	12.66	1025.16
3	12.81	1037.97
4	12.97	1050.94

Example 1.1.4   **Payroll**

A flowchart that will produce a list of 45 employees' names and salaries, given each individual name, hourly pay rate, and the number of hours

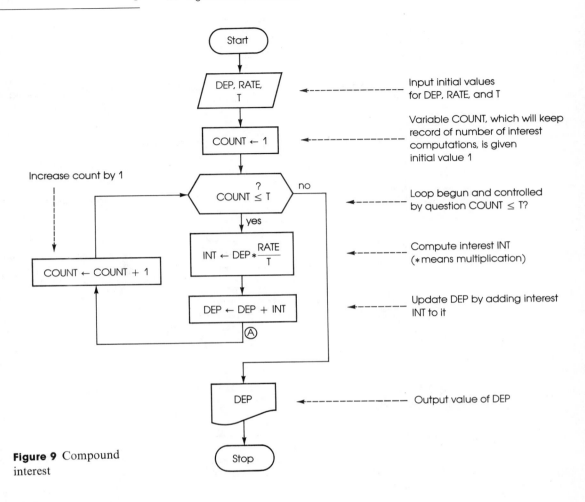

Figure 9 Compound interest

worked, is given in Figure 10. For instance, if

GALOIS, 32, 4.85

is the first set of input data, then

GALOIS, 155.20

will be the first line to be output. ∎

Example 1.1.5   **Age Category**

In Figure 11 there is a flowchart that, given the name and age of a person, will determine whether that person is a child (age ≤ 12), teenager (13 ≤ age ≤ 19), or an adult (age ≥ 20).

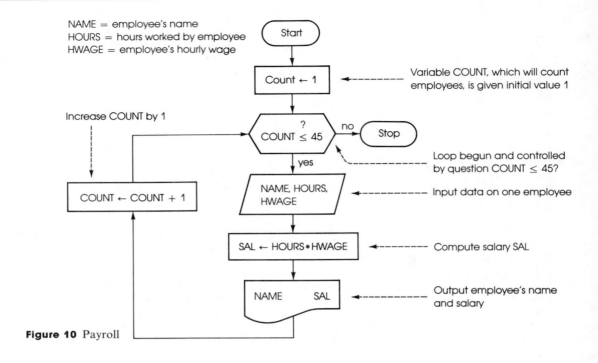

NAME = employee's name
HOURS = hours worked by employee
HWAGE = employee's hourly wage

Start

Count ← 1 ◄-------- Variable COUNT, which will count employees, is given initial value 1

Increase COUNT by 1

?
COUNT ≤ 45   no → Stop

yes

COUNT ← COUNT + 1

NAME, HOURS, HWAGE ◄-------- Input data on one employee

Loop begun and controlled by question COUNT ≤ 45?

SAL ← HOURS * HWAGE ◄-------- Compute salary SAL

NAME   SAL ◄-------- Output employee's name and salary

Figure 10 Payroll

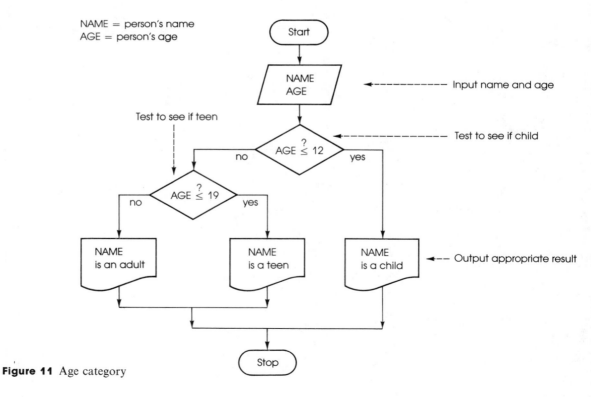

NAME = person's name
AGE = person's age

Start

NAME
AGE ◄-------- Input name and age

Test to see if teen

?
AGE ≤ 12   ◄-------- Test to see if child

no   yes

?
AGE ≤ 19

no   yes

NAME
is an adult

NAME
is a teen

NAME
is a child ◄-- Output appropriate result

Stop

Figure 11 Age category

Sample data will check the accuracy of this algorithm. For example, inputting

CLIBURN, 15

will produce the output

CLIBURN is a teen ∎

Example 1.1.6 **Oldest of Three**

Three people are known to be of different ages. Figure 12 gives a flowchart that will determine which is the oldest. Again, the reader is urged to test the validity of this flowchart by working through it using sample ages. ∎

The examples of this section illustrate the three basic components of flowcharting. The first example illustrates *sequential* flow, the second *repetitive* flow, and the third and fourth are combinations of the two. The last two examples exhibit *conditional* flow.

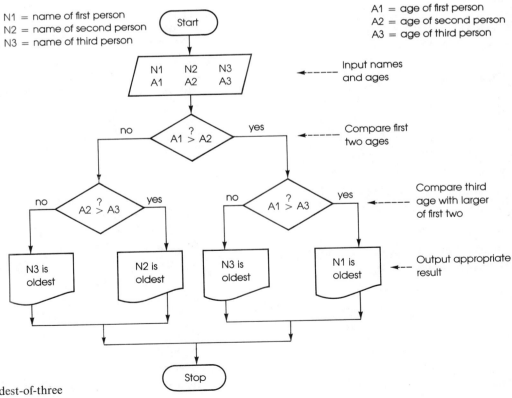

Figure 12 Oldest-of-three

As mentioned in the introduction to this chapter, we shall use flowcharts to illustrate algorithms, but you are not assumed to have skill in constructing them. For that reason, in selecting exercises at the end of this section and the next, it is not necessary to master those involving the construction of flowcharts. They are presented only as stimulation for those who may be interested.

EXERCISES **1.1**

1. Work through the flowchart of Figure 9 for the following sample data:

DEP = 2000   RATE = .06   T = 2

and construct a trace at point Ⓐ that includes values for COUNT, INT, DEP.

2. Work through the flowchart of Figure 9 for the following sample data:

DEP = 6500   RATE = .10   T = 5

3. If   ALBRIGHT, 35, 5
            DEMITRAS, 40, 4.50

are the first two sets of input data for the flowchart of Figure 10, then

a. do a trace (for this input only) of the variables NAME, HOURS, SAL, and COUNT at the point following the print operation.
b. What are the first two lines of output?

4. If   BURKE, 38, 5.25

is the first set of input data for the flowchart of Figure 10, then

a. do a trace (for this input only) of the variables NAME, HOURS, SAL, and COUNT at the point following COUNT ← COUNT + 1;
b. What is the first line of output?

5. Work through the flowchart of Figure 11 with input

KIM, 12

and determine the output.

6. Work through the flowchart of Figure 11 with input

MARK, 48

and determine the output.

7. Work through the flowchart of Figure 12 with input

N1 = JILL   A1 = 15

N2 = ALLIE   A2 = 16

N3 = SARAH   A3 = 12

and determine the output.

8. Work through the flowchart of Figure 12 with input

N1 = MATT   A1 = 17

N2 = DON   A2 = 40

N3 = SUE   A3 = 41

and determine the output.

Exercises 9–22 involve the construction of flowcharts and are optional.

9. Construct a strictly *sequential* flowchart that describes how to fill a gallon can using a one-quart measuring cup.

10. Construct a flowchart with repetitive flow that describes how to fill a gallon can using a one-quart measuring cup.

11. Without using the loop-controller hexagon symbol, construct a flowchart that shows how to mix black paint with white in order to attain a given shade of gray.

12. Using the hexagon symbol, construct a flowchart that shows how to mix black paint with white in order to attain a given shade of gray.

In Exercises 13 through 20, draw a flowchart that accomplishes the described task.

13. Determine the numbers of quarters, dimes, nickels, and pennies received in change when a one-dollar bill is offered as payment for a purchase.

14. Scratch one's back until the itch is gone.

15. Determine the average exam grade for a class of 35 students, given the 35 grades.

16. Generalize Exercise 15 to a class of N students, where N is a fixed positive integer.

17. Find the larger of two given unequal numbers.

18. Find the largest of four given numbers, no two of which are equal.

19. Given a family name, the number of children, and

the family income, determine if that family has more than two children and less than $25,000 in income.

20. Change a flat tire on a wheel having five lug nuts.

Definition: *A **sentinel** is a selected piece of data placed at the end of the input data to signal the end of the input (data) stream.*

21. Redo the flowchart in Example 1.1.4 of the text, but instead of 45 employees, use the sentinel XYZ, 0, 0 to indicate the end of the input data.

22. Redo Exercise 15 for a class of unknown size, but using the sentinel −5 to indicate the end of the exam grades.

1.2

Additional Examples of Flowcharts

Often the solution to a problem involves a combination of the three types of flow—sequential, repetitive, and conditional—described in the preceding section. The following examples were chosen to illustrate this.

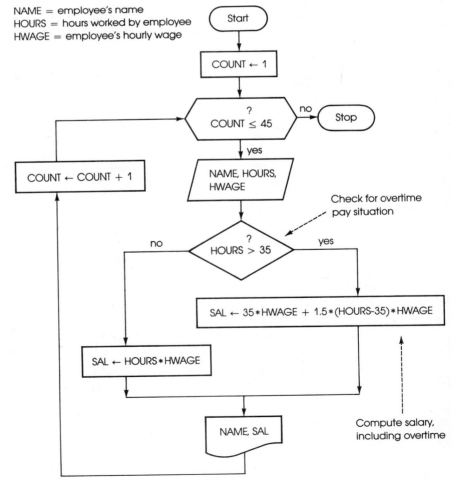

Figure 13 Payroll with overtime

Example 1.2.1 **Payroll with Overtime**

Figure 13 is a flowchart that will produce a list of 45 employees' names and salaries, given each individual name, hourly pay rate, and number of hours worked (see Example 1.1.4). However, here we take into account that a worker receives time-and-a-half pay for each hour beyond 35. ■

Example 1.2.2 **Balloting**

Suppose that 1785 valid ballots are cast in a two-person election. A flowchart that will produce the final vote tallies is given in Figure 14. In this example, the variables TALLY1 and TALLY2 are called *accumulators*. ■

Example 1.2.3 **Largest Number**

A flowchart that will produce the largest of 500 given numbers is drawn in Figure 15. ■

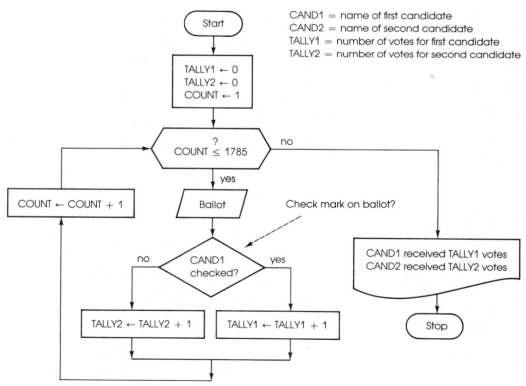

Figure 14 Balloting problem

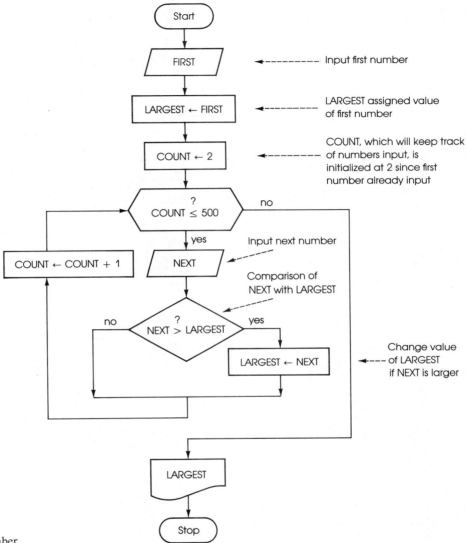

Figure 15 Largest number

Example 1.2.4 **Addition of Integers**

Recall the mechanical process used to add two positive integers. Figure 16 shows a flowchart that describes this process. ■

In each of the four preceding examples, one question was asked (conditional flow) within a loop (repetitive flow). The next example involves two questions within a loop.

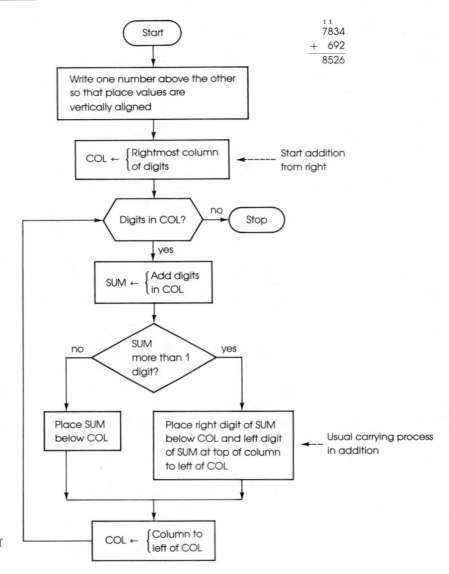

Figure 16 Addition of integers

Example **1.2.5**   Age Category

A flowchart is drawn in Figure 17 that, given the names and ages of 2000 people, will determine whether each person is a child (age ≤ 12), teenager (13 ≤ age ≤ 19), or adult (age ≥ 20). (Compare this with Example 1.1.5.)   ■

As always, the reader is urged to check the validity of the flowcharts in this section by working through each with sample data.

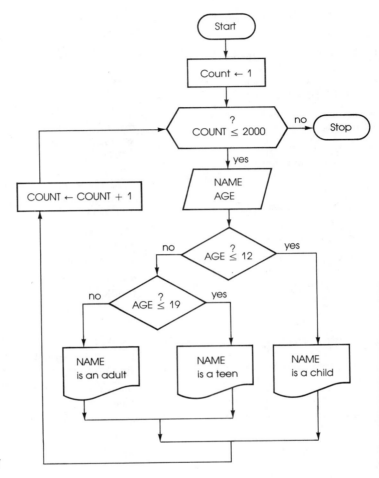

Figure 17 Age category

EXERCISES **1.2**

1. If   ALBRIGHT, 35, 5
       DEMITRAS, 40, 4.50

are the first two sets of input data for the flowchart in Figure 13, what are the first two lines of output?

2. If   BURKE, 38, 5.25
       ELLIS, 32, 4.65

are the first two sets of input data for the flowchart in Figure 13, what are the first two lines of output?

3. Work through the flowchart of Figure 16 for the sum of 6504 and 1687.

4. Work through the flowchart of Figure 16 for the sum of 999 and 101.

Exercises 5–16 involve the construction of flowcharts and are optional.

5. A certain business pays its sales force according to the following formula:

Salary = $250 + Commission

where

$$\text{Commission} = \begin{cases} 20\% \text{ of weekly sales if weekly} \\ \quad \text{sales} \leq \$1000 \\ 25\% \text{ of weekly sales if weekly} \\ \quad \text{sales} > \$1000 \end{cases}$$

Given the name and weekly sales for each of the 42

salespersons, draw a flowchart that will produce the name and salary of each salesperson.

6. Revise the flowchart in Example 1.2.2 to allow for the possibility that some of the 1785 ballots may be invalidly marked, and have the number of invalid ballots printed along with the candidate tallies.

7. Generalize Example 1.2.3 to handle N numbers, where N is any positive integer.

8. Construct a flowchart that describes how to fill a gallon container with a jar of unknown capacity without overfilling the container. (See Exercise 9 of Section 1.1 for comparison.)

9. Given the names and ages of 3500 adults, devise a flowchart that will produce the names of those eligible for Medicare coverage (that is, age $\geq$ 65).

10. Given the names and numerical exam scores of 54 students in a class, draw a flowchart that will produce the name and appropriate letter grade for each of the students, where

$$A \geq 90 \qquad 80 \leq B < 90 \qquad 70 \leq C < 80$$
$$60 \leq D < 70 \qquad F < 60$$

11. One can compute a college grade point average (GPA) as follows:

    a. For each course, multiply the number of course credits by the number of "quality points" earned by your final course grade (A = 4, B = 3, C = 2, D = 1, F = 0).

    b. Add the numbers obtained in part a for each course you took.

    c. Divide the number obtained in part b by the sum of the course credits attempted.

Devise a flowchart that will compute the GPA for any given number of courses.

12. One way of finding a path through a maze is to follow the rule "Move right when you can, left when you must." Assuming that (a) you can move *only* forward, one step at a time, (b) you can turn either right or left, and (c) you can discern a wall one step ahead, draw a flowchart for getting through a maze successfully.

13. Construct a flowchart that will determine the difference of two positive integers using the familiar mechanical process of subtraction.

14. Generalize the flowchart in Exercise 9 so that it can handle an unspecified number of people by using the *sentinel* data XYZ, -9 (See Exercises 1.1 for the definition of *sentinel*.)

15. A *prime number* is a positive integer greater than 1 whose only divisors are 1 and itself. Draw a flowchart that will determine if a given integer N is prime.

16. A *flag* is a variable that usually takes on only two values, like 0 and 1. Can the flowchart in Exercise 15 be made more efficient with the addition of a flag variable?

1.3

Decimal Numbers: Fixed and Floating Point

In problem solving, we often need to work sample data through a flowchart to determine if the output meets our expectations. If the flowchart is to be translated into a computer program, then we must know the form of data that the computer accepts and understand the type of data that it produces.

    The numbers we encounter in everyday life are **base 10** numbers, built from the ten **digits** 0, 1, 2, 3, 4, 5, 6, 7, 8, and 9. We learn in our early days of schooling that each digit in a number like

    73.5

has a particular **place value**:

    7 is in the tens place,
    3 is in the ones place, and
    5 is in the tenths place.

This allows us, if needed, to express base 10 numbers in **expanded form**.

Example 1.3.1 **Expanded Form**

$$73.5 = 7*10 + 3*1 + 5*.1$$
$$1762.903 = 1*1000 + 7*100 + 6*10 + 2*1 + 9*.1 + 0*.01 + 3*.001$$
$$.0504 = 5*.01 + 0*.001 + 4*.0001 \qquad \blacksquare$$

Note the importance of the zeroes in numbers like 1762.903 and .0504; the zeroes serve as *placeholders* for certain place values. Also, the place values in our base 10 system are powers of 10:

$$10^0 = 1 \qquad 10^1 = 10 \qquad 10^2 = 100 \qquad 10^3 = 1000 \qquad \text{etc.}$$
$$10^{-1} = .1 \qquad 10^{-2} = .01 \qquad 10^{-3} = .001 \qquad \text{etc.}$$

The three numbers in the preceding example are said to be in *fixed-point form*, which distinguishes them from another form we are about to present.

Computers are electronic devices that operate at amazing speed (approaching the speed of light) and are becoming increasingly compact (we are in the midst of a microelectronics revolution!). Nonetheless, computers have limitations. For instance, a computer contains only a finite number of memory cells and registers (special-purpose storage devices), each of which can store only a certain number of characters. Suppose a certain register was built to hold at most eight characters, as illustrated below:

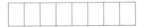

Then neither of the numbers

27,520,000,000   nor   .000000302

could be stored in their present form in this register, for each contains more than eight characters. However, we can rewrite these numbers as follows:

$$27,520,000,000 = .2752 * 10^{11}$$
$$.000000302 = .302 * 10^{-6}$$

In working with computers, it is important to express data using *just one line of type*, without resorting to subscript or superscript positions. Using the letter E (for exponent), we can further rewrite these numbers on one line as

$$27,520,000,000 = .2752E + 11$$
$$.000000302 = .302E - 6$$

The expressions on the right sides of these equations are examples of what is called the *floating-point form* of a number; the decimals .2752 and .302 are called *mantissas*, and the integers 11 and −6 are called *exponents*. Note that a positive exponent indicates that the decimal point of the mantissa is moved to the right; a negative exponent, to the left. In either case, zeroes can be added as needed.

If one now allots the five leftmost cells and the three rightmost cells of the register to hold the mantissa and exponent, respectively, then the eight-character register can hold the floating-point forms of the given numbers:

$$27{,}520{,}000{,}000 = .2752E + 11 \rightarrow \boxed{+ \;|\; 2 \;|\; 7 \;|\; 5 \;|\; 2 \;|\; + \;|\; 1 \;|\; 1}$$

Mantissa — Exponent

$$.000000302 = .302E - 6 \rightarrow \boxed{+ \;|\; 3 \;|\; 0 \;|\; 2 \;|\; 0 \;|\; - \;|\; 0 \;|\; 6}$$

Mantissa — Exponent

Example 1.3.2   Conversion from Fixed Point to Floating Point

To find the mantissa, move the decimal point to the left of the leftmost nonzero digit; to find the exponent, determine the direction and number of places the relocated decimal point must be moved so as to equal the original number. See the examples displayed below.

	Mantissa	Exponent
$26.9 = .269E + 2$	.269	+2
$-5402.78 = -.540278E + 4$	−.540278	+4
$.0173 = 0.173E - 1$	.173	−1
$-.00045 = -000.45E - 3$	−.45	−3
$.428 = .428E0$	.428	0

∎

Example 1.3.3   Conversion from Floating Point to Fixed Point

To convert from floating-point to fixed-point form, move the decimal point and the number of spaces in the direction indicated by the exponent, adding zeroes if needed.

$$.6903E + 3 = \quad 690.3 \quad = \quad 690.3$$

$$.52E - 2 \quad = \quad .0052 \quad = \quad .0052$$

$$-.481E + 5 \quad = \quad -48100. \quad = \quad -48{,}100$$

$$-.702E - 3 \quad = \quad -.000702 \quad = \quad -.000702$$

$$-.9E0 \qquad = \qquad\qquad -.9$$

$$.1E + 1 \quad = \quad 1. \quad = \quad 1$$

∎

Actually, floating-point notation can take many possible forms, depending on where you wish to place the decimal point in the mantissa.

Example 1.3.4   Different Forms of Floating Point[†]

$$729 = .729E + 3 = 7.29E + 2 = 72.9E + 1 = 729E0$$

$$.052 = .52E - 1 \ = 5.2E - 2 \ = 52E - 3$$

∎

[†] Whenever floating-point form is referred to later in this text, it is the form used in Examples 1.3.2 and 1.3.3.

The type of floating point that places the decimal point *to the right* of the first nonzero digit in the mantissa has been used for many years by laboratory scientists who work with numbers that are very large (such as astronomical data) or very small (such as nuclear physics measurements). Consequently, this type of floating point has become known as *scientific notation*.

Example 1.3.5   **Scientific Notation**

The expressions on the right sides of the following equations are the scientific notation forms of the numbers on the left:

$$4902 = 4.902E+3 \qquad .04902 = 4.902E-2$$
$$490.2 = 4.902E+2 \qquad .004902 = 4.902E-3$$
$$49.02 = 4.902E+1 \qquad 1 = 1E0$$
$$4.902 = 4.902E0 \qquad 16 = 1.6E+1$$
$$.4902 = 4.902E-1 \qquad -723.842 = -7.23842E+2$$

■

The two forms of floating point most commonly used by computers are the first form introduced (see Examples 1.3.2 and 1.3.3) and scientific notation. For example, the DEC PDP series of minicomputers use the former, while the Radio Shack TRS-80 microcomputer uses the latter. Even though computers at the machine level perform all their work in binary notation (which we shall discuss in Chapter 2), we shall conclude this section with illustrations of the mechanics of arithmetic computations in floating point.

Example 1.3.6   **Addition in Floating Point**

The steps in the following three computations are described in the flowchart in Figure 18.

$$
\left.\begin{array}{r} .803E+6 \\ +.19\ E+4 \end{array}\right\} \longrightarrow \begin{array}{r} .803\ E+6 \\ .0019E+6 \\ \hline \boxed{.8049E+6} \end{array}
$$

$$
\left.\begin{array}{r} .329E-7 \\ +.751E-8 \end{array}\right\} \longrightarrow \begin{array}{r} .329\ E-7 \\ .0751E-7 \\ \hline \boxed{.4041E-7} \end{array}
$$

$$
\left.\begin{array}{r} .449E-3 \\ +.962E-2 \end{array}\right\} \longrightarrow \begin{array}{r} .0449\ E-2 \\ .962\ \ E-2 \\ \hline 1.0069\ E-2 \\ = \boxed{.10069E-1} \end{array}
$$

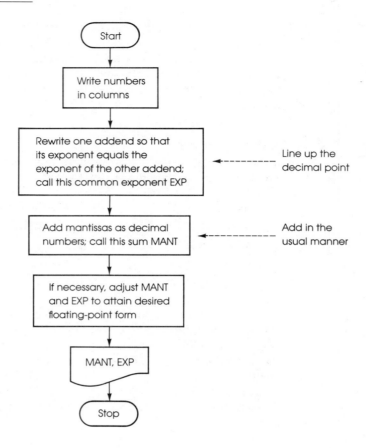

Figure 18 Floating-point addition

Example 1.3.7   **Subtraction in Floating Point**

The mechanics here are the same as addition except that one subtracts the mantissas.

$$\begin{array}{c} .803E+6 \\ -.19\ E+4 \end{array} \Biggr\} \longrightarrow \begin{array}{c} .803\ \ E+6 \\ -.0019E+6 \end{array}$$

$$\boxed{.8011E+6}$$

$$\begin{array}{c} .329E-7 \\ -.751E-8 \end{array} \Biggr\} \longrightarrow \begin{array}{c} .329\ \ E-7 \\ -.0751E-7 \end{array}$$

$$\boxed{.2539E-7}$$

$$\begin{array}{c} .1491E-2 \\ -.962\ E-1 \end{array} \Biggr\} \longrightarrow \begin{array}{c} .01491E-1 \\ -.962\ \ \ E-1 \end{array}$$

$$\boxed{-.94709E-1}$$

Example 1.3.8

Multiplication in Floating Point

Figure 19 exhibits the steps used in the next three computations.

$$(.803E+6) * (.1992E+3) = (.803 * .1992)E(6+3)$$
$$= \boxed{.1599576E+9}$$

$$(.329E-7) * (.751E-8) = (.329 * .751)E(-7-8)$$
$$= \boxed{.247079E-15}$$

$$(.149E+1) * (.592E-2) = (.149 * .592)E(1-2)$$
$$= .088208E-1$$
$$= \boxed{.88208E-2}$$

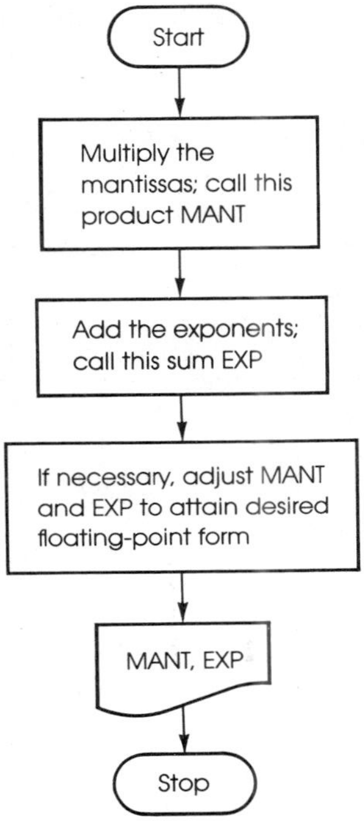

```
            Start

  Multiply the
  mantissas; call this
  product MANT

  Add the exponents;
  call this sum EXP

  If necessary, adjust MANT
  and EXP to attain desired
  floating-point form

       MANT, EXP

            Stop
```

Figure 19 Floating-point multiplication

Example 1.3.9

Division in Floating Point

The mechanics here are similar to multiplication except that we divide the mantissas and subtract the exponents.

$$(.803E+6)/(.1992E+3) = (.803/.1992)E(6-3)$$
$$= 4.03E+3$$
$$= \boxed{.403E+4}$$

$$(.329E-7)/(.751E-8) = (.329/.751)E(-7+8)$$
$$= \boxed{.438E+1}$$

$$(.149E+1)/(.592E-2) = (.149/.592)E(1+2)$$
$$= \boxed{.252E+3}$$

$$(-.592E-2)/(.149E+1) = (-.592/.149)E(-2-1)$$
$$= -3.973E-3$$
$$= \boxed{-.3973E-2}$$

In several instances in this last example, the equal sign was used in the sense of "is approximately equal to" rather than "is exactly equal to." This is often the case when using a computer to perform numerical computations, and it is the main topic of the next section.

EXERCISES **1.3**

1. In each of the following, determine the place value of the underlined digit:

  a. 7462   **b.** 682.09   **c.** 96,000,000
  d. .00013   **e.** 76.6   **f.** −1403.1

2. In each of the following, determine the place value of the underlined digit:

  a. 849   **b.** 202.6   **c.** 403007.51
  d. .0028   **e.** 509.5   **f.** −17.0542

3. Write the expanded form of each of the base 10 numbers in Exercise 1.

4. Write the expanded form of each of the base 10 numbers in Exercise 2.

Note: In the exercises below, the floating point referred to is that of Examples 1.3.2 and 1.3.3.

5. Convert each of the following fixed-point numbers to floating-point:

  a. 179.2   **b.** 450000   **c.** 1.865
  d. .902   **e.** .0571   **f.** .0001001
  g. −102   **h.** −1

6. Convert each of the following fixed-point numbers to floating-point:

  a. 42.43   **b.** 1982000   **c.** 7.063
  d. .101   **e.** −.00995   **f.** .0000001
  g. −14.9   **h.** −6

7. Convert each of the following floating-point numbers to fixed-point:

  a. .241E+2   **b.** .9E+5   **c.** .5026E+1
  d. .5E0   **e.** .62E−2   **f.** .40189E−6
  g. −.72E+3   **h.** −.5E−4

8. Convert each of the following floating-point numbers to fixed-point:

  a. .52E+3   **b.** .8071E+7   **c.** .112E0
  d. .4E−1   **e.** .762E−3   **f.** −.498E+1
  g. −.1E−3   **h.** −.1E+1

9. Convert each of the fixed-point numbers in Exercise 5 to scientific notation.

10. Convert each of the fixed-point numbers in Exercise 6 to scientific notation.

11. Complete the table below.

Fixed point	Floating point	Scientific notation
5		
1742.69		
63,029,001		
.42		
.00601		
	.6E+4	
	.6E−4	
	.4112E+6	
		1.02E−3
		−6.9987E+7

12. Complete the table below.

Fixed point	Floating point	Scientific notation
29		
1416		
462,004,692		
.0618		
.00000002		
	.15E+3	
	.15E−3	
	.698E+5	
		5.980E+5
		−2.37E−4

13. In each of the following, convert the given pair of numbers to floating point; then add, subtract, multiply, and divide them, giving your answers in floating point.

  a. 521 billion, 721 million
  b. .0000001012, .0000000987

14. In each of the following, convert the given pair of numbers to floating point; then add, subtract, multiply, and divide them, giving your answers in floating point.

  a. 902 million, 67 thousand
  b. .0000815, .00000609

15. Redo Exercise 13 using scientific notation rather than floating point.

16. Redo Exercise 14 using scientific notation rather than floating point.

17. Compute the product and quotient of 1,697,000,000 and .00000142, expressing your answer in fixed point.

18. Compute the product and quotient of 51,062,000,000 and .0000402, expressing your answer in fixed point.

19. Explain why, at machine level, (digital) computers deal with rational numbers and *not* irrational numbers.

1.4

Approximation and Error

In the preceding section, we discussed the importance of floating-point notation relative to computer operation. For instance, given a register

that can hold only eight characters, the fixed-point number

726,000,000,000

cannot be stored in that form; whereas that same register can hold the floating-point form

| + | 7 | 2 | 6 | 0 | + | 1 | 2 |

But even the use of floating point has its limitations, for the above register cannot hold any number larger than

| + | 9 | 9 | 9 | 9 | + | 9 | 9 | $\rightarrow .9999 * 10^{99}$

or smaller than

| − | 9 | 9 | 9 | 9 | + | 9 | 9 | $\rightarrow -.9999 * 10^{99}$

and it cannot hold in *its exact form* the number

$$5268.47 = .526847E + 4$$

since the mantissa has more than four digits. When faced with a floating-point number with more digits in the exponent part than its register can accommodate, a situation called ***overflow***, most computers are built to send an ***error message*** to the user (for instance, "floating point error" is the message on the DEC PDP series) indicating the problem. Having more digits in the mantissa than the register can accommodate is a less serious problem, and computers are generally built to respond by either truncating or rounding, which we shall now describe. In the remainder of this section, some examples will be done in fixed-point notation, and others in floating-point. It would be a good idea to think of these latter examples in relation to a computer whose registers store only four mantissa digits (excluding sign) like the register above.

> **Truncation**
>
> *To truncate* a number at a particular decimal place, delete all the digits to the right of that place.

Example 1.4.1

Truncation: Fixed Point

Number	Number truncated at units	Number truncated at second decimal
43.6574	43	43.65
7.42	7	7.42
.59241	0	.59
.00995	0	.00

In effect, the truncation to the units place in this example produces the integer parts of the given numbers. Since this capability is sometimes useful, many computer languages have a built-in function available to perform this type of truncation. (In BASIC, FORTRAN, PL/C, and Pascal, the functions are INT, INT, FLOOR, and TRUNC, respectively.)

Example 1.4.2

Truncation: Floating Point

In each case, the mantissa is truncated to four digits.

Number	Truncated number
.52984E + 2	.5298E + 2
.123456E + 12	.1234E + 12
.40251E − 3	.4025E − 3
.6100926E − 5	.6100E − 5

Truncation can also be used to find the remainder when two integers are divided on a computer or calculator.

Example 1.4.3

Application of Truncation

Using computer operations, find the remainder when 3479 is divided by 8.

Step 1: Do the division. $3479/8 = 434.875$

Step 2: Truncate at units place. 434

Step 3: Multiply by 8. $434 * 8 = 3472$

Step 4: Subtract. $3479 - 3472 = 7$

The remainder is 7.

Rounding

To round off a number to a particular decimal place, the next digit to the right is considered. If this digit is:

1. 0, 1, 2, 3, or 4, then the original decimal place is left unchanged and all following digits are deleted;
2. 5, 6, 7, 8, or 9, then the original decimal place is increased by 1 (carrying if necessary) and all following digits are deleted.[†]

Example 1.4.4   **Rounding: Floating Point**

In each case, the mantissa is rounded to four digits.

Number	Rounded number
.52984E+2	.5298E+2
.123456E+12	.1235E+12
.40251E−3	.4025E−3
.6100926E−5	.6101E−5

■

Example 1.4.5   **Rounding**

The fascinating number e appears in many important applications. Expressed to 15 places, it equals

   2.718281828459045

To 13 places $e = 2.7182818284590$

To 10 places $e = 2.7182818285$

To 7 places $e = 2.7182818$

To 5 places $e = 2.71828$

To 2 places $e = 2.72$

■

As mentioned after Example 1.4.1, many high-level programming languages have a built-in function that yields the integer part of any positive number. For instance, if TRD (for "truncate decimal") is such a function, then

   $TRD(7.2) = 7$
   $TRD(27.062) = 27$
   $TRD(0.82) = 0$
   $TRD(5) = 5$

Such a function can be used within a computer program to round off positive numbers.

[†] Some texts use the "odd-add" rule when the test digit is 5: If the test digit is 5 followed by all zeroes, the preceding digit is left unchanged if even, but increased by one if odd.

Example 1.4.6   **Rounding Using Truncation**

Assuming the availability of the TRD function and the usual computer arithmetic operations, round 37.825 and 9.4721 to the first decimal place.

Step 1:	Multiply by 10	378.25	94.721
Step 2:	Add .5	378.75	95.221
Step 3:	Apply TRD function	378	95
Step 4:	Divide by 10	37.8	9.5

The answers are 37.8 and 9.5, respectively. ∎

Example 1.4.7   **Rounding Using Truncation**

Assuming TRD and the usual arithmetic operations, round 37.825 and 9.4721 to the second decimal place.

Step 1:	Multiply by 100	3782.5	947.21
Step 2:	Add .5	3783.0	947.71
Step 3:	Apply TRD function	3783	947
Step 4:	Divide by 100	37.83	9.47

The answers are 37.83 and 9.47 respectively. ∎

If you're guessing that this procedure can be generalized to handle any decimal place, you're right! This will be dealt with in certain exercises at the end of this section.

The methods of truncation and rounding yield numbers that are *approximations* to the original ones. Computers are built to use one of these methods (usually rounding) when faced with storing a floating-point number whose mantissa is too long to fit. In so doing, the approximation then differs from the original number, and an *error* is introduced.

The topic of errors generated in computer usage is both extensive and difficult. We shall be content here to introduce the notions of *absolute error* and *relative error*, and apply them to a few examples of common computer errors.

Definition

Let $\begin{cases} \text{NUM} = \text{original number} \\ \text{APP} = \text{approximation to NUM} \end{cases}$

Then we define *absolute error* by

$$\text{ABS} = \text{NUM} - \text{APP}$$

and *relative error* by

$$\text{REL} = \text{ABS/NUM}$$

Usually REL is expressed as a percentage.

Of these two quantities, the relative error is generally the better measure of the approximation. For instance, if 1000 is approximated by 999 and 10 is approximated by 9, then the absolute error in each case is 1. But the relative error in the first is

$$\frac{1}{1000} = .001 = .1\%$$

whereas the relative error of the second is a much larger

$$\frac{1}{10} = .1 = 10\%$$

Example 1.4.8   **Truncation Error: Fixed Point**

If 5.637 is truncated at the second decimal place, then

$$
\begin{aligned}
\text{NUM} &= 5.637\\
-\text{APP} &= 5.63\\
\hline
\text{ABS} &= \boxed{.007}
\end{aligned}
$$

$$\text{REL} = \frac{.007}{5.637} = .00124 = \boxed{.124\%}$$

Example 1.4.9   **Rounding Error: Fixed Point**

If 5.637 is rounded to the second decimal place, then

$$
\begin{aligned}
\text{NUM} &= 5.637\\
-\text{APP} &= 5.64\\
\hline
\text{ABS} &= \boxed{-.003}
\end{aligned}
$$

$$\text{REL} = \frac{-.003}{5.637} = -.000532 = \boxed{-.0532\%}$$

Example 1.4.10   **Truncation Error: Floating Point**

If .49208E+3 is truncated at the fourth digit, then

$$
\begin{aligned}
\text{NUM} &= .49208\text{E}+3\\
-\text{APP} &= .4920\text{E}+3\\
\hline
\text{ABS} &= \boxed{.00008\text{E}+3}
\end{aligned}
$$

$$\text{REL} = \frac{.00008\text{E}+3}{.49208\text{E}+3}$$

$$= \left(\frac{.00008}{.49208}\right)\text{E}(+3-3)$$

$$= .00016\text{E}0 = \boxed{.016\%}$$

Example 1.4.11   **Rounding Error: Floating Point**

If .49208E + 3 is rounded to the fourth digit, then

$$
\begin{aligned}
\text{NUM} &= \quad .49208\text{E} + 3 \\
-\text{APP} &= \quad .4921\ \text{E} + 3 \\
\hline
&= \boxed{-.00002\text{E} + 3}
\end{aligned}
$$

$$
\begin{aligned}
\text{REL} &= \frac{-.00002\text{E} + 3}{.49208\text{E} + 3} \\[2mm]
&= \left(\frac{-.00002}{.49208}\right)\text{E}(+3 - 3) \\[2mm]
&= .000041\text{E}0 = \boxed{.0041\%}
\end{aligned}
$$

■

In the previous four examples, observe that the magnitudes of both the absolute and relative errors are smaller for rounding than for truncating.

Example 1.4.12   **Conversion Error: Floating Point**

If the fraction $\frac{1}{3} = .333\ldots$ is stored as a floating-point number with a mantissa truncated to four digits, then

$$
\begin{aligned}
\text{NUM} &= \tfrac{1}{3} \\
\text{APP} &= \quad .3333\text{E}0
\end{aligned}
$$

$$
\begin{aligned}
3 * \text{NUM} &= 1 \qquad \text{E}0 \\
-3 * \quad \text{APP} &= .9999\text{E}0 \\
\hline
3 * \text{ABS} &= .0001\text{E}0
\end{aligned}
$$

$$
\text{ABS} = \frac{.0001\text{E}0}{3} = \frac{.0001}{3} = \boxed{.0000333\ldots}
$$

$$
\begin{aligned}
3 * \text{REL} &= 3 * \frac{\text{ABS}}{\text{NUM}} \\[2mm]
&= \frac{.0001\text{E}0}{\frac{1}{3}} = 3 * .0001 = .0003
\end{aligned}
$$

$$
\text{REL} = \frac{.0003}{3}
$$

$$
= .0001 = \boxed{.01\%}
$$

■

While the conversion error in the preceding example is quite small, as the approximation is used repeatedly in calculations, the resultant error becomes larger.

Example 1.4.13   **Accumulating Error: Floating Point**

Let's compute the sum of six terms, each the floating-point form of $\frac{1}{3}$:

$$
\begin{array}{r}
.3333\text{E}0 \\
+.3333\text{E}0 \\
\hline
.6666\text{E}0 \\
+.3333\text{E}0 \\
\hline
.9999\text{E}0 \\
+.3333\text{E}0 \\
\hline
1.333\ \text{E}0 \\
+.3333\text{E}0 \\
\hline
1.666\ \text{E}0 \\
+.3333\text{E}0 \\
\hline
1.999\ \text{E}0
\end{array}
$$

Remember: The mantissa is truncated to only four digits.

Then

$$\text{NUM} = 6 * \frac{1}{3} = 2$$

$$\text{APP} = 1.999\text{E}0$$

$$\text{ABS} = 2 - 1.999 = \boxed{.001}$$

$$\text{REL} = \frac{.001}{2} = .0005 = \boxed{.05\%}$$

Notice how rapidly the error has increased, for REL here is five times that of the previous example! Obviously, this type of error can be of great concern in a program performing many such calculations.

Also, if two numbers that are very close in value are approximated, then the relative error of the difference of these approximations can be quite large.

Example 1.4.14   **Subtraction Error: Floating Point**

Consider the two numbers

$$.62194\text{E}+4 \quad \text{and} \quad .62186\text{E}+4$$

The actual difference is

$$
\begin{array}{r}
.62194\text{E}+4 \\
-.62186\text{E}+4 \\
\hline
.00008\text{E}+4
\end{array}
$$

whereas if the computer truncates to four mantissa digits, then it computes

$$
\begin{array}{r}
.6219E+4 \\
-.6218E+4 \\
\hline
.0001E+4
\end{array}
$$

Thus,

$$
\begin{array}{rl}
\text{NUM} = & .00008E+4 \\
-\ \text{APP} = & .0001\ \ E+4 \\
\hline
\text{ABS} = & \boxed{-.00002E+4} \\
\end{array}
$$

$$
\begin{aligned}
\text{REL} &= \frac{-.00002E+4}{.00008E+4} \\
&= \left(\frac{-.00002}{.00008}\right)E(+4-4) \\
&= -.25E0 \\
&= \boxed{-25\%}
\end{aligned}
$$

The relative error here is a whopping 25%! ■

The notion of *significant digit* relates directly to numerical data arising from measurements made in the physical world; hence, it has been a tool of the laboratory scientist for many years. Rather than attempt to describe significant digits in this broader setting, we shall restrict the context to computer science.

Significant Digits

1. Any nonzero digit is significant.
2. A zero between nonzero digits is significant.
3. Zeroes on the left are not significant.
4. Zeroes on the right are not significant unless otherwise specified.

For numbers in floating-point form, these rules are applied only to the mantissa.

Example 1.4.15   **Significant Digits: Fixed Point**

Number	Significant digits
34.56	3, 4, 5, 6
4.802	4, 8, 0, 2
.00351	3, 5, 1 (The zeroes on the left merely locate the decimal point.)
5.070	5, 0, 7 (See comment after the next example.)

■

Example 1.4.16 **Significant Digits: Floating Point**

Number	Significant digits
.6254E + 3	6, 2, 5, 4
.1002E − 4	1, 0, 0, 2
.76E − 5	7, 6
.7600E + 12	7, 6

∎

The last parts of the two preceding examples deserve comment. If one recorded

5.070   and   .7600E + 12

as measurements in a laboratory experiment, then the rightmost zeroes in each number would be considered significant, for there would be no other reason to write them. However, if these same numbers were part of computer output, these trailing zeroes may or may not be meaningful. So unless we have additional information to the contrary, these rightmost zeroes are considered as *not* significant.

Example 1.4.17 **Significant Figures: Rounding**

Round each given number to three significant figures (counting from the left).

Number	Rounded number
487600	488000
23.042	23.0
.003855	.00386

EXERCISES 1.4

1. Truncate each of the following numbers at the third decimal place and then at the first decimal place:

a. 273.55137   **b.** −3.274501
c. 62.0329   **d.** .0000427

2. Truncate each of the following numbers at the third decimal place and then at the first decimal place.

a. 49.5238   **b.** −572.40955
c. 8.0612   **d.** .00097

3. A computer truncates the mantissa of floating-point numbers to six digits. In what form will each of the following be stored?

a. .6950432E + 1   **b.** .420055782E − 4
c. .112233E0   **d.** .5000009E − 11

4. A computer truncates the mantissa of floating-point numbers to six digits. In what form will each of the following be stored?

a. .54908541E + 5   **b.** .7039278E + 12
c. .998877E − 2   **d.** .1040005E − 6

5. Redo Exercise 1 but round rather than truncate.

6. Redo Exercise 2 but round rather than truncate.

7. Redo Exercise 3 but round rather than truncate.

8. Redo Exercise 4 but round rather than truncate.

9. Given the numbers 7.156 and 9.27:

a. Multiply and round the answer to two decimal places.
b. Round each number to one decimal place and multiply.
c. Multiply and truncate the answer to two decimal places.
d. Truncate each number to one decimal place and multiply.

10. Given the numbers 6.35 and 11.064:

a. Multiply and round the answer to two decimal places.
b. Round each number to one decimal place and multiply.

c. Multiply and truncate the answer to two decimal places.

d. Truncate each number to one decimal place and multiply.

11. Draw a flowchart for the application of truncation described in Example 1.4.3, where

DIV = dividend   DOR = divisor

REM = remainder

Then test your flowchart on

DIV = 748   DOR = 7

12. Use the method of Example 1.4.3 to find the remainder when 1049 is divided by 12.

13. Using Examples 1.4.6 and 1.4.7 as models, draw a flowchart that will round a given number NUM to the nth decimal place, where n is a fixed positive integer. Then use this flowchart to round NUM = 4.60983 to the
  a. third decimal place   **b.** fourth decimal place

14. Generalizing on the technique shown in Examples 1.4.6 and 1.4.7, round 12.90854 to the
  a. third decimal place   **b.** fourth decimal place

15. Compute the absolute and relative errors if 7.064 is
  a. truncated to the first decimal place
  b. rounded to the first decimal place

16. Compute the absolute and relative errors if 4.251 is
  a. truncated to the first decimal place
  b. rounded to the first decimal place

17. Compute the absolute and relative errors if the mantissa of .4235E+4 is
  a. truncated to three digits
  b. rounded to three digits

18. Compute the absolute and relative errors if the mantissa of .6949E−6 is
  a. truncated to three digits
  b. rounded to three digits

19. Find the absolute and relative errors produced when $\frac{1}{6}$ is stored in floating-point form in a computer that truncates the mantissa at
  a. four digits   **b.** six digits

20. Find the absolute and relative errors produced when $\frac{15}{99}$ is stored in floating-point form in a computer that truncates the mantissa at
  a. four digits   **b.** six digits

21. Consider the sum of six terms, each being the floating-point form of $\frac{1}{6}$. Find the absolute and relative errors produced in a computer that truncates the mantissa at
  a. four digits   **b.** six digits

22. Consider the sum of six terms, each being the floating-point form of $\frac{15}{99}$. Find the absolute and relative errors produced in a computer that truncates the mantissa at
  a. four digits   **b.** six digits

23. Find the absolute and relative errors produced when the difference

$$.4035171E+7$$
$$-.4035169E+7$$

is performed by a computer that truncates mantissas to six digits.

24. Find the absolute and relative errors produced when the difference

$$.9817513$$
$$-.9817504$$

is performed by a computer that truncates mantissas to six digits.

25. Find the absolute and relative errors produced in the computation of
  a. Exercise 9b   **b.** Exercise 9d

26. Find the absolute and relative errors produced in the computation of
  a. Exercise 10b   **b.** Exercise 10d

27. If a computer stores in floating-point form and truncates to three digits in the mantissa, how many times can $\frac{75}{8}$ be added to itself before the relative error exceeds .5%?

28. If a computer stores in floating-point form and truncates to three digits in the mantissa, how many times can $\frac{43}{8}$ be added to itself before the relative error exceeds .5%?

29. For each of the following expressions, determine the number of significant digits:

425.16	.83129E−12
63.025	.543E+3
.07020	.71902E0
1.0	.1E+15
1	.100E−3

30. For each of the following expressions, determine the number of significant digits:

5116.1	.29842E + 5
203.89	.36E − 2
0.0056	.60920E + 1
.0520	.40001E + 12
3	.70E − 1

31. Round each of the numbers in Exercise 29 to
   a. two significant digits if possible
   b. one significant digit.

32. Round each of the numbers in Exercise 30 to
   a. two significant digits if possible
   b. one significant digit.

1.5

Binary Operations

The operations on numbers that we have seen in previous sections are formally called *binary operations*.

Definition

> A *binary operation* is an operation that takes exactly two inputs and produces exactly one output.

Whenever we use the word *operation* in this text, we shall mean "binary operation." The ordinary arithmetic operations $+$, $-$, $*$, and $/$ are familiar examples:

Two inputs		One output
↓ ↓		↓
5 + 3	=	8
7 − 9	=	−2
6 * 4	=	24
−9 / 6	=	−1.5

A less familiar example, but one that is important in computer application, is the operation of *concatenation*, or joining together, which makes sense for nonnumeric inputs. If we use the symbol ◯ for the operation of concatenation,[†] then some examples are as follows:

Two inputs		One output
↓ ↓		↓
EUC ◯ LID	=	EUCLID
WORK ◯ MAN	=	WORKMAN
7 ◯ 2	=	72

However, mathematical expressions often involve several of these operations acting on several numbers. For instance, the expression

$$4 + 3 * 5$$

[†] In BASIC, FORTRAN, and PL/C, the symbols for concatenation are +, ‖, and ‖, respectively.

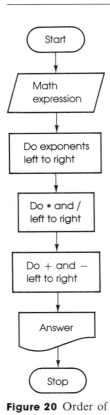

Figure 20 Order of operations

Example 1.5.1

involves two operations and three numbers. The problem with such expressions is that the answer obtained depends on the order in which the operations are performed. Here, if the $+$ is done first, one obtains

$$7 * 5 = 35$$

However, if the $*$ is done first, the result is

$$4 + 15 = 19$$

To eliminate this ambiguity, mathematicians have established the following convention.

Order of Operations (Figure 20)

   1. All exponentiations[†] are done from left to right.
   2. All multiplications and divisions are done from left to right.
   3. All additions and subtractions are done from left to right.

If an expression is encountered that involves several operations, the above order is followed. In this way there is no ambiguity in the expression, and there is only one correct answer. Our example above, $4 + 3 * 5$, contains both addition and multiplication. Since multiplication has a higher priority, it is done first, and the correct answer is 19.

Order of Operations

The brace symbol $\smile$ is used to indicate which operation is performed at a particular stage.

$$2 + \underbrace{4 * 3} - 3 * 7 * 2 + 14 \qquad \underbrace{4 ** 3} * 5 - 2/5/2$$

$$= 2 + \quad 12 \quad - \underbrace{3 * 7} * 2 + 14 \qquad = \underbrace{64 \ * 5} - 2/5/2$$

$$= 2 + \quad 12 \quad - \underbrace{21 \ * 2} + 14 \qquad = \quad 320 \quad - \underbrace{2/5/2}$$

$$= \underbrace{2 + \ 12} \quad - \quad 42 \quad + 14 \qquad = \quad 320 \quad - \underbrace{.4 \ /2}$$

$$= \quad \underbrace{14 \quad - \quad 42} \quad + 14 \qquad = \underbrace{320 \quad - \quad .2}$$

$$= \quad \underbrace{-28 \quad\quad + 14} \qquad = \quad \boxed{319.8}$$

$$= \quad\quad \boxed{-14} \qquad\qquad\qquad\qquad \blacksquare$$

[†] The symbol we shall use for exponentiation, or raising to a power, is $**$. For example, $2 ** 3 = 2 * 2 * 2 = 8$.

Suppose in the example

$$4 + 3 * 5$$

that we want to do the addition first. Then we must devise a way to preempt the order of operations as stated, and this is done by using **parentheses** as grouping symbols. The expression is thus written as

$$(4 + 3) * 5$$

where the parentheses indicate that the addition is to be done first. Here the result is

$$7 * 5 = 35$$

Several sets of parentheses are involved in the expression

$$(5 * (-2)) * (3 + (3 * 7))$$

but there is a standard, natural way of proceeding. Let us call a set of parentheses **atomic** if it contains no other parentheses, except possibly those around a single number. For instance, in the above expression the sets

$$(5 * (-2)) \quad \text{and} \quad (3 * 7)$$

are atomic, whereas $(3 + (3 * 7))$ is not atomic. The rules for evaluating such an expression are stated in the following algorithm, which has a repetitive part to it.

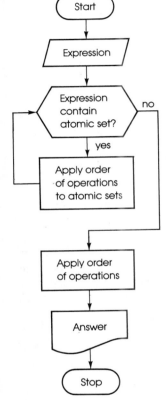

Figure 21 Evaluation of expressions

Evaluation of Expressions with Parentheses (Figure 21)

　　1. Evaluate all atomic sets using the order of operations.
　　2. Repeat step 1 until no parentheses remain.
　　3. Evaluate the result using the order of operations.

Applying this algorithm to the above expression gives

$$
\underbrace{(5 * (-2))}_{\text{atomic}} * (3 + \underbrace{(3 * 7)}_{\text{atomic}})
$$

$$
= \quad \downarrow \quad\quad\quad\quad\quad \downarrow
$$

$$
= \quad -10 \quad * (3 \quad + 21)
$$

$$
\underbrace{}_{\text{atomic}}
$$

$$
\downarrow
$$

$$
= \quad -10 \quad * \quad 24
$$

$$
= \quad\quad\quad -240
$$

Example 1.5.2   **Expressions with Parentheses**

The brace ⌣ is used to indicate atomic sets of parentheses at each stage.

$$2 * (3 + \underbrace{(3 * (-7))})$$
$$= 2 * \underbrace{(3 + (-21))}$$
$$= 2 * \quad (-18)$$
$$= \quad \boxed{-36}$$

$$5 * ((\underbrace{21 - 5 * 3}) + (7/\underbrace{(6 + 2)}))$$
$$= 5 * \quad (6 \quad + \underbrace{(7/\ 8)})$$
$$= 5 * \quad \underbrace{(6 \quad + \ .875)}$$
$$= 5 * \quad 6.875$$
$$= \quad \boxed{34.375}$$

$$4 * (3 + (6/4 - \underbrace{(4 + 3)} * 4) + 1.5)$$
$$= 4 * (3 + \underbrace{(6/4 - \quad 7 \quad * 4)} + 1.5)$$
$$= 4 * \underbrace{(3 + \quad (-26.5) \quad + 1.5)}$$
$$= 4 * \quad (-22)$$
$$= \quad \boxed{-88}$$

The correct use of parentheses is very important whenever one works with expressions containing different operations because parentheses allow you to execute the operations in the intended order. They can also be used to insure absolute clarity. For example, in

$$4 + (5/4) = 5.25$$

the parentheses are not really needed if one follows the order of operations convention, but they can be included to insure that an uninformed person executes the operations in the correct sequence.

Happily, a computer, just like us, can perform only one binary operation at a time, although the speed at which it does so may give the false impression that it can execute many operations simultaneously. That being the case, computers must be instructed on how to execute multioperation expressions. Most computers, from the large mainframe down to the pocket microcomputers, use the order of operations and rules for handling parentheses discussed in this section, as do many hand calculators that use algebraic notation. A computer can give accurate results *only* when it has been correctly programmed, and that requires that all formulas be carefully written. One should always read the manual that comes with a particular machine to be sure of the correct notational procedures.

There is a method of writing expressions of the type just discussed *without* using parentheses. It is done by using **Polish notation**, named in honor of the Polish mathematician J. Lukasiewicz, who introduced it. Here we shall use Polish notation in **prefix form**, where the operation symbol ∘ is placed to the *left* of the operands.

Example 1.5.3   **Converting to Polish Notation: Constants**

The key here is to write each operation

$$a \circ b \qquad \text{in the form} \qquad \circ\, a\, b$$

working first with atomic sets of parentheses, if any.

$$4 + \underbrace{(3 * 5)}$$

$$= 4 + \quad *3,5$$

$$= \overset{\downarrow}{+4} \quad *3,5$$

$$= \boxed{+4 * 3,5}$$

$$\underbrace{(4 + 3)} * 5$$

$$= \quad +4,3 \quad * 5$$

$$= \overset{\downarrow}{*} + 4,3 \; 5$$

$$= \boxed{* + 4,3,5}$$

$$\underbrace{(6 * (-3))} * (4 + \underbrace{(5/7)})$$

$$= (*6,-3) \quad * \quad \underbrace{(4 + /5,7)}$$

$$= (*6,-3) \quad * \quad (+4/5,7)$$

$$= \overset{\downarrow}{*} * 6,-3 \qquad + 4/5,7$$

$$= \boxed{* * 6,-3 + 4/5,7}$$

The commas are used here to separate numbers so that, for instance 3,5 is not read as thirty-five. These commas are unnecessary when single-letter variables are used. ■

Example 1.5.4 | **Converting to Polish Notation: Variables**

$$a - \underbrace{(b/c)}$$

$$= a - \quad /bc$$

$$= \overset{\downarrow}{-}a \;\; /bc$$

$$= \boxed{-a/bc}$$

$$\underbrace{(a * b)} - \underbrace{(c * d)}$$

$$= \quad *ab \quad - \quad *cd$$

$$= \overset{\downarrow}{-}*ab \qquad *cd$$

$$= \boxed{-*ab*cd}$$

$$\underbrace{((a + b)/(c * d))} - \underbrace{(e - f)}$$

$$= \quad \underbrace{(+ab/*cd)} \quad - (-ef)$$

$$= (/ + ab * cd) \quad - (-ef)$$

$$= \overset{\swarrow}{-}/ + ab * cd \qquad -ef$$

$$= \boxed{-/+ab*cd-ef}$$ ■

Example 1.5.5 | **Converting from Polish Notation: Variables**

Here, work from left to right, converting each operation that is immediately followed by *two* operands (that is, variables or numbers) to standard form; repeat this process as necessary.

$$\underbrace{*ab}/c$$

$$= (a*b)/c$$

$$= \boxed{(a*b)/c}$$

$$-\underbrace{/ab} + \underbrace{cd}$$

$$= -(a/b)(c+d)$$

$$= (a/b)\overset{\frown}{-}(c+d)$$

$$= \boxed{(a/b)-(c+d)}$$

$$-*\underbrace{+ab}/cd + ef$$

$$= -*(a+b)(c/d)(e+f)$$

$$= -((a+b)*(c/d))(e+f)$$

$$= ((a+b)*(c/d))\overset{\frown}{-}(e+f)$$

$$= \boxed{((a+b)*(c/d))-(e+f)}$$

This type of notation has wide application in an area of computer science called **compiling**. Also, Polish notation can be expressed in **postfix form**, where the operation symbol ∘ is placed to the *right* of the operands. When using Polish notation for expressions that involve exponents, to avoid ambiguity it is preferable to use a single symbol like ↑ to represent exponentiation rather than the double asterisk **. ■

EXERCISES 1.5

1. Calculate the value of each expression.

a. $4 * 5 + 7$ **b.** $4 * (5 + 7)$
c. $(4 * 5) + 7$ **d.** $2 * (3 ** 2 + (5 - 7) * (-4))$
e. $3 / 5 * 2$ **f.** $3 / (5 * 2)$

2. Calculate the value of each expression.

a. $5 + 4 - 3 * 2 / 1$ **b.** $5 / 4 + 3 - 2 * 1$
c. $5 * 4 / 3 + 2 - 1$ **d.** $5 - 4 * 3 / 2 + 1$

3. Calculate.

a. $2 + (3 * 2.55 - (7 + 4 * (3 ** 3 - 18.3)) + 4) / 8$
b. $(3 + 5) / 5 * 11 + 16 / (2 * 3 - 4)$

4. Calculate.

a. $2 ** 4 * 3.125 / (25 - (2 + 3 * 6) + (-12))$
b. $(3 + (1 / (3 + (1 / (3 + (1 / (3 + (1 / 3))))))))$

5. Determine which parentheses are unnecessary.

a. $(4 + 2) / (6 + 4)$ **b.** $5 * (6 ** 2)$
c. $(5 * 6) ** 2$ **d.** $4 - (5) / (6 + (3 ** 2))$

6. Determine which parentheses are unnecessary.

a. $(5 - 4) / 3$ **b.** $(5 * 4) / 3$
c. $7 - (6 - 5)$ **d.** $(5 ** 2) / (2 + (3 + 4))$

7. True or false?

a. $6 + 4 / 2 = (6 + 4) / 2$
b. $6 + (4 / 2) = 6 + 4 / 2$

c. $5 * 2 ** 3 = 5 * (2 ** 3)$
d. $5 * 2 ** 3 = (5 * 2) ** 3$

8. True or false?

a. $7 - 2 - 3 = 7 - (2 - 3)$
b. $7 - 2 - 3 = (7 - 2) - 3$
c. $9 / 8 - 2 = (9 / 8) - 2$
d. $9 / 8 - 2 = 9 / (8 - 2)$

9. Convert each of the following expressions into Polish notation:

a. $(7 - 3) / 4$ **b.** $a * (b - c) * d$
c. $(x - y - z) / (w + v)$

10. Convert each of the following expressions into Polish notation:

a. $(5 + 6) / (-4 * (10 / 3))$ **b.** $a / b * (c + d)$

11. Convert each of the following expressions from Polish notation into standard notation:

a. $*6 - 4,2$
b. $/ + abc$
c. $* + uv + - / wxy * yz$

12. Convert each of the following expressions from Polish notation into standard notation:

a. $*/10,3 - -4,5,6$
b. $- * / abcd$

CHAPTER 2

Computer-Related Arithmetic

The memory and processing circuits of a digital computer are made of electronic devices called *flip-flops*, each of which is in one of two states, on or off, much like a light switch in a room. Thus, it is natural to associate two symbols to these states; traditionally, the digit 1 has been assigned to *on*, and 0 to *off*. Thus, at machine level, the processing and storage of data involve the manipulation of long strings of the symbols 0 and 1, called *binary digits*, or bits.

The computer performs all mathematical calculations using binary notation and binary arithmetic. While this type of notation is often longer and more cumbersome than our familiar decimal notation (as we shall see in Section 2.1), the fantastic speed at which the machine can process binary strings more than compensates for the notational disadvantage. Both numeric and alphabetic information is usually stored within a computer in binary-coded form. Table 1 gives a partial listing of one frequently used eight-bit binary code that we'll discuss in Section 2.5.

For these reasons, we shall study binary notation, arithmetic, and codes in this chapter. Material relating to octal and hexadecimal numbers, which are useful to machine and assembly language programmers, is also presented.

2.1

Binary Numbers

Before we delve into the specifics of binary numbers (and octal and hexadecimal numbers later in this chapter), we should pause and review some facts about our familiar decimal numbers.

The *decimal number system* (the prefix *deci-* means "ten") is built from the ten digits 0, 1, 2, 3, 4, 5, 6, 7, 8, and 9. It is a *place-value system* because the position of a digit within a number is important; for example,

$$209 = 2*100 + 0*10 + 9*1$$
$$= 2*10^2 + 0*10^1 + 9*10^0$$

Table 1 Representation of Characters

8-bit code	Hexadecimal	Decimal	Punched card code	BCD characters	EBCDIC characters
1100 0001	C1	193	12–1	A	A
1100 0010	C2	194	12–2	B	B
1100 0011	C3	195	12–3	C	C
1100 0100	C4	196	12–4	D	D
1100 0101	C5	197	12–5	E	E
1100 0110	C6	198	12–6	F	F
1100 0111	C7	199	12–7	G	G
1100 1000	C8	200	12–8	H	H
1100 1001	C9	201	12–9	I	I
1101 0000	D0	208	11–0	!	
1101 0001	D1	209	11–1	J	J
1101 0010	D2	210	11–2	K	K
1101 0011	D3	211	11–3	L	L
1101 0100	D4	212	11–4	M	M
1101 0101	D5	213	11–5	N	N
1101 0110	D6	214	11–6	O	O
1101 0111	D7	215	11–7	P	P
1101 1000	D8	216	11–8	Q	Q
1101 1001	D9	217	11–9	R	R
1110 0010	E2	226	0–2	S	S
1110 0011	E3	227	0–3	T	T
1110 0100	E4	228	0–4	U	U
1110 0101	E5	229	0–5	V	V
1110 0110	E6	230	0–6	W	W
1110 0111	E7	231	0–7	X	X
1110 1000	E8	232	0–8	Y	Y
1110 1001	E9	233	0–9	Z	Z
1111 0000	F0	240	0	0	0
1111 0001	F1	241	1	1	1
1111 0010	F2	242	2	2	2
1111 0011	F3	243	3	3	3
1111 0100	F4	244	4	4	4
1111 0101	F5	245	5	5	5
1111 0110	F6	246	6	6	6
1111 0111	F7	247	7	7	7
1111 1000	F8	248	8	8	8
1111 1001	F9	249	9	9	9

$$47.83 = 4*10 \ + 7*1 \ \ + 8*.1 \ \ \ + 3*.01$$
$$= 4*10^1 + 7*10^0 + 8*10^{-1} + 3*10^{-2}$$

In the decimal system *the place values are powers of 10*. Usually near the beginning of our study of math, we are taught the mechanics of adding, subtracting, multiplying, and dividing such numbers. One such algorithm was expressed in flowchart form in Example 1.2.4.

Now let's turn to another system. The prefix *bi-* means "two," and the **binary number system** is built from the two digits

    0 and 1

called **binary digits**, or **bits**. To become familiar with counting in this new system, let's imagine we have just bought a new car. In a standard car, the odometer has six digits

0	0	0	0	0	0

and behind the dashboard the entire mechanism would appear as shown in Figure 1. On each of the six disks there are the ten decimal digits; as the car moves, the rightmost disk, which measures tenths of a mile, turns. These disks are linked together in such a way that whenever any one disk turns from 9 to 0, then the disk to its left moves one-tenth of a turn to the next digit. Figure 2 shows what happens in three different cases when a car is driven one-tenth of a mile, thus causing the rightmost disk to increase by 1.

Figure 1

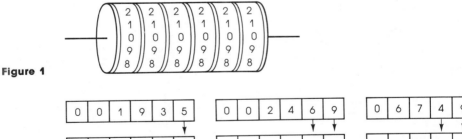

Figure 2

Well, this fancy new car we have purchased (or the bank has purchased for us) is equipped with a binary odometer. It still has six digits, which is not very realistic (see exercise 5 in Exercises 2.1) if we plan to do much driving!

0	0	0	0	0	0

This odometer has three important differences compared with the usual odometer discussed above. First, each of the six disks contains only the two binary digits, 0 and 1, as indicated in Figure 3. Second, the linkage between disks is such that whenever any one disk turns from 1 to 0, then the disk to

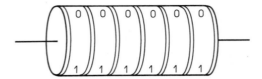

Figure 3

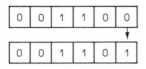

Figure 4

the left moves one-half of a turn to the next digit. Third, the rightmost disk moves to the next digit after the car has covered one full mile. In Figure 4 we see three different instances involving a car that is driven one mile, so that the rightmost disk of the odometer increases by 1.

Finally, starting with an odometer reading of all zeroes and successively adding 1, we can count using binary notation; in Figure 5 we count from 0

Odometer reading						Associated binary number	Miles driven (decimal notation)
0	0	0	0	0	0	0	0
0	0	0	0	0	1	1	1
0	0	0	0	1	0	10	2
0	0	0	0	1	1	11	3
0	0	0	1	0	0	100	4
0	0	0	1	0	1	101	5
0	0	0	1	1	0	110	6
0	0	0	1	1	1	111	7
0	0	1	0	0	0	1000	8
0	0	1	0	0	1	1001	9
0	0	1	0	1	0	1010	10
0	0	1	0	1	1	1011	11
0	0	1	1	0	0	1100	12
0	0	1	1	0	1	1101	13

Figure 5

to 13 (decimal) miles. As you can see in the figure, the decimal counting

0, 1, 2, 3, 4, 5, 6, 7, 8, . . .

is equivalent to binary counting

0, 1, 10, 11, 100, 101, 110, 111, 1000, . . .

Notice that like the decimal system, the binary system is a *place-value system.*

Example 2.1.1   **Binary Notation: Expanded Form**

a. $10 = 1*2 \ + 0*1$
$\quad\quad = 1*2^1 + 0*2^0$

b. $110 = 1*4 \ + 1*2 \ + 0*1$
$\quad\quad\ = 1*2^2 + 1*2^1 + 0*2^0$

c. $1101 = 1*8 \ + 1*4 \ + 0*2 \ + 1*1$
$\quad\quad\ \ = 1*2^3 + 1*2^2 + 0*2^1 + 1*2^0$

d. $10.11 = 1*2^1 + 0*2^0 + 1*2^{-1} + 1*2^{-2}$

The place values here are powers of 2; 0 may sometimes play the role of a placeholder, as it does with the numbers in the above example. Using the expanded form makes the process of converting numbers from binary to decimal rather straightforward.   ■

Example 2.1.2   **Conversion: Binary to Decimal**[†]

a. $1101_{bin} = 1*2^3 + 1*2^2 + 0*2^1 + 1*2^0$
$\quad\quad\quad\ = \ 8 \ + \ 4 \ + \ 0 \ + \ 1$
$\quad\quad\quad\ = 13_{dec}$

b. $10110110_{bin} = 1*2^7 + 0*2^6 + 1*2^5 + 1*2^4 + 0*2^3 + 1*2^2 + 1*2^1 + 0*2^0$
$\quad\quad\quad\quad\ = 128 \ + \ 0 \ + 32 \ + 16 \ + \ 0 \ + \ 4 \ + \ 2 \ + \ 0$
$\quad\quad\quad\quad\ = 182_{dec}$

c. $101.1_{bin} = 1*2^2 + 0*2^1 + 1*2^0 + 1*2^{-1}$
$\quad\quad\quad\ = \ 4 \ + \ 0 \ + \ 1 \ + \ \frac{1}{2}$
$\quad\quad\quad\ = 5.5_{dec}$

d. $1.1001_{bin} = 1*2^0 + 1*2^{-1} + 0*2^{-2} + 0*2^{-3} + 1*2^{-4}$
$\quad\quad\quad\quad = \ 1 \ + \ \frac{1}{2} \ + \ 0 \ + \ 0 \ + \ \frac{1}{16}$
$\quad\quad\quad\quad = 1.5625_{dec}$ ■

[†] To distinguish numbers expressed in different systems, we shall use the appropriate subscript when necessary.

A *binary integer* is a binary number that does not require the use of a binary point in its expression, like the numbers in (a) and (b) above. A *decimal integer* is defined similarly.

Example 2.1.3   Algorithm: Conversion of Binary Integer to Decimal

The flowchart in Figure 6 takes in a binary integer BININT and produces the decimal integer equivalent DECINT, where

VAL   = particular place value

MAX   = number of digits in BININT

POS   = counter of digits in BININT

BIT_{POS} = digit of BININT in position POS from the right

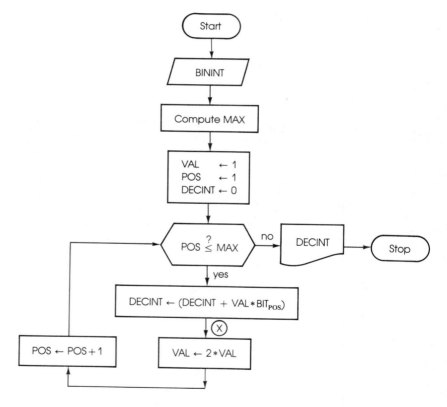

Figure 6

For instance, if BININT = 11001, then MAX = 5 and $BIT_1 = 1$, $BIT_2 = 0$, $BIT_3 = 0$, $BIT_4 = 1$, and $BIT_5 = 1$. Given this initial value, Table 2 traces the flowchart of Figure 6 at point $\circledX$ for variables POS, VAL, BIT_{POS}, and DECINT. From this trace we see that $11001_{bin} = 25_{dec}$. ∎

To illustrate conversion methods from decimal to binary, we shall deal separately with integers and nonintegers (those decimal numbers requiring a decimal point). For each case, two conversion techniques will be given. In

Table 2

POS	VAL	BIT$_{POS}$	DECINT
1	1	1	1
2	2	0	1
3	4	0	1
4	8	1	9
5	16	1	25

this discussion, you should be familiar with the following smaller powers of 2 (Figure 7).

$2^0 = 1$ $2^6 = 64$ $2^{-1} = \frac{1}{2} = .5$

$2^1 = 2$ $2^7 = 128$ $2^{-2} = \frac{1}{4} = .25$

$2^2 = 4$ $2^8 = 256$ $2^{-3} = \frac{1}{8} = .125$

$2^3 = 8$ $2^9 = 512$ $2^{-4} = \frac{1}{16} = .0625$

$2^4 = 16$ $2^{10} = 1024$ $2^{-5} = \frac{1}{32} = .03125$

$2^5 = 32$ $\qquad\qquad$ $2^{-6} = \frac{1}{64} = .015625$

Figure 7 $2^{-7} = \frac{1}{128} = .0078125$

Example 2.1.4   **Conversion: Decimal Integer to Binary**

Method 1A: Successive subtraction of the highest possible power of 2 and use of expanded form.

a. Convert 98_{dec} to binary.

First,

$$
\begin{array}{r}
98 \\
-64 \\
\hline
34 \\
-32 \\
\hline
2 \\
-2 \\
\hline
0
\end{array}
$$

As a result of these successive subtractions, we can write:

$98_{dec} = 64 + 32 + 2$

$\quad = 2^6 + 2^5 + 2^1$

$\quad = 1*2^6 + 1*2^5 + 0*2^4 + 0*2^3 + 0*2^2 + 1*2^1$

$\quad\quad + 0*2^0$

$\quad = 1100010_{bin}$

b. Convert 709_{dec} to binary:

First,

$$
\begin{array}{r}
709 \\
-512 \\
\hline
197 \\
-128 \\
\hline
69 \\
-64 \\
\hline
5 \\
-4 \\
\hline
1 \\
-1 \\
\hline
0
\end{array}
$$

Then we have

$709_{dec} = 512 + 128 + 64 + 4 + 1$

$\quad = 2^9 + 2^7 + 2^6 + 2^2 + 2^0$

$\quad = 1*2^9 + 0*2^8 + 1*2^7 + 1*2^6 + 0*2^5 + 0*2^4$

$\quad\quad + 0*2^3 + 1*2^2 + 0*2^1 + 1*2^0$

$\quad = 1011000101_{bin}$

Method 2A: Successive divisions by 2, specifying the remainder at each stage.

a. Convert 98_{dec} to binary.
First,

Remainder

$$
\begin{array}{r|l}
2\,)\,98 & 0 \\
2\,)\,49 & 1 \\
2\,)\,24 & 0 \\
2\,)\,12 & 0 \\
2\,)\,6 & 0 \\
2\,)\,3 & 1 \\
2\,)\,1 & 1 \\
0 &
\end{array}
$$

Now, reading the remainders from *bottom to top*, we have

$$1100010_{bin}$$

b. Convert 709_{dec} to binary.
First,

Remainder

$$
\begin{array}{r|l}
2\,)\,709 & 1 \\
2\,)\,354 & 0 \\
2\,)\,177 & 1 \\
2\,)\,88 & 0 \\
2\,)\,44 & 0 \\
2\,)\,22 & 0 \\
2\,)\,11 & 1 \\
2\,)\,5 & 1 \\
2\,)\,2 & 0 \\
2\,)\,1 & 1 \\
0 &
\end{array}
$$

So, reading again from *bottom to top*, the answer is

$$1011000101_{bin} \qquad \blacksquare$$

In the next example, we shall deal with decimals having zero to the left of the decimal point.

Example 2.1.5 **Conversion: Decimal Noninteger to Binary**

Method 1B: Successive subtraction of the highest possible power of 2 and use of expanded form.

a. Convert $.6875_{dec}$ to binary.

First,

$$
\begin{array}{r}
.6875 \\
-.5 \\
\hline
.1875 \\
-.125 \\
\hline
.0625 \\
-.0625 \\
\hline
0
\end{array}
$$

Because of the subtractions on the left, we can say

$$
\begin{aligned}
.6875_{dec} &= .5 + .125 + .0625 \\
&= 2^{-1} + 2^{-3} + 2^{-4} \\
&= 1*2^{-1} + 0*2^{-2} + 1*2^{-3} + 1*2^{-4} \\
&= .1011_{bin}
\end{aligned}
$$

b. Convert $.40625_{dec}$ to binary.
First,

$$
\begin{array}{r}
.40625 \\
-.25 \\
\hline
.15625 \\
-.125 \\
\hline
.03125 \\
-.03125 \\
\hline
0
\end{array}
$$

So we have

$$
\begin{aligned}
.40625_{dec} &= .25 + .125 + .03125 \\
&= 2^{-2} + 2^{-3} + 2^{-5} \\
&= 0*2^{-1} + 1*2^{-2} + 1*2^{-3} \\
&\quad + 0*2^{-4} + 1*2^{-5} \\
&= .01101_{bin}
\end{aligned}
$$

Method 2B: Successive multiplications by 2, removing the integer part at each stage.

a. Convert $.6875_{dec}$ to binary.
First.

	Integer Part
$2*.6875 = 1.375$	1
$2*.375 = 0.75$	0
$2*.75 = 1.5$	1
$2*.5 = 1.0$	1

Now, reading the integer parts from *top to bottom* gives the answer

$$.1011_{bin}$$

b. Convert $.40625_{dec}$ to binary.
First,

	Integer Part
$2*.40625 = 0.8125$	0
$2*.8125 = 1.625$	1
$2*.625 = 1.25$	1
$2*.25 = 0.5$	0
$2*.5 = 1.0$	1

Again, reading from *top to bottom*, we get

$$.01101_{bin}$$ ∎

There is one big difference between the techniques for converting decimal integers (Methods 1A and 2A) and those for converting decimal non-integers (Methods 1B and 2B). The former always terminate in a finite number of steps, whereas the latter may not. The nonterminating situation is more easily discovered with Method 2B than with Method 1B; thus, in general, Method 2B is preferred to Method 1B. The next example deals with this case.

Example 2.1.6   **Conversion to Nonterminating Binary Number**

Convert $.6_{dec}$ to binary.
Using the preferred technique, Method 2B, we have:

	Integer Part
$2*.6 = 1.2$	1
$2*.2 = 0.4$	0
$2*.4 = 0.8$	0
$2*.8 = 1.6$	1
$2*.6 = 1.2$	1
STOP!	

Since our fifth line of work duplicates the first, this work would continue forever, producing the binary digits 1001 over and over again.

Therefore, we conclude that

$$.6_{dec} = .100110011001\ldots$$
$$= .\overline{1001}_{bin}$$

where the overbar identifies those digits that repeat *ad infinitum*.   ∎

The next example summarizes the results of the preceding three examples.

Example 2.1.7   **Conversion: Decimal to Binary**

a. Convert 18.75_{dec} to binary.
The procedure here is to separate the given number into integer and noninteger parts and to convert each part separately.

Remainder		Integer Part	
2)18 0		$2*.75 = 1.5$ 1	
2) 9 1		$2*.5\ = 1.0$ 1	
2) 4 0			
2) 2 0			
2) 1 1			
0			

So, $18_{dec} = 10010_{bin}$ and $.75_{dec} = .11_{bin}$. Bringing these two results together,

$$18.75_{dec} = 10010.11_{bin}$$

b. Convert 283.1_{dec} to binary.

Again, working on the two parts individually:

	Remainder		Integer Part
2)283	1	$2*.1 = 0.2$	0
2)141	1	$2*.2 = 0.4$	0
2) 70	0	$2*.4 = 0.8$	0
2) 35	1	$2*.8 = 1.6$	1
2) 17	1	$2*.6 = 1.2$	1
2) 8	0	$2*.2 = 0.4$	0
2) 4	0	STOP!	
2) 2	0		
2) 1	1		
0			

Because line 6 repeats line 2, we have $.1_{dec} = .0\overline{0011}_{bin}$.

Thus, $283_{dec} = 100011011_{bin}$

Table 3

DECINT	POS	REM_{POS}
115	1	1
57	2	1
28	3	0
14	4	0
7	5	1
3	6	1
1	7	1

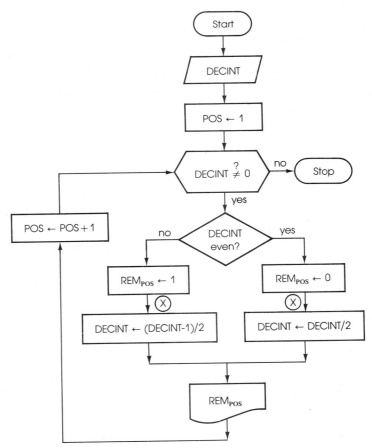

Figure 8

As a result,

$$283.1_{dec} = 100011011.0\overline{0011}_{bin}$$ ■

Some knowledge of convergence and geometric series is required to convert to binary a decimal number with an infinitely repeating noninteger part, or to convert to decimal a binary number with an infinitely repeating noninteger part. We shall not deal with that type of conversion here.

Example 2.1.8 **Algorithm: Conversion of Decimal Integer to Binary**

Figure 8 shows a flowchart for Method 2A that converts a given positive decimal integer DECINT into a binary integer by successive divisions by 2. The variable POS counts the digits REM_{POS} of the answer. A trace of this flowchart at point $\bigotimes$ for variables DECINT, POS, REM_{POS}, and initial value DECINT = 115 is given in Table 3. From it we can conclude that

$$115_{dec} = 1110011_{bin}$$ ■

EXERCISES 2.1

1. Write the expanded form for each of the following:
  a. Decimal numbers: 79, 40003, 91.085, 1011.1
  b. Binary numbers: 11, 110010, .001, 100.1

2. Write the expanded form for each of the following:
  a. Decimal numbers: 405, 1010, .008, 732.09
  b. Binary numbers: 111, 1010, 10.001, 1100.01

3. Extend Figure 5 to include 14 through 20 miles driven.

4. Extend Figure 5 to include 21 through 25 miles driven.

5. How many miles (in decimal notation) can you drive before the six-digit binary odometer described in the text returns to 000000?

6. If we add two digits to the six-digit binary odometer described in the text, how many miles (in decimal notation) can you drive before the odometer returns to 00000000?

7. Convert each of the following binary numbers to decimal numbers:
  a. 101 **b.** 11010 **c.** 10110101
  d. 1111011110 **e.** 110.1 **f.** 10.001011

8. Convert each of the following binary numbers to decimal numbers:
  a. 110 **b.** 10011 **c.** 1101010
  d. 100010001 **e.** 101.01 **f.** 1.11001

9. Convert each of the following decimal numbers to binary numbers by using two different methods:
  a. 70 **b.** 1425 **c.** .625 **d.** .2890625

10. Convert each of the following decimal numbers to binary numbers by using two different methods:
  a. 89 **b.** 912 **c.** .5625 **d.** .203125

11. Convert each of the following decimal numbers to binary form:
  a. 147.5 **b.** 1.125 **c.** 942.65625
  d. 303.2 **e.** 2.7 **f.** 55.875

12. Convert each of the following decimal numbers to binary form:
  a. 96.25 **b.** 1.0625 **c.** 503.4
  d. 881.4375 **e.** 13.3 **f.** 45.15625

13. Do a trace at point $\bigotimes$ of the flowchart in Figure 6 for the variables POS, BIT_{POS}, and DECINT, given the initial value BININT = 1011000. From this trace, what is the decimal form of 1011000?

14. Do a trace at point $\bigotimes$ of the flowchart in Figure 8 for the variables DECINT, POS, and REM_{POS}, given the initial value DECINT = 99. From this trace determine the binary form of 99.

15. a. Construct a flowchart for the procedure described in Method 1A of the text, which converts a decimal integer to a binary integer.

b. Test this flowchart by entering the initial value 57 and working through to produce the binary form of 57.

16. a. Construct a flowchart for the procedure described in Method 2B of the text, assuming that the given decimal number with no integer part can be converted to a terminating binary number (that is, a binary number with finitely many digits).

b. Test this flowchart by entering the initial value .3125 and working through to produce the binary form of .3125.

17. Here is another method for converting binary integers to decimal integers:

Given a binary integer, double the leftmost digit and add it to the digit to the right. Double this sum

and add it to the next digit to the right. Repeat this process until the rightmost digit has been added. The resulting sum is the decimal form of the original binary integer.

a. Use this method to find the decimal forms of 101, 11010, and 1101010.

b. Draw a flowchart to describe this algorithm.

18. a. Is it possible that a decimal number with finitely many digits converts to a binary number with infinitely many digits?

b. Is it possible that a binary number with finitely many digits converts to a decimal number with infinitely many digits?

2.2

Binary Arithmetic

As mentioned in the introduction to this chapter, the computer performs all numeric calculations—in fact, all of its processing work—using binary notation, though not necessarily the "straight" binary notation of the preceding section. (Binary computer codes are the topic of Section 2.5.) So here we shall illustrate techniques for performing binary addition, subtraction, multiplication, and division. First, however, you must clearly have in mind the mechanics involved in computing the sum, difference, product, and quotient of two decimal numbers, because we will mimic these mechanics for binary numbers. Those who are a little rusty with the mechanics of these computations should spend a few moments working through some review exercises.

$$0 + 0 = 0 \qquad 0 + 0 + 0 = 0$$
$$1 + 0 = 1 \qquad 1 + 0 + 0 = 1$$
$$1 + 1 = 10 \qquad 1 + 1 + 0 = 10$$
$$\qquad\qquad 1 + 1 + 1 = 11$$

Figure 9 Binary addition facts

To find the sum of two binary numbers, we vertically align the place values and then add columns, starting from the rightmost, using the binary addition facts of Figure 9. Whenever a column sum has two digits, you must carry the left digit (always a 1).

Example 2.2.1   Binary Addition: Two Addends

a.

```
   11 ◄────── carrying row ──────►  111
  1011                            1001100
 + 110                          + 1011101
 ─────                          ─────────
 10001                           10101001
```

b.

```
  1 1 1 ◄───── carrying row ─────►  1 1 1      1
 11.011                            1101.101
 + 1.11                          +  11.0011
 ──────                          ──────────
 101.001                          10000.1101
```

■

Example 2.2.2   **Binary Addition: More Than Two Addends**

Compute the sum

$$
\begin{array}{r}
10100 \\
11110 \\
10101 \\
+\,11011 \\
\hline
\end{array}
$$

Method 1: Add columns as in the preceding example. Now the carrying process becomes more complicated, since we may have to carry more than one digit. To do this, you must know binary sums of more than three digits; these digit sums are easily computed by using the "counting list":

$$
\begin{array}{ccccccccc}
+1 & +1 & +1 & +1 & +1 & +1 & +1 & +1 & +1 \\
0, & 1, & 10, & 11, & 100, & 101, & 110, & 111, & 1000, & 1001, \ldots
\end{array}
$$

$$
\left.\begin{array}{r}
1 \\
1\,0 \\
1\,1
\end{array}\right\} \quad \text{carrying rows}
$$

$$
\left.\begin{array}{r}
10100 \\
11110 \\
10101 \\
+\,11011
\end{array}\right\} \quad \text{original addends}
$$

$$
\begin{array}{r}
\hline
1100010 \quad \text{answer}
\end{array}
$$

Method 2: Repeat the technique of Example 2.2.1 three times, adding precisely two binary numbers at each of the three steps.

Step 1:

$$
\begin{array}{rl}
1\,1 & \\
10100 & \text{first addend} \\
+\,11110 & \text{second addend} \\
\hline
110010 &
\end{array}
$$

Step 2:

$$
\begin{array}{rl}
1 & \\
110010 & \\
+\,10101 & \text{third addend} \\
\hline
1000111 &
\end{array}
$$

Step 3:

$$
\begin{array}{rl}
1\,1\,1\,1\,1 & \\
1000111 & \\
+\,11011 & \text{fourth addend} \\
\hline
1100010 & \text{answer}
\end{array}
$$

Note that at each step there is at most one carrying row.

The next example, a precise description of the addition procedure illustrated above, is extremely important. For it presents an algorithm that not only computes the sum of two integers but also, as we shall see later in this section, it plays a central role in computing the difference, product, and quotient of two binary integers.

Example 2.2.3

Algorithm: Addition of Two Binary Integers

In Example 1.2.4 we stated a wordy algorithm to describe the mechanics of adding two decimal integers. If you review the associated flowchart now, you will see that it also applies to the sum of two binary integers. Here in Figure 10 we refine that flowchart by using several variables:

a, b = two given positive binary integers to be added

MAX = 1 + number of digits in longer of a and b

POS = counter of columns in addition process

a_{POS} = digit of a in position POS from right

b_{POS} = digit of b in position POS from right

c_{POS} = carry digit in position POS from right

SUM = sum of a_{POS}, b_{POS}, c_{POS}

ANS = sum of a, b

S_{POS} = digit of ANS in position POS from the right

For instance, if

$$a = 11110 \quad \text{and} \quad b = 1101$$

then MAX = 6 and

$a_1 = 0$	$a_2 = 1$	$a_3 = 1$	$a_4 = 1$	$a_5 = 1$	$a_6 = 0$
$b_1 = 1$	$b_2 = 0$	$b_3 = 1$	$b_4 = 1$	$b_5 = 0$	$b_6 = 0$

Table 4 gives a trace of the flowchart at point $\widehat{X}$ for several of the variables. From the rightmost column of this trace, we can read off the correct sum

$$\text{ANS} = 101011$$

of $a = 11110$ and $b = 1101$.

In Chapter 4 we shall discuss an electronic device called a ***full adder***, which is used in computers to execute the inside of the loop in Figure 10. ∎

Now let us describe the mechanics of the other three arithmetic operations. To find the difference of two binary numbers, we again align the place values vertically and then, working column by column from right to left, subtract the bottom digit from the top digit, borrowing when necessary. Borrowing is the tricky part, especially if you must skip over several zeroes

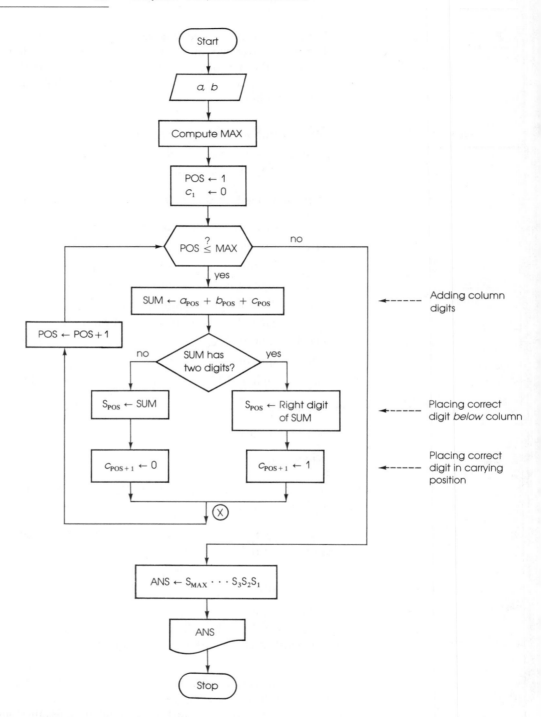

Figure 10 Addition algorithm for two binary integers

Table 4 Trace for $a = 11110$ and $b = 1101$

POS	a_{POS}	b_{POS}	c_{POS}	c_{POS+1}	SUM	S_{POS}
1	0	1	0	0	① → 1
2	1	0	0	0	① → 1
3	1	1	0	1	①⓪ → 0
4	1	1	1	1	①① → 1
5	1	0	1	1	①⓪ → 0
6	0	0	1	0	① → 1

$0 - 0 = 0$
$1 - 0 = 1$
$1 - 1 = 0$
$10 - 1 = 1$

Figure 11 Binary subtraction facts

to borrow from a digit several spaces to the left. The subtraction facts are given in Figure 11.

Example 2.2.4

Binary Subtraction

a.
```
   0 10              10          0 10 0 10  }←── borrowing rows
                   0 ̷0 10
  1 ̷1 ̷0 1         ̷1 ̷1 ̷0 1       10 ̷1 0 ̷1 ̷0
  − 1 0           − 1  1 0       − 1 0 0 1 0 1
  ───────         ─────────      ─────────────
  1 0 1 1          1  1 1          1 0 1 ←── answers
```

b.
```
    0 10            0 10    0 1 ̷0 10     0 1 1 1 1  10  }←── borrowing rows
                            0 1 ̷0 ̷1 ̷0
  1 ̷1 . ̷0 1 1     1 ̷1 ̷0 1 . ̷1 ̷0 ̷1 ̷0    ̷1 ̷0 ̷0 ̷0 ̷0 . ̷0 1
  − 1 . 1 1       − 1 1 . 0 0 1 1       − . 1
  ───────────     ─────────────────    ──────────────
  1 . 1 0 1        1 0 1 0 . 0 1 1 1     1 1 1 1 . 1 1 ←─────── answers    ∎
```

To multiply two binary numbers, we place one above the other; however, vertically aligning the place values is not necessary. Working from right to left on the bottom factor, we then multiply each digit d with the top factor, placing each product underneath the two given factors so that its rightmost digit is vertically aligned with d. Finally, we add these resulting products. The simple multiplication facts needed are listed in Figure 12.

$0 * 0 = 0$
$0 * 1 = 0$
$1 * 1 = 1$

Figure 12 Binary multiplication facts

Example 2.2.5

Binary Multiplication

a.
```
     110                      110110
    *10                      *1011
    ─────                    ─────────
     000                      110110
  + 110                       110110
  ──────                      000000
    1100                    + 110110
                            ──────────
                            1001010010
```

The reader should verify the second result, placing the omitted carry digits where appropriate. Immediately we realize that the addition is the hard part of this multiplication procedure.

b. One can simplify the above mechanics somewhat by omitting zero rows to be added. Thus:

```
    1101              10001
   *101             *10001
   ----             ------
    1101              10001
  +1101            +10001
  -------          --------
 1000001          100100001
```

c. When multiplying numbers that involve binary points, remember that the number of binary places in the answer is the sum of the binary places in the two factors.

```
  100.1  one place        10.11  two places
 *1.11   two places      *1.01   two places
 ------                  -----
   1001                   1011
   1001                 +1011
 +1001                  ------
 -------                11.0111  four places
111.111  three places
```

One can divide two binary numbers by mimicking the familiar division algorithm used for decimal numbers. This process involves repeated multiplications and subtractions until the remainder is less than the quotient.

Example 2.2.6   **Binary Division**

a.

```
          110                  101                     1111
     11 ) 10010          11 ) 1111           101 ) 1001101
        - 11                - 11                  - 101
        ----                ----                  -----
          11                  01                   1001
        - 11                - 00                  - 101
        ----                ----                  -----
           0                  11                   1000
                           - 11                  - 101
                           ----                  -----
                              0                    111
                                                 - 101
                                                 -----
                                                   10   remainder
```

b. When dividing numbers with binary points, first eliminate the binary point of the divisor by moving it as far to the right as necessary and then move the binary point of the dividend the same number of places to the right, adding zeroes if necessary.

```
             10.11                            110.
    1.01. )  11.10.11            10.001.)  1101.100.
          - 10 1                        - 1000 1
          -------                       --------
            1 00 1                        101 00
          - 10 1                        - 100 01
          -------                       --------
            10 01                          110
          - 1 01                         - 000
          -------                       -------
            1 00   remainder               110   remainder
```

While the procedure demonstrated in Example 2.2.4 is a perfectly correct way to determine binary differences, it is not how computers subtract. Instead, a computer calculates a quantity called a **complement** and then performs two successive additions to obtain a desired binary difference. The following examples specify the steps of this process; you can learn why this process gives the correct answer by working through exercises 60 and 61 at the end of this section.

Definition

> The (ones) *complement of a binary number* is that binary number formed by replacing each 0 by 1 and each 1 by 0 in the original number.

Example 2.2.7   **Binary Complements**

Binary number	Complement
10	01
1100	0011
101101	010010
10.01	01.10
1.00101	0.11010

Example 2.2.8   **Subtraction by Complements**

Compute

$$\begin{array}{r} 10110 \\ -1101 \\ \hline \end{array}$$

First, if necessary, add zero(es) to the left of the bottom number so that it has as many digits as the top number:

$$\begin{array}{r} 10110 \\ -01101 \\ \hline \end{array}$$

Second, compute the complement of the bottom number:

01101   has complement   10010

Third, add this complement to the top number:

$$\begin{array}{r} 10110 \\ +10010 \\ \hline 101000 \end{array}$$

Last, take the leftmost 1 of this sum and add it to what remains:

$$\begin{array}{r} 01000 \\ +1 \\ \hline 1001 \end{array}$$

The reader can verify that 1001 is the desired difference.

Now let's do the same work in a more compact form.

Example 2.2.9   **Subtraction by Complements**

a.
$$
\begin{array}{ccc}
1101 & 1101 & 1101 \\
-\quad 10 \longrightarrow & -0010 \longrightarrow & +1101 \\
& & \overline{\text{①}1010} \\
& & \searrow +1 \\
& & \boxed{1011}
\end{array}
$$

b.
$$
\begin{array}{cc}
101010 & 101010 \\
-100101 \longrightarrow & +011010 \\
& \overline{\text{①}000100} \\
& \searrow +1 \\
& \boxed{101}
\end{array}
$$

c.
$$
\begin{array}{ccc}
1101 & 1101 & 1101 \\
-110 \longrightarrow & -0110 \longrightarrow & +1001 \\
& & \overline{\text{①}0110} \\
& & \searrow +1 \\
& & \boxed{111}
\end{array}
$$

d. The process can also be used on noninteger binary numbers as long as the place values are aligned and zeroes are added, if necessary, on both the right and left of the bottom number:

$$
\begin{array}{ccc}
11.01 & 11.01 & 11.01 \\
-1.1 \longrightarrow & -01.10 \longrightarrow & +10.01 \\
& & \overline{\text{①}01.10} \\
& & \searrow +1 \\
& & \boxed{1.11}
\end{array}
$$

■

The subtraction-by-complements method just described will give the correct answer only when the top number is larger than the bottom. The result obtained when the bottom is larger than the top will be interpreted in the last example of Section 2.6 (Modular Arithmetic).

Example 2.2.10   **Algorithm: Subtraction of Two Binary Integers**

By using the complement method, we can now construct a flowchart for subtraction that includes two applications of the addition algorithm given in Example 2.2.3. Recall from the flowchart for addition drawn in Figure 10 that the two numbers to be added are input using variables a and b, and their sum $a + b$ is produced with variable name ANS.

For the subtraction flowchart constructed in Figure 13, we use the following variables:

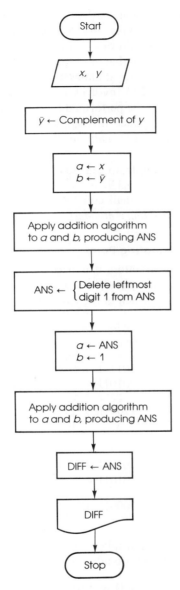

Figure 13

x = given positive binary integer

y = given positive binary integer to be subtracted from x where $x > y$

$\bar{y}$ = complement of y

$DIFF$ = difference $x - y$

This flowchart will take in two positive binary integers x and y where $x > y$ and produce the difference $x - y$. The reader is urged to run through this diagram with sample values for x and y. ∎

Now we can justify our earlier assertion about the significance of the addition algorithm. We have just delineated a subtraction algorithm that entails two successive additions. From our work with multiplication (Example 2.2.5), it is evident that multiplication reduces to several successive additions (with correct vertical alignment) of one of the factors. The division procedure (Example 2.2.6) involves repeated multiplications and subtractions, each of which can be performed by successive additions. Therefore, we have a result of importance in computer design: *all four basic arithmetic operations can be reduced to addition(s)!*

In Section 1.3 we introduced floating-point notation and gave several examples (1.3.6 through 1.3.9) of computations with decimal numbers in floating-point form. We shall conclude this section with computations involving *floating-point binary numbers.*

Example 2.2.11

Binary Arithmetic: Floating Point

For addition and subtraction, it is necessary to align place values (that is, make the exponent parts equal) before adding or subtracting the mantissas. It makes no difference which of the two exponents is adjusted, as long as the work is done correctly.

Addition

$$
\begin{array}{ll}
.1011\mathrm{E}+2 & .1011\ \mathrm{E}+2 \\
+.0101\mathrm{E}+1 \longrightarrow & +.00101\mathrm{E}+2 \\
\hline
& \boxed{.11011\mathrm{E}+2}
\end{array}
\qquad
\begin{array}{ll}
.1011\mathrm{E}-3 & .01011\mathrm{E}-2 \\
+.0101\mathrm{E}-2 \longrightarrow & +.0101\ \mathrm{E}-2 \\
\hline
& \boxed{.10101\mathrm{E}-2}
\end{array}
$$

Subtraction

$$
\begin{array}{ll}
.1011\mathrm{E}+2 & .1011\ \mathrm{E}+2 \\
-.0101\mathrm{E}+1 \longrightarrow & -.00101\mathrm{E}+2 \\
\hline
& \boxed{.10001\mathrm{E}+2}
\end{array}
\qquad
\begin{array}{ll}
.1011\mathrm{E}-3 & .01011\mathrm{E}-2 \\
-.0101\mathrm{E}-2 \longrightarrow & -.0101\ \mathrm{E}-2 \\
\hline
& \boxed{.00001\mathrm{E}-2}
\end{array}
$$

Multiplication: When multiplying floating-point forms, the mantissas are multiplied and exponents are added.

$$(.1011\mathrm{E}+2)*(.0101\mathrm{E}+1) = (.1011*.0101)\mathrm{E}(2+1)$$
$$= \boxed{.00110111\mathrm{E}+3}$$

$$(.1011\mathrm{E}-3)*(.0101\mathrm{E}-2) = (.1011*0101)\mathrm{E}(-3-2)$$
$$= \boxed{.00110111\mathrm{E}-5}$$

Division: When dividing floating-point forms, the mantissas are divided and the exponents are subtracted.

$$(.1011E+2)/(.0101E+1) = (.1011/.0101)E(2-1)$$
$$= \boxed{10.0011E+1}$$

$$(.1011E-3)/(.0101E-2) = (.1011/.0101)E(-3+2)$$
$$= \boxed{10.0011E-1}$$

The reader should verify the accuracy of the answers for the two computations

$$.1011*.0101 \quad \text{and} \quad .1011/.0101$$

Actually, the quotient is a repeating binary, so the last two answers above are only approximations. ■

EXERCISES **2.2**

Compute the following binary sums:

1. 101
   $+10$

2. 101
   $+11$

3. 1101
   $+110$

4. 1011
   $+101$

5. 10001
   $+11$

6. 10110
   $+100$

7. 11111
   $+1$

8. 1111111
   $+10$

9. 1.001
   $+.01$

10. 1101.101
   $+111.001$

11. 111
   1010
   $+101$

12. 1011
   110
   1101
   $+11$

13. 10001
   1110
   1001
   100
   $+111$

14. 111
   110
   101
   100
   $+111$

15. 1.1
   .01
   $+1.11$

16. 1.1
   1.1
   $+1.1$

17. .01
   .01
   $+.001$

18. 11
   .01
   $+1.11$

Use the method of Example 2.2.4 to compute the following binary differences:

19. 111
   -10

20. 101
   -11

21. 1101
   -111

22. 10100
   -1011

23. 11111
   -10101

24. 10101010
   -1010101

25. 1.0011
   $-.1010$

26. 110.01
   -11.1

27. 100.0
   $-.1$

28. 1000.01
   $-.11$

Find each of the following binary products:

29. 101
   $*10$

30. 111
   $*11$

31. 101101
   $*1001$

32. 10101
   $*10101$

33. 110.01
   $*1.1$

34. 1011.101
   $*10.01$

35. 11111
   $*.1$

36. 11.0010
   $*101$

Perform each of the following binary divisions and specify the integer remainder:

37. 11 $\overline{)\,11101}$

38. 10 $\overline{)\,11010}$

39. 110 $\overline{)\,1011011}$

40. 101 $\overline{)\,11010001}$

Compute each of the following divisions to at most two binary places:

41. 10 $\overline{)\,101}$

42. 11 $\overline{)\,101}$

43. 1.1 $\overline{)\,1.01}$

44. 1.1 $\overline{)\,1.001}$

45. .001 $\overline{)\,1.01}$

46. 10.101 $\overline{)\,1011.1}$

47. Find the binary (ones) complement of each of the following:

a. 11
b. 101
c. 11010
d. 100000
e. 1.101
f. 101.001

48. To each of the numbers in exercise 47, add zero(es) to the left if necessary so that it is a six-digit binary number. Then find the (ones) complement of this six-digit number.

49. Use the subtraction-by-complements method (see Examples 2.2.8 and 2.2.9) to compute the differences in exercises 19, 21, 23, 25, 27.

50. Write each of the following floating-point binary numbers in fixed-point form:

   a. .1011E$+$2   **b.** .110E$-$3
   c. .1E$+$5   **d.** .101E$-$4

51. Write each of the following floating-point binary numbers in fixed-point form:

   a. .110E$+$4   **b.** .1011E$-$2
   c. .111E0   **d.** 1001E$-$1

52. Convert each of the following fixed-point numbers into floating-point form with the binary point leftmost:

   a. 1   **b.** 1101
   c. .1010   **d.** 110.1

53. Convert each of the following fixed-point numbers into floating-point form with the binary point leftmost:

   a. 10   **b.** 101
   c. 1.01   **d.** .0011

54. Add, subtract, multiply, and divide each of the following pairs of floating-point binary numbers:

   a. .1101E$+$3, .1010E$+$1
   b. .1101E$-$2, .1010E$-$4

55. Add, subtract, multiply, and divide each of the following pairs of floating-point binary numbers:

   a. .1111E$+$1, .101E$-$1   **b.** .1111E$-$1, .101E$-$2

In the next two exercises, for the given initial values of a and b, do a complete trace at point $\textcircled{X}$ of the flowchart in Figure 10 for the variables POS, a_{POS}, b_{POS}, c_{POS}, c_{POS+1}, SUM, and S_{POS}. From the trace, read off the value of $a + b$.

56. $a = 1011, b = 110$

57. $a = 110110, b = 10010$

58. Construct a flowchart for the algorithm illustrated in Example 2.2.4 for binary subtraction, when applied to binary integers.

59. Construct a flowchart for the algorithm presented in Example 2.2.5 for binary multiplication, when applied to binary integers.

60. Justify the method of subtraction by complements used in Example 2.2.8 by giving a mathematical reason for each of the steps below:

$$10110 - 1101$$
$$= 10110 - 1101 + 100000 \quad - 100000$$
$$= 10110 - 1101 + 11111 + 1 - 100000$$
$$= 10110 + 11111 - 1101 \quad - 100000 + 1$$
$$= 10110 + 10010 \quad - 100000 + 1$$
$$= \quad 101000 \quad - 100000 + 1$$
$$= \quad 1000 \quad + 1$$
$$= \quad 1001$$

Compare the steps of this procedure with the steps in the example cited.

61. Rewrite each of the four subtractions (using complements) in Example 2.2.9 in the form illustrated in exercise 60.

62. The *twos complement of a binary number* is that binary number formed by adding 1 to the (ones) complement of the given number.

   a. Find the twos complement of each binary number in exercise 47.

   b. Review the use of the (ones) complement in Example 2.2.8 and modify this process so as to use the twos complement.

   c. Use the twos complement to compute $1101 - 10$

2.3

Octal Numbers

In this section and the next we present two number systems that are naturally related to the binary system. The importance of these systems will become apparent in our discussion of binary codes (Section 2.5).

The prefix *oct-* means "eight," and the ***octal number system*** is built from the eight digits

0, 1, 2, 3, 4, 5, 6, and 7

called ***octal digits.*** Recall our discussion of the binary odometer in Section 2.1. Now we wish to adapt that device so that it becomes an octal odometer, which we do as follows. First, each of the six odometer disks should contain the octal digits, as shown in Figure 14. Second, the linkage between the disks is such that whenever any one disk turns from 7 to 0, then the disk to the left moves one-eighth of a turn to the next digit. Third, the rightmost disk moves to the next digit after the car has traveled one full mile. In Figure 15, we are given three examples of a car that is driven one mile, thus increasing the rightmost digit of the odometer by 1.

Figure 14

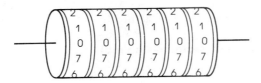

Figure 15

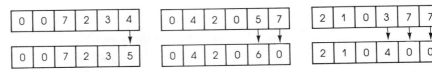

We can use this device to count in octal numbers by starting with an all-zero odometer reading and successively adding 1. In Table 5 we count up to 33 (decimal) miles, where for simplicity's sake leftmost zeroes on the octal numbers are omitted.

Table 5

Odometer reading (Octal)	Miles driven (Decimal)	Odometer reading (Octal)	Miles driven (Decimal)	Odometer reading (Octal)	Miles driven (Decimal)
0	0	14	12	27	23
1	1	15	13	30	24
2	2	16	14	31	25
3	3	17	15	32	26
4	4	20	16	33	27
5	5	21	17	34	28
6	6	22	18	35	29
7	7	23	19	36	30
10	8	24	20	37	31
11	9	25	21	40	32
12	10	26	22	41	33
13	11				

Like the binary and decimal systems, the octal numbers form a *place-value system*.

Example 2.3.1 **Octal Notation: Expanded Form**

a. $32_{oct} = 3*8 \ + 2*1$
$= 3*8^1 + 2*8^0$

b. $574_{oct} = 5*64 + 7*8 \ + 4*1$
$= 5*8^2 + 7*8^1 + 4*8^0$

c. $1011_{oct} = 1*8^3 + 0*8^2 + 1*8^1 \ + 1*8^0$

d. $12.67_{oct} = 1*8^1 + 2*8^0 + 6*8^{-1} + 7*8^{-2}$ ∎

Here *the place values are powers of* 8; the digit 0 may play the role of a place-holder, as it does in (c) above. Using the expanded form, conversion from octal to decimal form is direct.

Example 2.3.2 **Conversion: Octal to Decimal**

a. $26_{oct} = 2*8^1 + 6*8^0$
$= \ 16 \ + \ 6$
$= \ \ \ \ 22_{dec}$

b. $551_{oct} = 5*8^2 + 5*8^1 + 1*8^0$
$= 5*64 + 5*8 \ + 1*1$
$= \ 320 \ + \ 40 \ + \ 1$
$= 361_{dec}$

c. $40007_{oct} = 4*8^4 \ \ \ + 0*8^3 + 0*8^2 + 0*8^1 + 7*8^0$
$= 4*4096 + \ \ 0 \ \ + \ \ 0 \ \ + \ \ 0 \ \ + 7*1$
$= 16391_{dec}$

d. $30.42_{oct} = 3*8^1 + 0*8^0 + 4*8^{-1} + 2*8^{-2}$
$= \ 24 \ + \ 0 \ + \ .5 \ + .03125$
$= 24.53125_{dec}$ ∎

To be consistent in our use of the word *integer*, an octal number that can be written *without* an octal point, such as in (a), (b), and (c) of the preceding example, is called an **octal integer**. The flowchart in the next example is a modification of the one in Figure 6, which is related to Example 2.1.3.

Example 2.3.3 **Algorithm: Conversion of Octal Integer to Decimal**

The flowchart constructed in Figure 16 will take in a given octal integer OCTINT and print out the equivalent decimal integer DECINT, where

VAL   = particular place value
MAX   = number of digits in OCTINT

Table 6

POS	VAL	OCT_{POS}	DECINT
1	1	5	5
2	8	0	5
3	64	1	69
4	512	7	3653

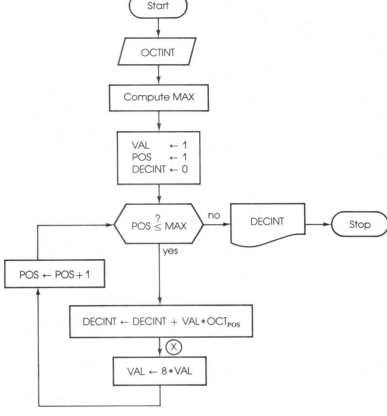

Figure 16

POS   = counter of digits in OCTINT

OCT_{POS} = digit of OCTINT in position POS from right

For example, if OCTINT = 7105, then MAX = 4 and $OCT_1 = 5$, $OCT_2 = 0$, $OCT_3 = 1$, and $OCT_4 = 7$. For this initial value of OCTINT, in Table 6 there is a trace of the flowchart at point $\textcircled{X}$ for variables POS, VAL, OCT_{POS}, and DECINT, from which we can conclude that

$$7105_{oct} = 3653_{dec}$$ ∎

At this point we could present several techniques for converting in the opposite direction, that is from decimal to octal; however, it is always possible, and often easier, to convert from decimal to binary to octal. For this reason we shall not linger on direct decimal-to-octal conversion; instead, we shall illustrate a technique that's not entirely new to us and then proceed to the methods of octal–binary conversion.

Example 2.3.4 **Conversion: Decimal Integer to Octal**

The procedure here is a variation of Method 2A given in Example 2.1.4. Here we successively divide by 8 and specify the remainder at each stage.

a. Convert 156_{dec} to octal.
First,

<table>
<tr><td></td><td>Remainder</td></tr>
<tr><td>8)156</td><td>4</td></tr>
<tr><td>8) 19</td><td>3</td></tr>
<tr><td>8) 2</td><td>2</td></tr>
<tr><td>0</td><td></td></tr>
</table>

Second, reading the remainders from *bottom to top*, we have

$$234_{oct}$$

b. Convert 9085_{dec} to octal.

<table>
<tr><td></td><td>Remainder</td></tr>
<tr><td>8)9085</td><td>5</td></tr>
<tr><td>8)1135</td><td>7</td></tr>
<tr><td>8) 141</td><td>5</td></tr>
<tr><td>8) 17</td><td>1</td></tr>
<tr><td>8 2</td><td>2</td></tr>
<tr><td>0</td><td></td></tr>
</table>

So the answer is

$$21575_{oct}$$

∎

Suppose we wish to convert an octal number, say 65, to binary. Using the expanded-form notation in both systems, we have

$$65_{oct} = 6*8^1 + 5*8^0$$
$$= 6*8 + 5*1$$
$$= 6 *2^3 + 5 *2^0$$
$$= \overbrace{(4+2)}*2^3 + \overbrace{(4+1)}*2^0$$
$$= (1*2^2 + 1*2^1 + 0*2^0)*2^3 + (1*2^2 + 0*2^1 + 1*2^0)*2^0$$
$$= 1*2^2*2^3 + 1*2^1*2^3 + 0*2^0*2^3 + 1*2^2*2^0 + 0*2^1*2^0 + 1*2^0*2^0$$
$$= 1*2^5 + 1*2^4 + 0*2^3 + 1*2^2 + 0*2^1 + 1*2^0$$
$$= 110101_{bin}$$

But observe that the answer here could have been obtained mechanically by converting individually the octal digits 6 and 5 to their binary equivalents using Table 7 and placing the conversions side by side. This procedure is valid in general, even for numbers with octal points. We shall use this abbreviated version in the next example.

Table 7

Octal digit	0	1	2	3	4	5	6	7
Binary equivalent	000	001	010	011	100	101	110	111

Example 2.3.5   **Conversion: Octal to Binary**

a.

b.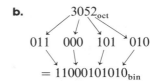

c.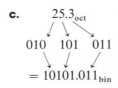

d.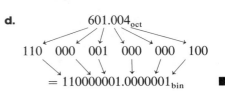

Now suppose we want to convert the binary number 11011101 into octal. Again, using the expanded forms

$$11011101_{bin} = \underbrace{1*2^7 + 1*2^6}_{} + \underbrace{0*2^5 + 1*2^4 + 1*2^3}_{} + \underbrace{1*2^2 + 0*2^1 + 1*2^0}_{}$$

$$= \underbrace{(1*2^1 + 1*2^0)*2^6}_{} + \underbrace{(0*2^2 + 1*2^1 + 1*2^0)*2^3}_{} + \underbrace{(1*2^2 + 0*2^1 + 1*2^0)}_{}$$

$$= \quad\quad 3*64 \quad\quad + \quad\quad 3*8 \quad\quad + \quad\quad 5$$

$$= \quad\quad 3*8^2 \quad\quad + \quad\quad 3*8^1 \quad\quad + \quad\quad 5*8^0$$

$$= 335_{oct}$$

This result can be attained mechanically by taking the given number and, starting at the rightmost digit and working leftward, converting each triple of consecutive binary digits into its octal equivalent, using Table 7 if necessary. Let's illustrate the method.

Example 2.3.6   **Conversion: Binary to Octal**

a.

b.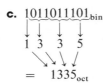

c.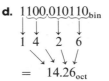

Note that in (d) above you must work to the left and right separately, *starting at the binary point*. At the right side, you may have to add zero(es) to complete a triple. ∎

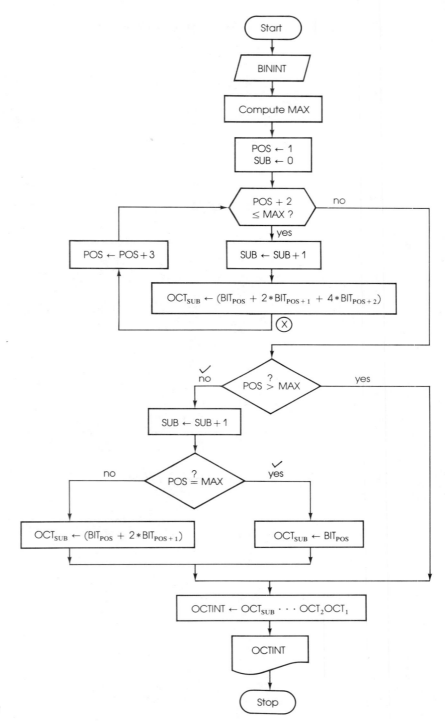

Figure 17

Example 2.3.7

Algorithm: Conversion of Binary Integer to Octal

In Figure 17 we have drawn a flowchart for the algorithm of the preceding example when applied to binary integers, where

$$\begin{aligned}
\text{BININT} &= \text{given binary integer} \\
\text{MAX} &= \text{number of digits in BININT} \\
\text{POS} &= \text{counter of digits in BININT} \\
\text{BIT}_{\text{POS}} &= \text{digit of BININT in position POS from right} \\
\text{OCTINT} &= \text{octal equivalent of BININT} \\
\text{SUB} &= \text{counter of digits in OCTINT} \\
\text{OCT}_{\text{SUB}} &= \text{digit of OCTINT in position SUB from right}
\end{aligned}$$

The bottom half of this flowchart is needed to handle the conversion of the leftmost digit(s) of BININT. A trace of the flowchart at point $\widehat{X}$ when the initial value of BININT is 1011101 (so MAX = 7) is given in Table 8. For this initial value, those bottom parts of the diagram marked with a check ($\checkmark$) are traversed, yielding $\text{OCT}_3 = 1$. Combining this with the information $\text{OCT}_1 = 5$ and $\text{OCT}_2 = 3$ from the trace, we have OCTINT = 135. ∎

Table 8

POS	POS + 3	BIT_{POS}	$\text{BIT}_{\text{POS}+1}$	$\text{BIT}_{\text{POS}+2}$	SUB	OCT_{SUB}
1	4	1	0	1	1	5
4	7	1	1	0	2	3

We did not spend much effort earlier in this section converting decimal numbers to octal, because the simple binary–octal conversion technique can be used once a decimal number is converted to binary.

Example 2.3.8

Conversion: Decimal to Octal via Binary

Convert 345.625_{dec} to octal.

First, convert the integer and noninteger parts to binary:

	Remainder		Integer Part
2)345	1	$2*.625 = 1.25$	1
2)172	0	$2*.25 = 0.5$	0
2) 86	0	$2*.5 = 1.0$	1
2) 43	1		
2) 21	1	$.625_{\text{dec}} = .101_{\text{bin}}$	
2) 10	0		
2) 5	1		
2) 2	0		
2) 1	1		
0			

$$345_{dec} = 101011001_{bin}$$

So,

$$345.625_{dec} = \underline{101011001}.\underline{101}_{bin}$$
$$= \quad 5 \quad 3 \quad 1 \;.\; 5_{oct}$$

We conclude this section with examples of octal addition and subtraction, since these operations can be useful to computer personnel who work with machine language. As with decimal and binary forms, to add two octal numbers we must vertically align place values and then add columns of digits, starting with the rightmost. Here, however, we must use the addition table for the octal digits given in Table 9. Whenever a column sum has two digits, the left digit is carried in the usual fashion.

Table 9 Octal Addition Table

+	0	1	2	3	4	5	6	7
0	0	1	2	3	4	5	6	7
1	1	2	3	4	5	6	7	10
2	2	3	4	5	6	7	10	11
3	3	4	5	6	7	10	11	12
4	4	5	6	7	10	11	12	13
5	5	6	7	10	11	12	13	14
6	6	7	10	11	12	13	14	15
7	7	10	11	12	13	14	15	16

Example 2.3.9 **Octal Addition**

```
      1 ←————— carrying row ————→ 111
    4036                         57771
  + 251                       + 4032
   ————                        ——————
   4307                         64023
```

```
   1   1 ←———— carrying row ————→ 11
   17.12                        2007.425
 + 6.264                      + 107.61
  ——————                       ————————
  25.404                        2117.235
```

Table 9 also helps in performing subtractions, where borrowing is done as usual when needed. Remember that $10 - 1 = 7$.

Example 2.3.10 **Octal Subtraction**

```
   1 13 ←——————— borrowing row ————————→ 3 7 7  7 15
   3 2̸ 3 5                              4̸ 0 0 0 5̸
 −   1 4 2                            −   1 6 2 7
   ————————                            ——————————
   3 0 7 3                              3 6 1 5 6
```

$$2\ 12 \longleftarrow \text{borrowing rows} \longrightarrow \begin{cases}1\ 1\ 12 \\ 1\ \ 2\ 14 \quad 6\ 10\end{cases}$$

$$\begin{array}{r} 29.3\,2\,1 \\ -16.0\,4 \\ \hline 13.2\,6\,1 \end{array} \qquad \begin{array}{r} 1\,2\,3\,4.5\,6\,7\,0 \\ -\ 7\,6\,5.4\,3\,2\,1 \\ \hline 2\,4\,7.1\,3\,4\,7 \ \blacksquare \end{array}$$

EXERCISES **2.3**

1. Extend Table 5 to include 34 through 55 miles driven.

2. Extend Table 5 to include 56 through 75 miles driven.

3. List the first 20 even octal integers.

4. List the first 20 odd octal integers.

5. Write the expanded forms of the following octal numbers:

   a. 50   **b.** 111   **c.** .24
   d. 4007.1   **e.** 26.7706

6. Write the expanded forms of the following octal numbers:

   a. 71   **b.** 206   **c.** .71
   d. 7710.5   **e.** 4.00007

7. How many miles (in decimal notation) can you drive before the six-digit octal odometer described in the text returns to 000000?

8. If we add two digits to the six-digit octal odometer described in the text, how many miles (in decimal notation) can you drive before the odometer returns to 00000000?

9. Convert each of the following octal numbers to decimal form:

   a. 15   **b.** 706   **c.** 2437
   d. .01   **e.** 77.7   **f.** 1111.111

10. Convert each of the following octal numbers to decimal form:

   a. 66   **b.** 351   **c.** 7654
   d. .007   **e.** 12.1   **f.** 700000.432

11. Use the method of Example 2.3.4 to convert each of the following decimal integers to octal form:

   a. 49   **b.** 243   **c.** 6002   **d.** 98765

12. Use the method of Example 2.3.4 to convert each of the following decimal integers to octal form:

   a. 98   **b.** 407   **c.** 1049   **d.** 87654

13. Convert each of the following octal numbers to binary form:

   a. 71   **b.** 432   **c.** 10774
   d. .25   **e.** 12.6   **f.** 507.0063

14. Convert each of the following octal numbers to binary form:

   a. 26   **b.** 675   **c.** 71003
   d. .7   **e.** 74.1   **f.** 1004.777

15. Convert each of the following binary numbers to octal form:

   a. 11010   **b.** 10110101   **c.** 1111011110
   d. .011   **e.** 1110.1   **f.** 10.001011

16. Convert each of the following binary numbers to octal form:

   a. 10011   **b.** 1101010   **c.** 100010001
   d. .1011   **e.** 101.01   **f.** 1.11001

17. Convert each of the following decimal numbers to octal form:

   a. 82.5   **b.** 639.875   **c.** 9402.125   **d.** 111.1

18. Convert each of the following decimal numbers to octal form:

   a. 59.25   **b.** 801.5
   c. 40007.4375   **d.** 2222.6

19. Complete the table below so that the numbers in each column are equivalent.

Binary	101.1	____	____
Octal	____	101.1	____
Decimal	____	____	4716

20. Complete the table below so that the numbers in each column are equivalent.

Binary	1101	____	____
Octal	____	47.02	____
Decimal	____	____	11111

21. Use the octal counting list $0, 1, \ldots, 16$ to verify the rows beginning with 1, 3, 5, and 7 in the octal addition table (Table 9).

22. Use the octal counting list $0, 1, \ldots, 16$ to verify the rows beginning with 0, 2, 4, and 6 in the octal addition table (Table 9).

Compute the octal sums.

23. 351
 $+134$

24. 726
 $+61$

25. 4472
 $+406$

26. 60345
 $+16235$

27. 12.7
 $+3.42$

28. 100.015
 $+72.77$

29. 5246.123
 $+6666.666$

30. 4075.67
 $+105.214$

Compute the octal differences.

31. 425
 -64

32. 306
 -21

33. 5320
 -411

34. 2043
 -252

35. .423
 $-.333$

36. .726
 $-.15$

37. 617.001
 -47.52

38. 7654.321
 -123.4567

39. Construct an octal multiplication table for the octal digits 0, 1, 2, 3, 4, 5, 6, and 7.

40. Use the table constructed in the preceding exercise and the usual multiplication algorithm (illustrated for binary numbers in Example 2.2.5) to compute each of the following octal products:

a. 75
 $*5$

b. 63
 $*42$

c. 517
 $*603$

41. The usual "long division" algorithm is illustrated for binary numbers in Example 2.2.6. Use it to perform the following octal divisions:

a. $7 \overline{)425}$ **b.** $23 \overline{)604}$ **c.** $502 \overline{)47251}$

42. For the value OCTINT = 20374, construct a trace at point $\textcircled{X}$ of the flowchart in Figure 16 for the variables VAL, POS, DECINT, OCT$_{POS}$. What is the decimal form of 20374_{oct}?

43. A mathematical way of stating the *Division Algorithm* used in Example 2.2.6 and exercise 41 above is as follows: For any given integers a and b (where $b > 0$), there exist integers q and r (called the quotient and remainder, respectively) with the property that $a = b * q + r$ where $0 \le r < b$. Regarding the mechanical procedure illustrated in Example 2.3.4:

a. Justify it by applying the Division Algorithm stated above. (*Hint:* Take $b = 8$.)
b. Construct a flowchart for this procedure that takes in a decimal integer DECINT and produces its octal equivalent OCTINT.

44. For the given value BININT = 11100101110:

a. Do a trace at point $\textcircled{X}$ of the flowchart in Figure 17 for the variables listed in Table 8.
b. After performing the trace in (a), complete the bottom half of the flowchart in Figure 17.
c. Use the results of (a) and (b) to write the octal form of the given BININT.

45. Do exercise 44 for the initial value
BININT = 10011001

46. Construct a flowchart that will take in any binary number *beginning with a binary point* (for example, .01101) and produce the octal equivalent. (*Hint:* Use the flowchart of Figure 17 as a model and make the necessary changes.)

47. Construct a flowchart that will take in any octal *integer* and produce its binary equivalent.

48. Do the same as in exercise 47 but for any octal number *beginning with an octal point* (for example, .7025).

49. a. When will an octal number with infinitely many repeating octal digits (for example, $72.6\overline{32}$) convert to a binary number with infinitely many repeating binary digits?
b. When will a binary number with infinitely many repeating binary digits (for example, $1.0\overline{0001}$) convert to an octal number with infinitely many repeating octal digits?

2.4

Hexadecimal Numbers

The prefix *hex-* means "six," and the **hexadecimal number system** is constructed from 16 symbols called digits. Usually these digits are represented by the ten digits of the decimal system and the first six letters of the alpha-

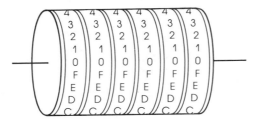

Figure 18

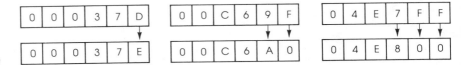

Figure 19

bet; so

0, 1, 2, 3, 4, 5, 6, 7, 8, 9, A, B, C, D, E, and F

are the ***hexadecimal digits***.

To demonstrate how we count in this system, we again employ the notion of an auto odometer. This time we are interested in building a hexadecimal odometer. First, each of the odometer's six disks must contain the sixteen hexadecimal digits, as indicated in Figure 18. Second, the disks are linked so that whenever any one disk turns from F to 0, then the disk to its left moves $\frac{1}{16}$ of a turn to the next digit. Last, the rightmost disk moves to the next digit after the car has gone one full mile. Several examples of a car that travels one mile, causing the rightmost disk to move $\frac{1}{16}$ of a turn to the next digit, are given in Figure 19. In Table 10 we use the odometer to count miles by starting with an all-zero reading and repeatedly adding one. For simplicity, leftmost zeroes are not included in the odometer reading. You should examine Table 10 long enough to memorize the decimal equivalents of the 16 hexadecimal digits.

Table 10

Odometer reading (Hexadecimal)	Miles driven (Decimal)	Odometer reading (Hexadecimal)	Miles driven (Decimal)	Odometer reading (Hexadecimal)	Miles driven (Decimal)
0	0	C	12	17	23
1	1	D	13	18	24
2	2	E	14	19	25
3	3	F	15	1A	26
4	4	10	16	1B	27
5	5	11	17	1C	28
6	6	12	18	1D	29
7	7	13	19	1E	30
8	8	14	20	1F	31
9	9	15	21	20	32
A	10	16	22	21	33
B	11				

Just like the decimal, binary, and octal systems previously discussed, the hexadecimal numbers constitute a *place-value system.*

Example 2.4.1   **Hexadecimal Notation: Expanded Form**

a. $32_{hex} = 3*16 + 2*1$
$= 3*16^1 + 2*16^0$

b. $A7_{hex} = A*16 + 7*1$
$= A*16^1 + 7*16^0$

c. $F20B_{hex} = F*16^3 + 2*16^2 + 0*16^1 + B*16^0$

d. $9C.E4D_{hex} = 9*16^1 + C*16^0 + E*16^{-1} + 4*16^{-2} + D*16^{-3}$ ■

In this system, the *place values are powers of 16*, and 0 plays the usual role of placeholder. Conversion from hexadecimal (or "hex," for short) to decimal form is direct if we use the expanded form and the decimal equivalents of the hex digits.

Example 2.4.2   **Conversion: Hex to Decimal**

a. $26_{hex} = 2*16^1 + 6*16^0$
$= 32 + 6$
$= 38_{dec}$

b. $B2_{hex} = B*16^1 + 2*16^0$
$= 11*16 + 2*1$
$= 176 + 2$
$= 178_{dec}$

c. $AEB_{hex} = A*16^2 + E*16^1 + B*16^0$
$= 10*256 + 14*16 + 11*1$
$= 2560 + 224 + 11$
$= 2795_{dec}$

d. $.10C_{hex} = 1*16^{-1} + 0*16^{-2} + C*16^{-3}$
$= 1*\frac{1}{16} + 0 + 12*(\frac{1}{16})^3$
$= .0625 + 0 + .00292969$
$= .06542969_{dec}$

e. $4F.D5_{hex} = 4*16^1 + F*16^0 + D*16^{-1} + 5*16^{-2}$
$= 4*16 + 15*1 + 13*\frac{1}{16} + 5*(\frac{1}{16})^2$
$= 64 + 15 + .8125 + .01953$
$= 79.83203_{dec}$ ■

A hex number that can be expressed without the need of a hexadecimal point, as in (a), (b), and (c) above, is called a ***hexadecimal integer***. The next example is a modification of Example 2.3.3.

Example 2.4.3   **Algorithm: Conversion of Hex Integer to Decimal**

Figure 20 gives a flowchart that receives a hex integer HEXINT and produces the equivalent decimal integer DECINT, where

VAL　　　 = particular place value

MAX　　　 = number of digits in HEXINT

POS　　　 = counter of digits in HEXINT

HEX_{POS} 　 = digit of HEXINT in position POS from right

$DHEX_{POS}$ = decimal equivalent of HEX_{POS}

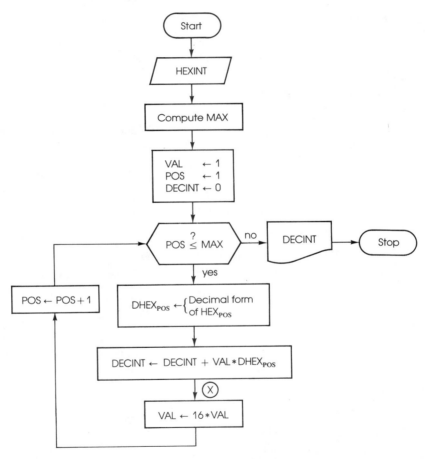

Figure 20

For instance, if HEXINT = B9E, then MAX = 3 and HEX_1 = E, HEX_2 = 9, and HEX_3 = B. A trace of this flowchart at point $\bigotimes$ for initial value HEXINT = B9E is given in Table 11. From this trace we determine that

$$B9E_{hex} = 2974_{dec}$$ ■

Right now we could present several methods for converting decimal numbers to hex form. However, as with decimal–octal conversion, it is

Table 11

POS	VAL	HEX$_{POS}$	DHEX$_{POS}$	DECINT
1	1	E	14	14
2	16	9	9	158
3	256	B	11	2974

always possible, and often easier, to convert from decimal to binary to hex. So we shall pause here only to present one procedure for converting decimal *integers* to hex, and then move on to hex–binary conversions.

Example 2.4.4   **Conversion: Decimal Integer to Hex**

As with previous conversions, we repeatedly divide by 16 and specify the remainder at each stage. The remainders read last to first give the desired hex equivalent.

a. Convert 156_{dec} to hex.
First,

$$
\begin{array}{r|c}
 & \text{Remainder} \\
\hline
16\,)156 & C \\
16\,)\ \ \ 9 & 9 \\
\hline
0 &
\end{array}
$$

So, reading from *bottom to top*, the answer is

$$\boxed{9C_{hex}}$$

b. Convert 9085_{dec} to hex.

$$
\begin{array}{r|c}
 & \text{Remainder} \\
\hline
16\,)9085 & D \\
16\,)\ 567 & 7 \\
16\,)\ \ 35 & 3 \\
16\,)\ \ \ 2 & 2 \\
\hline
0 &
\end{array}
$$

Again, reading from *bottom to top*, we have

$$\boxed{237D_{hex}}$$ ■

Turning to hex–binary conversion, suppose we want to find the binary equivalent of $B5_{hex}$. Employing expanded-form notation gives:

$$
\begin{aligned}
B5_{hex} &= B*16^1 + 5*16^0 \\
&= 11*16 + 5*1 \\
&= \quad (8 + 2 + 1)*16 \qquad\qquad + \qquad\qquad (4+1)*1 \\
&= \quad (2^3 + 2^1 + 2^0) \qquad *2^4 + \qquad (2^2 + 2^0) \qquad\qquad *2^0 \\
&= (1*2^3 + 0*2^2 + 1*2^1 + 1*2^0)*2^4 + (0*2^3 + 1*2^2 + 0*2^1 + 1*2^0)*2^0 \\
&= (1*2^3*2^4 + 0*2^2*2^4 + 1*2^1*2^4 + 1*2^0*2^4) + (0*2^3*2^0 + 1*2^2*2^0 + 0*2^1*2^0 + 1*2^0*2^0) \\
&= 1*2^7 + 0*2^6 + 1*2^5 + 1*2^4 + 0*2^3 + 1*2^2 + 0*2^1 + 1*2^0 \\
&= 10110101_{bin}
\end{aligned}
$$

However, note that one can bypass the intermediate work and write the correct answer by converting individually the hex digits B and 5 to their four-digit binary equivalents using Table 12 and placing these conversions side by side. This short-cut mechanical process can be shown to be valid in general; it is illustrated in the next example.

Table 12

Hexadecimal digit	0	1	2	3	4	5	6	7	8	9	A	B	C	D	E	F
Binary equivalent	0000	0001	0010	0011	0100	0101	0110	0111	1000	1001	1010	1011	1100	1101	1110	1111

Example 2.4.5 **Conversion: Hex to Binary**

a. $2D_{hex}$

  0010 1101

  $= 101101_{bin}$

b. $F1E6_{hex}$

 1111 0001 1110 0110

 $= 1111000111100110_{bin}$

c. $7.A_{hex}$

  0111 1010

  $= 111.101_{bin}$

d. $C8.B5_{hex}$

 1100 1000 1011 0101

 $= 11001000.10110101_{bin}$ ∎

Now let's consider the opposite conversion. If 111010010 is a given binary number, we'll use expanded-form notation in converting it to hex form:

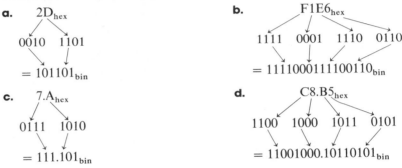

$$111010010_{bin} = 1*2^8 + \underbrace{1*2^7 + 1*2^6 + 0*2^5 + 1*2^4} \quad + \underbrace{0*2^3 + 0*2^2 + 1*2^1 + 0*2^0}$$

$$= 1*2^8 + \underbrace{(1*2^3 + 1*2^2 + 0*2^1 + 1*2^0)*2^4} + \underbrace{0*2^3 + 0*2^2 + 1*2^1 + 0*2^0}$$

$$= 1*256 + \qquad\qquad 13*16 + \qquad\qquad 2$$
$$= 1*16^2 + \qquad\qquad D*16^1 + \qquad\qquad 2*16^0$$
$$= 1D2_{hex}$$

This answer can be obtained mechanically by taking the given number and, starting at the rightmost digit and working to the left, converting each four-bit string into its hex equivalent by using Table 12. If the number involves a binary point, one must *start at the binary point* and work separately to the left and right, adding zero(es) at the right end if necessary.

Example 2.4.6 **Conversion: Binary to Hex**

a. 11000110_{bin}

  C 6

 $= C6_{hex}$

b. 101011_{bin}

  2 B

 $= 2B_{hex}$

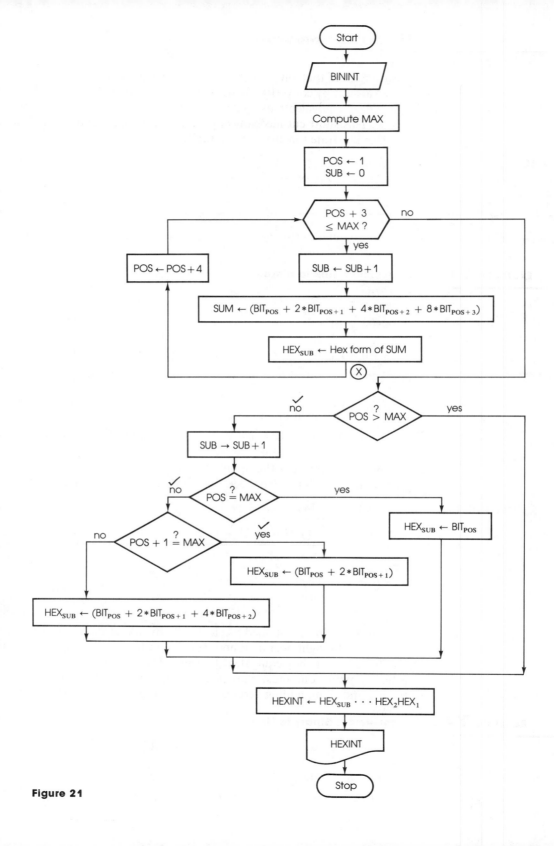

Figure 21

c. $\underbrace{11001010111}_{}{}_{bin}$

657

$= 657_{hex}$

d. $\underbrace{11101.10001100}_{}{}_{bin}$

$1D8C$

$= 1D.8C_{hex}$ ∎

Example 2.4.7

Algorithm: Conversion of Binary Integer to Hex

The flowchart drawn in Figure 21 is a modification of the one in Figure 17 (Section 2.3) of this chapter. It accepts a binary integer BININT and produces its hexadecimal equivalent HEXINT, where

MAX $=$ number of digits in BININT

POS $=$ counter of digits in BININT

BIT$_{POS}$ $= $ digit of BININT in position POS from right

SUB $=$ counter of digits in HEXINT

HEX$_{SUB}$ $=$ digit of HEXINT in position POS from right

As before, the bottom half of this flowchart is needed to handle the conversion of the leftmost digit(s) of BININT. For initial value BININT = 1011100111 (MAX = 10), a trace of this flowchart at point $\widehat{X}$ is given in Table 13. For this initial value, those bottom parts marked with a check apply, giving HEX$_3$ = 2. Combined with the information HEX$_1$ = 7 and HEX$_2$ = E from the trace, we have

HEXINT = 2E7 ∎

Table 13

POS	POS + 4	BIT$_{POS}$	BIT$_{POS+1}$	BIT$_{POS+2}$	BIT$_{POS+3}$	SUM	SUB	HEX$_{SUB}$
1	5	1	1	1	0	7	1	7
5	9	0	1	1	1	14	2	E

With the short binary–hex conversion procedure in hand, we now can easily convert any decimal number to hex form by first changing to binary and then to hex.

Example 2.4.8

Conversion: Decimal to Hex via Binary

Convert 291.75_{dec} to hex.

First, change separately to binary the integer and noninteger parts:

Remainder$$Integer Part

$2\,)\,\overline{291}12*.75 = 1.51\,|$

$2\,)\,\overline{145}12*.5 = 1.01\,\downarrow$

$2\,)\,\overline{72}0$

$2\,)\,\overline{36}0.75_{dec} = .11_{bin}$

$2\,)\,\overline{18}0$

$2\,)\,\overline{9}1$

$2\,)\,\overline{4}0$

$2\,)\,\overline{2}0\,\uparrow$

$2\,)\,\overline{1}1\,|$

0

$291_{dec} = 100100011_{bin}$

So,

$$291.75_{dec} = \underbrace{1001}\underbrace{0001}\underbrace{1}.\underbrace{1100}_{bin}$$
$$\begin{array}{ccccc} \downarrow & \downarrow & \downarrow & & \downarrow \\ 1 & 2 & 3 & . & C_{hex} \end{array}$$

∎

As in the preceding section on octal numbers, we shall conclude this section with examples of addition and subtraction. The steps in the process of adding hex numbers are the same as before: Align place values, add columns from the right to the left using the hexadecimal addition table (see Table 14) if necessary, and carry the left digit whenever a column sum has two digits.

Table 14 Hex Addition Table

+	0	1	2	3	4	5	6	7	8	9	A	B	C	D	E	F
0	0	1	2	3	4	5	6	7	8	9	A	B	C	D	E	F
1	1	2	3	4	5	6	7	8	9	A	B	C	D	E	F	10
2	2	3	4	5	6	7	8	9	A	B	C	D	E	F	10	11
3	3	4	5	6	7	8	9	A	B	C	D	E	F	10	11	12
4	4	5	6	7	8	9	A	B	C	D	E	F	10	11	12	13
5	5	6	7	8	9	A	B	C	D	E	F	10	11	12	13	14
6	6	7	8	9	A	B	C	D	E	F	10	11	12	13	14	15
7	7	8	9	A	B	C	D	E	F	10	11	12	13	14	15	16
8	8	9	A	B	C	D	E	F	10	11	12	13	14	15	16	17
9	9	A	B	C	D	E	F	10	11	12	13	14	15	16	17	18
A	A	B	C	D	E	F	10	11	12	13	14	15	16	17	18	19
B	B	C	D	E	F	10	11	12	13	14	15	16	17	18	19	1A
C	C	D	E	F	10	11	12	13	14	15	16	17	18	19	1A	1B
D	D	E	F	10	11	12	13	14	15	16	17	18	19	1A	1B	1C
E	E	F	10	11	12	13	14	15	16	17	18	19	1A	1B	1C	1D
F	F	10	11	12	13	14	15	16	17	18	19	1A	1B	1C	1D	1E

Example 2.4.9   **Hex Addition**

$$\begin{array}{r} {}^{1}\leftarrow \text{carrying row} \rightarrow {}^{1} \\ \begin{array}{r} 497 \\ +86 \\ \hline 51D \end{array} \qquad \begin{array}{r} \text{A.BC} \\ +7.8 \\ \hline 12.3C \end{array} \end{array}$$

$$\begin{array}{r} {}^{111}\leftarrow \text{carrying row} \rightarrow {}^{11} \\ \begin{array}{r} C2764 \\ +A89E \\ \hline CD002 \end{array} \qquad \begin{array}{r} 98.F \\ +E.15 \\ \hline A7.05 \end{array} \end{array}$$

∎

For subtraction, use the addition table and remember that $10 - 1 = F$.

Example 2.4.10 **Hex Subtraction**

$$
\begin{array}{r}
\text{borrowing row} \longrightarrow \quad {}^{C\ 15} \\
\end{array}
$$

$$
\begin{array}{r}
B7 \\
-A4 \\
\hline
13
\end{array}
\qquad
\begin{array}{r}
C\cancel{D}\,\cancel{5} \\
-\ 7\,A \\
\hline
C\,5\,B
\end{array}
$$

$$
\begin{array}{r}
{}^{5\ 10\ 8\ \ 11}\longleftarrow \text{borrowing row} \longrightarrow \quad {}^{8\ F\ F\ F\ \ 13}
\end{array}
$$

$$
\begin{array}{r}
6\cancel{0}\,9.\cancel{1}\,2 \\
-\ E\,0.B \\
\hline
5\,2\,8.6\,2
\end{array}
\qquad
\begin{array}{r}
\cancel{9}\,\cancel{0}\,\cancel{0}\,\cancel{0}.3 \\
-\ 4\,C.8 \\
\hline
8\,F\,B\,3.B
\end{array}
\qquad \blacksquare
$$

EXERCISES 2.4

1. Extend Table 10 to include 34 through 55 miles driven.

2. Extend Table 10 to include 56 through 75 miles driven.

3. List the first 20 even hex integers.

4. List the first 20 odd hex integers.

5. Write the expanded forms of the following hex numbers:

   a. 50   **b.** 4D1   **c.** .2B
   d. D.07   **e.** F6.90E

6. Write the expanded forms of the following hex numbers:

   a. 111   **b.** AB   **c.** .0D
   d. 9.C   **e.** E0.00F

7. How many miles (in decimal form) can you drive before the six-digit hex odometer described in the text returns to 000000?

8. If we add two digits to the six-digit hex odometer described in the text, how many miles (in decimal form) can you drive before the odometer returns to 00000000?

9. Convert each of the following hex numbers to decimal form:

   a. 17   **b.** 8E0   **c.** ABCD
   d. .01   **e.** F.F   **f.** 111.0C1

10. Convert each of the following hex numbers to decimal form:

   a. 99   **b.** EFG   **c.** 789A
   d. .007   **e.** 1.AA   **f.** 600.1B

11. Use the method of Example 2.4.4 to convert each of the following decimal integers to hex form:

   a. 32   **b.** 99   **c.** 253   **d.** 1000

12. Use the method of Example 2.4.4 to convert each of the following decimal integers to hex form:

   a. 45   **b.** 100   **c.** 441   **d.** 999

13. Convert each of the following hex numbers to binary form:

   a. 9B   **b.** 432   **c.** E0A5
   d. .B   **e.** D.19   **f.** 80F.0C

14. Convert each of the following hex numbers to binary form:

   a. 66   **b.** DEF   **c.** 7A30
   d. .0F   **e.** 1.B   **f.** CC.F09

15. Convert each of the following binary numbers to hex notation:

   a. 11010   **b.** 10110101   **c.** 111011110
   d. .0111   **e.** 101.101   **f.** 10.001011

16. Convert each of the following binary numbers to hex notation:

   a. 10010   **b.** 1101010   **c.** 1000100011
   d. .011   **e.** 1101.01   **f.** 1.11001

17. Convert each of the following decimal numbers to hex form:

   a. 82.5   **b.** 639.875
   c. 9402.125   **d.** 111.1

18. Convert each of the following decimal numbers to hex form:

   a. 59.25   **b.** 801.5
   c. 40007.4375   **d.** 2222.6

19. Complete the table below so that the numbers in each column are equivalent.

Binary	11.01	——	——
Decimal	——	1907	——
Hex	——	——	2D.1

20. Complete the table below so that the numbers in each column are equivalent.

Binary	1000.1	——	——
Decimal	——	1111	——
Hex	——	——	F0E

21. Use the hex counting list $0, 1, \ldots, 1E$ to verify the rows beginning with $5, 9, C,$ and F in the hex addition table (Table 14).

22. Use the hex counting list $0, 1, \ldots, 1E$ to verify the rows beginning with 4, 8, and D in the hex addition table (Table 14).

Compute the hex sums.

23. 9730
+ 78

24. 14A
+ 137

25. C9E3
+ 395B

26. D.2A
+ 7.E

27. 12.5
+ F.09

28. FFF.FF
+ 123.45

Compute the hex differences.

29. E7
− 61

30. C5
− A7

31. D000A
− 10B

32. 84.0E
− F.9

33. ABC.D
− EF.E1

34. F000.25
− C62.26

35. Construct a hex multiplication table for the hex digits $0, 1, \ldots, F$.

36. Use the table constructed in the preceding exercise and the usual multiplication algorithm (illustrated for binary numbers in Example 2.2.5) to compute each of the following hex products:

a. 72
∗ 5

b. 6B
∗ 25

c. F19
∗ A0A

37. The usual "long division" algorithm is illustrated for binary numbers in Example 2.2.6. Use it to perform the following hex divisions:

a. $7\overline{)425}$ **b.** $A\overline{)3F5}$ **c.** $4E\overline{)7C2D}$

38. For the value HEXINT = 10C8B, construct a trace at point $\overline{X}$ of the flowchart in Figure 20 for the variables POS, VAL, HEX_{POS}, $DHEX_{POS}$, and DECINT. What is the decimal form of $10C8B_{hex}$?

39. The Division Algorithm was stated in exercise 43 of Exercises 2.3. In regard to the procedure given in Example 2.4.4:

a. Justify it by applying the Division Algorithm. (*Hint:* Take $b = 16$.)
b. Construct a flowchart for this procedure that takes in a decimal integer DECINT and produces its hex equivalent HEXINT.

40. For the given value BININT = 1101011:

a. Do a trace at point $\overline{X}$ of the flowchart in Figure 21 for the variables listed in Table 13.
b. After completing the trace in (a), complete the bottom half of the flowchart in Figure 21.
c. Use the results of (a) and (b) to write the hex form of the given BININT.

41. Do exercise 40 for the initial value BININT = 11111000101.

42. Construct a flowchart that will take in any binary number *beginning with a binary point* (for example, .01101) and produce the hex equivalent. (*Hint:* Use the flowchart of Figure 21 as a model and make the necessary changes.)

43. Construct a flowchart that will take in any hex *integer* and produce its binary equivalent.

44. Do the same as in exercise 43 but for any hex number *beginning with a hex point* (for example, .B4E).

45. a. When will a hex number with infinitely many repeating hex digits (for example, $B7.\overline{A12}$) convert to a binary number with infinitely many repeating binary digits?
b. When will a binary number with infinitely many repeating binary digits (for example, $1.\overline{0001}$) convert to a hex number with infinitely many repeating hex digits?

2.5

Binary Codes

Computers can be built to perform arithmetic operations employing the binary notation of the preceding sections, called *straight (or pure) binary coding*. But the typical computer is not limited to this type of coding. For instance, a computer can execute the following addition of decimal numbers:

$$\begin{array}{r} 52 \\ +27 \\ \hline \end{array} \xrightarrow[\text{Convert to binary}]{} \begin{array}{r} 110100 \\ +11011 \\ \hline 1001111 \end{array} \xrightarrow[]{\text{Convert to decimal}} 79$$

Note that in this method there are not separate additions of the ones digits 7 and 2, and the tens digits 2 and 5. The identity of these digits is lost in the conversion to pure binary coding. However, in many applications, such as the processing of financial data, it is desirable to maintain the identity of each decimal digit. For this reason, most computers can encode each decimal digit separately into a binary notation called *binary-coded decimal (BCD) notation*, of which there are several types.

Example 2.5.1

Four-Bit BCD Notation

One type of four-bit BCD coding is given in Table 15. It encodes only the ten decimal digits and is directly related to straight binary code insofar as each decimal digit is coded by its four-bit binary equivalent, adding zeroes to the left if necessary.

Decimal number	Straight binary equivalent[†]	Four-bit BCD form[†]
5	101	0101
26	1101	0010 0110
193	1100 0001	0001 1001 0011
1408	101 1000 0000	0001 0100 0000 1000

■

Example 2.5.2

Addition Using Four-Bit BCD

a. Remember that BCD retains the identity of each decimal digit, so when adding two such coded numbers, the four-bit strings representing the same place value must be vertically aligned and added:

$$\begin{array}{r} 52 \longrightarrow \\ +27 \longrightarrow \\ \end{array} \begin{array}{r} 0101\ 0010 \\ +0010\ 0111 \\ \hline 0111\ 1001 \end{array} \longrightarrow \boxed{79}$$

b. When the sum of two four-bit strings exceeds the decimal digit 9, then 10 (= 1010 in binary) must be subtracted and 1 (= 0001 in binary) carried to the four-bit string to the left.

[†] Whenever it enhances readability, we shall leave spaces within long numeric strings.

Table 15

Character	Four-bit BCD	Six-bit BCD	Octal Equivalent of 6-bit BCD	EBCDIC	Hexadecimal Equivalent of EBCDIC	ASCII-8	Hexadecimal Equivalent of ASCII-8
0	0000	001010	12	11110000	F0	10110000	B0
1	0001	000001	01	11110001	F1	10110001	B1
2	0010	000010	02	11110010	F2	10110010	B2
3	0011	000011	03	11110011	F3	10110011	B3
4	0100	000100	04	11110100	F4	10110100	B4
5	0101	000101	05	11110101	F5	10110101	B5
6	0110	000110	06	11110110	F6	10110110	B6
7	0111	000111	07	11110111	F7	10110111	B7
8	1000	001000	10	11111000	F8	10111000	B8
9	1001	001001	11	11111001	F9	10111001	B9
A		110001	61	11000001	C1	11000001	C1
B		110010	62	11000010	C2	11000010	C2
C		110011	63	11000011	C3	11000011	C3
D		110100	64	11000100	C4	11000100	C4
E		110101	65	11000101	C5	11000101	C5
F		110110	66	11000110	C6	11000110	C6
G		110111	67	11000111	C7	11000111	C7
H		111000	70	11001000	C8	11001000	C8
I		111001	71	11001001	C9	11001001	C9
J		100001	41	11010001	D1	11001010	CA
K		100010	42	11010010	D2	11001011	CB
L		100011	43	11010011	D3	11001100	CC
M		100100	44	11010100	D4	11001101	CD
N		100101	45	11010101	D5	11001110	CE
O		100110	46	11010110	D6	11001111	CF
P		100111	47	11010111	D7	11010000	D0
Q		101000	50	11011000	D8	11010001	D1
R		101001	51	11011001	D9	11010010	D2
S		010010	22	11100010	E2	11010011	D3
T		010011	23	11100011	E3	11010100	D4
U		010100	24	11100100	E4	11010101	D5
V		010101	25	11100101	E5	11010110	D6
W		010110	26	11100110	E6	11010111	D7
X		010111	27	11100111	E7	11011000	D8
Y		011000	30	11101000	E8	11011001	D9
Z		011001	31	11101001	E9	11011010	DA
+		010000	20	01001110	4E	10101011	AB
−		100000	40	01101101	6D	11011111	DF
,		011011	33	01101011	6B	10101100	AC
.		111011	73	01001011	4B	10101110	AE
(Space)		000000	00	01000000	40	10100000	A0

$$
\begin{array}{r}
57 \longrightarrow \quad 0101 \quad 0111 \\
+36 \longrightarrow \; +0011 \quad 0110 \\
\hline
1000 \quad 1101 \\
+0001 - 1010 \\
\hline
1001 \quad 0011 \longrightarrow \boxed{93}
\end{array}
$$

—— Four bits on right exceed ten so subtract ten and add one to the left four-bit string

$$
\begin{array}{r}
69 \longrightarrow \quad 0110 \quad 1001 \\
+18 \longrightarrow \; +0001 \quad 1000 \\
\hline
0111 \quad 10001 \\
+0001 - 1010 \\
\hline
1000 \quad 0111 \longrightarrow \boxed{87}
\end{array}
$$

—— Four bits on right exceed ten so subtract ten and add one to the left four-bit string

■

Ordinarily, computers are designed to process data that is not entirely numeric, for instance, the names and addresses of customers. In this case, the four-bit BCD code just described is insufficient, since it only allows for the coding of at most $2^4 = 16$ distinct symbols—thus the need for codes with more than four bits.

Table 15 also contains an example of a six-bit code; the four bits on the right are called **numeric** (they are used to encode the decimal digits in straight binary form), and the two bits on the left are called **zone bits**. This allows us to encode a maximum of $2^6 = 64$ distinct characters; each such character is represented by a combination of zone and numeric bits. Recall that each binary triple has an octal-digit equivalent; so each six-bit string has a two-digit octal representation. These octal equivalents are also given in Table 15.

Example **2.5.3**   Six-Bit BCD Notation

Datum	6-bit BCD form	Octal equivalent
35	000011 000101	03 05
HELLO	111000 110101 100011 100011 100110	70 65 43 43 46
J. SMITH	100001 111011 000000 010010 100100 111001 010011 111000	41 73 00 22 44 71 23 70
9 MAIN ST.	001001 000000 100100 110001 111001 100101 000000 010010 010011 111011	11 00 44 51 71 45 00 22 23 73

■

For several reasons, including the need to encode more than 64 characters, the computer industry has developed codes with more than six bits. The seven-bit ASCII-7 code (the acronym ASCII stands for the American Standard Code for Information Interchange) was developed to improve compatibility among various intersystem communications devices. The two most widely used eight-bit BCD codes are ASCII-8 and EBCDIC (for Extended Binary Coded Decimal Interchange Code). Portions of these two codes with their hexadecimal equivalents are also listed in Table 15. Each can encode up to $2^8 = 256$ characters. The 4 bits on the right are called *numeric bits*; those on the left are *zone bits*.

Example 2.5.4 **Eight-Bit BCD Notation**

In each case we list a piece of data, the ASCII-8 equivalent, the hex equivalent of ASCII-8, the EBCDIC equivalent, and the hex equivalent of EBCDIC, in that order.

a.
```
   3          5
1011 0011 1011 0101
  B    3    B    5
1111 0011 1111 0101
  F    3    F    5
```

b.
```
   H         E          L          L          O
1100 1000 1100 0101 1100 1100 1100 1100 1100 1111
  C    8    C    5    C    C    C    C    C    F
1100 1000 1100 0101 1101 0011 1101 0011 1101 0110
  C    8    C    5    D    3    D    3    D    6
```

c.
```
   J                    (Space)
1100 1010 1010 1110 1010 0000
  C    A    A    E    A    0
1101 0001 0100 1011 0100 0000
  D    1    4    B    4    0
```
```
             S         M          I          T          H
          1101 0011 1100 1101 1100 1001 1101 0100 1100 1000
            D    3    C    D    C    9    D    4    C    8
          1110 0010 1101 0100 1100 1001 1110 0011 1100 1000
            E    2    D    4    C    9    E    3    C    8     ■
```

Codes that use the four rightmost bits to distinguish numeric characters have an important advantage: When processing numeric data exclusively, two decimal digits can be packed into one eight-bit space. For example, the EBCDIC code for the decimal number 37 is

```
1111 0011 1111 0111
Zone  ↑   Zone  ↑

     Numeric   Numeric
```

By extracting only the numeric portions of this string and placing them side by side, as follows:

```
0011 0111
```

we have "packed" the two-digit number into one eight-bit space. This is called *packed decimal format*.

Example 2.5.5   **Addition Using EBCDIC: Packed Decimal Format**

a. 35 $\longrightarrow$ 1111 0011 1111 0101
   +14 $\longrightarrow$ +1111 0001 1111 0100

$\xrightarrow{\text{packed}}$ 0011 0101
          +0001 0100
          0100 1001 $\xrightarrow{\text{unpacked}}$ 1111 0100 1111 1001 → $\boxed{49}$

b. 59 $\longrightarrow$ 1111 0101 1111 1001
   +23 $\longrightarrow$ +1111 0010 1111 0011

$\xrightarrow{\text{packed}}$ 0101   1001
         +0010   0011
         0111   1100
     +0001 − 1010
      1000   0010 $\xrightarrow{\text{unpacked}}$ 1111 1000 1111 0010 → $\boxed{82}$ ∎

Having read through Examples 2.5.3 and 2.5.4 and having compared the long binary strings with their octal or hexadecimal equivalents, the reader may already have surmised one reason why the octal and hexadecimal systems are useful in computer science. The computer performs all of its work in binary notation. If a human must deal with the computer at that level (as when debugging a machine language program), it is much easier for the person to process the equivalent, yet shorter, octal or hex strings.

For example, suppose the computer has output the following list of EBCDIC binary strings

    1100 1000 1100 0010
    1111 0110 1110 0101
    1110 0111 1100 0101
    1111 1001 1100 0001
    1101 0100 1111 1000
    1101 0001 1101 0001
    1111 0010 1110 1001
    1100 1001 1100 0110
    1111 0101 1111 0110
    1110 0011 1110 0100

and you want to compare it with the following hand-held listing to determine if there is any discrepancy:

    1100 1000 1100 0010
    1111 0110 1110 0101
    1110 0111 1100 0101
    1111 1001 1100 0001
    1101 0100 1111 1000
    1101 0001 1101 0001
    1111 0010 1110 0101
    1100 1001 1100 0110
    1111 0101 1111 0110
    1110 0011 1110 0100

Rather than do the digit-by-digit, line-by-line comparison of these two lists, it is less tedious and less time consuming to compare the lists of hexadecimal equivalents.

First list	Second list
C8C2	C8C2
F6F5	F6F5
E7C5	E7C5
F9C1	F9C1
D4F8	D4F8
D1D1	D1D1
F2E9	F2E5
C9C6	C9C6
F5F6	F5F6
E3E4	E3E4

The list on the left corresponds to the machine output list, the one on the right to the hand-held list. There is a variation in the seventh lines here, and likewise on the seventh lines of the binary lists.

As we have seen in previous sections, the binary–octal and binary–hex conversions are quick and mechanical, so the computer can be programmed to efficiently produce these equivalent lists.

E X E R C I S E S **2.5**

In each of the first four exercises a decimal number is given. For each number determine (a) the straight binary equivalent, (b) the hex equivalent of (a), (c) the four-bit BCD code, (d) the hex equivalent of (c).

1. 49   **2.** 506   **3.** 1111   **4.** 7832

5. Perform the following additions using four-bit BCD code and the technique of Example 2.5.2:

   a. 25   **b.** 45   **c.** 187
        + 33   + 26   + 69

6. Perform the following additions using four-bit BCD code and the technique of Example 2.5.2:

   a. 43   **b.** 27   **c.** 162
        + 16   + 25   + 84

7. Translate each of the following into four-bit BCD code and mimic the usual subtraction algorithm to compute the difference:

   a. 75   **b.** 54
        − 42   − 36

8. Translate each of the following into four-bit BCD code and mimic the usual subtraction algorithm to compute the difference:

   a. 49   **b.** 92
        − 23   − 57

In exercises 9 through 12, a piece of data is given. For each of these find (a) the six-bit BCD code, (b) the octal equivalent of (a), (c) the ASCII-8 code, (d) the hex equivalent of (c), (e) the EBCDIC code, (f) the hex equivalent of (e).

9. ANNE   **10.** 16 NORTH RD.

11. 12.6 + 5   **12.** MAY 7, 1931

In each of the next four exercises, write the EBCDIC form of the given decimal number and then the packed decimal form.

13. 48   **14.** 92   **15.** 156   **16.** 2073

Translate each of the following into ASCII-8 code and perform the addition in packed decimal format as illustrated in Example 2.5.5.

17. 26   **18.** 47   **19.** 89   **20.** 75
       + 12   + 36   + 4   + 5

2.6

Modular Arithmetic

Any digital counter that operates on the principle of an automobile odometer is an application of a type of arithmetic called *modular arithmetic*. When a car's odometer turns 100,000 miles, the reading goes from 999999 to 000000 (remember, the rightmost digit counts tenths of a mile!), and the leftmost digit 1 is lost. Future odometer readings should be equated with that reading altered to include a 1 affixed to the left.

What happens in a calculating device or computer when such a situation occurs depends on how the device is built. One possibility is to have the machine send an *overflow signal* to the operator; another is to proceed without interruption, as with the car odometer. In this section, we present the fundamentals of modular arithmetic and relate it to the subtraction-by-complement method illustrated in Section 2.2. *Our work in this section deals exclusively with integers*; the first four examples use the decimal integer notation; the last three, binary integer notation.

The first definition is a general statement of a concept with which you may be acquainted.

Definition

> Let m be a positive integer. Integer a is called a *multiple of m* if there is some integer k with the property
>
> $$a = k * m$$

Example 2.6.1

Multiples

a. 20 is a multiple of 5 because $20 = 4 * 5$.
   (Using the definition, $m = 5$, $a = 20$, $k = 4$.)

b. 72 is a multiple of 6 because $72 = 12 * 6$.
   (Here, $m = 6$, $a = 72$, $k = 12$.)

c. 72 is a multiple of 12 because $72 = 6 * 12$.
   (Here, $m = 12$, $a = 72$, $k = 6$.)

d. -88 is a multiple of 11 because $-88 = (-8) * 11$.
   (Here, $m = 11$, $a = -88$, $k = -8$.)

e. 0 is a multiple of 7 because $0 = 0 * 7$.
   (Here, $m = 7$, $a = 0$, $k = 0$.)

f. -10 is a multiple of 10 because $-10 = (-1) * 10$.
   ($m = 10$, $a = -10$, and $k = -1$.) ∎

Definition

> Let m be a positive integer. Integers a and b are called *congruent modulo m* if $a - b$ is a multiple of m.
>
> **Notation:** $a \equiv b \pmod{m}$

Example 2.6.2   **Congruence mod *m***

a. For $m = 4$,

$$0 \equiv 4 \ (\text{mod } 4) \text{ since} \qquad 0 - 4 = \quad -4 \text{ is a multiple of } 4$$
$$1 \equiv 5 \ (\text{mod } 4) \text{ since} \qquad 1 - 5 = \quad -4 \text{ is a multiple of } 4$$
$$2 \equiv 6 \ (\text{mod } 4) \text{ since} \qquad 2 - 6 = \quad -4 \text{ is a multiple of } 4$$
$$3 \equiv 7 \ (\text{mod } 4) \text{ since} \qquad 3 - 7 = \quad -4 \text{ is a multiple of } 4$$
$$4 \equiv 8 \ (\text{mod } 4) \text{ since} \qquad 4 - 8 = \quad -4 \text{ is a multiple of } 4$$
$$100 \equiv 0 \ (\text{mod } 4) \text{ since} \quad 100 - 0 = \quad 100 \text{ is a multiple of } 4$$
$$-51 \equiv 1 \ (\text{mod } 4) \text{ since} \ -51 - 1 = \ -52 \text{ is a multiple of } 4$$
$$-77 \equiv 3 \ (\text{mod } 4) \text{ since} \ -77 - 3 = \ -80 \text{ is a multiple of } 4$$
$$222 \equiv 2 \ (\text{mod } 4) \text{ since} \quad 222 - 2 = \quad 220 \text{ is a multiple of } 4$$
$$105 \not\equiv 3 \ (\text{mod } 4) \text{ since} \quad 105 - 3 = \quad 102 \text{ is } not \text{ a multiple of } 4$$

By now you have probably realized that each integer, positive or negative, is congruent mod 4 to one of the integers 0, 1, 2, or 3; so these four integers are called the ***integers mod 4***. Let's compute a few sums involving these integers:

Table 16 Integers mod 4

+	0	1	2	3
0	0	1	2	3
1	1	2	3	0
2	2	3	0	1
3	3	0	1	2

$$1 + 1 = 2 \equiv 2 \quad (\text{mod } 4) \qquad 2 + 2 = 4 \equiv 0 \quad (\text{mod } 4)$$
$$2 + 3 = 5 \equiv 1 \quad (\text{mod } 4) \qquad 3 + 3 = 6 \equiv 2 \quad (\text{mod } 4)$$

The complete addition table for integers mod 4 is given in Table 16.

b. For $m = 7$,

$$0 \equiv \ 7 \ (\text{mod } 7) \text{ since} \quad 0 - \ 7 = -7 \text{ is a multiple of } 7$$
$$2 \equiv \ 9 \ (\text{mod } 7) \text{ since} \quad 2 - \ 9 = -7 \text{ is a multiple of } 7$$
$$5 \equiv 12 \ (\text{mod } 7) \text{ since} \quad 5 - 12 = -7 \text{ is a multiple of } 7$$
$$8 \equiv \ 1 \ (\text{mod } 7) \text{ since} \quad 8 - \ 1 = \ 7 \text{ is a multiple of } 7$$
$$243 \equiv \ 5 \ (\text{mod } 7) \text{ since} \ 243 - \ 5 = 238 \text{ is a multiple of } 7$$
$$94 \equiv \ 3 \ (\text{mod } 7) \text{ since} \quad 94 - \ 3 = \ 91 \text{ is a multiple of } 7$$
$$17 \not\equiv \ 5 \ (\text{mod } 7) \text{ since} \quad 17 - \ 5 = \ 12 \text{ is } not \text{ a multiple of } 7$$

Table 17 Integers mod 7

+	0	1	2	3	4	5	6
0	0	1	2	3	4	5	6
1	1	2	3	4	5	6	0
2	2	3	4	5	6	0	1
3	3	4	5	6	0	1	2
4	4	5	6	0	1	2	3
5	5	6	0	1	2	3	4
6	6	0	1	2	3	4	5

In fact, each integer is congruent mod 7 to one of the integers 0, 1, 2, 3, 4, 5, or 6, which are called the ***integers mod 7***. To add some of these:

$$2 + 3 = \quad 5 \equiv 5 \quad (\text{mod } 7) \qquad 3 + 4 = 7 \equiv 0 \quad (\text{mod } 7)$$
$$5 + 5 = 10 \equiv 3 \quad (\text{mod } 7) \qquad 6 + 2 = 8 \equiv 1 \quad (\text{mod } 7)$$

Table 17 gives the complete addition table for the integers mod 7.

c. For $m = 100$,

$$0 \equiv 100 \ (\text{mod } 100) \text{ since} \qquad 0 - 100 = -100 \text{ is a multiple of } 100$$
$$51 \equiv 151 \ (\text{mod } 100) \text{ since} \qquad 51 - 151 = -100 \text{ is a multiple of } 100$$
$$786 \equiv 86 \ (\text{mod } 100) \text{ since} \qquad 786 - 86 = 700 \text{ is a multiple of } 100$$
$$-432 \equiv 68 \ (\text{mod } 100) \text{ since} \ -432 - 68 = -500 \text{ is a multiple of } 100$$
$$5000 \equiv 0 \ (\text{mod } 100) \text{ since} \ 5000 - 0 = 5000 \text{ is a multiple of } 100$$
$$15 \not\equiv 45 \ (\text{mod } 100) \text{ since} \qquad 15 - 45 = -30 \text{ is } not \text{ a multiple of } 100$$

The integers $0, 1, \ldots, 99$ are called the ***integers mod 100***; each integer is congruent mod 100 to one of these. ∎

Example 2.6.3

Clock Arithmetic: Congruence mod 12

The familiar arithmetic involving clock hours is really arithmetic mod 12.

$$12 \equiv 0 \ (\text{mod } 12)$$
$$9 + \ 4 = \ 13 \equiv \ 1 \ (\text{mod } 12)$$
$$6 + 10 = \ 16 \equiv \ 4 \ (\text{mod } 12)$$
$$3 - \ 7 = -4 \equiv \ 8 \ (\text{mod } 12)$$
$$1 - \ 2 = -1 \equiv 11 \ (\text{mod } 12)$$

The reader should relate each of the last four operations with the appropriate hour hand movement shown in Figure 22. ∎

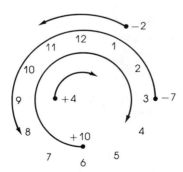

Figure 22

Example 2.6.4

Digital Counter

A museum attendant who oversees a popular exhibit area clicks his three-digit counter each time a person enters that particular location. If at the end of the day the counter reads

| 5 | 6 | 2 |

then the attendant knows that the number of people entering his area that day was 562, or 1562, or 2562, and so on. The counter is a mod 1000 device; note that

$$1000 = 10^3$$

and the counter has three digits. ∎

The preceding example can be generalized: Any digital counter with k digits that operates like an odometer gives readings modulo 10^k. Note that the conventional car odometer with six digits counts tenths of a mile, so it gives tenth-of-a-mile readings mod

$$10^6 = 1000000$$

or mile readings mod

$$10^5 = 100000$$

The remaining examples of this section involve integers written in binary notation.

Example 2.6.5 **Congruence: Binary Notation**

a. For $m = 10$,

$$10 \equiv 0 \,(\text{mod } 10) \text{ since } \quad 10 - 0 = \quad 10 \text{ is a multiple of } 10$$
$$11 \equiv 1 \,(\text{mod } 10) \text{ since } \quad 11 - 1 = \quad 10 \text{ is a multiple of } 10$$
$$100 \equiv 0 \,(\text{mod } 10) \text{ since } \quad 100 - 0 = \quad 100 \text{ is a multiple of } 10$$
$$1000 \equiv 0 \,(\text{mod } 10) \text{ since } \quad 1000 - 0 = 1000 \text{ is a multiple of } 10$$
$$10000 \not\equiv 1 \,(\text{mod } 10) \text{ since } 10000 - 1 = 1111 \text{ is } not \text{ a multiple of } 10$$

b. For $m = 100$,

$$100 \equiv 0 \,(\text{mod } 100) \text{ since } \quad 100 - 0 = \quad 100 \text{ is a multiple of } 100$$
$$101 \equiv 1 \,(\text{mod } 100) \text{ since } \quad 101 - 1 = \quad 100 \text{ is a multiple of } 100$$
$$110 \equiv 10 \,(\text{mod } 100) \text{ since } \quad 110 - 10 = \quad 100 \text{ is a multiple of } 100$$
$$111 \equiv 11 \,(\text{mod } 100) \text{ since } \quad 111 - 11 = \quad 100 \text{ is a multiple of } 100$$
$$1000 \equiv 0 \,(\text{mod } 100) \text{ since } \quad 1000 - 0 = \quad 1000 \text{ is a multiple of } 100$$
$$10011 \not\equiv 10 \,(\text{mod } 100) \text{ since } 10011 - 10 = 10001 \text{ is } not \text{ a multiple of } 100$$

∎

The next example illustrates how the subtraction-by-complements method is accomplished by digital registers that can only add and compute complements. The example uses two facts that can be verified directly from the definition of congruence; namely, for any integer a,

$$a \equiv a + m \,(\text{mod } m) \quad \text{and} \quad a \equiv a - m \,(\text{mod } m)$$

Example 2.6.6   **Congruence: Subtraction by Complements**

The binary numbers subtracted here are those of Example 2.2.8, so you may want to review that illustration first. The work on the right side below indicates how the appropriate steps would be handled by five-digit registers.

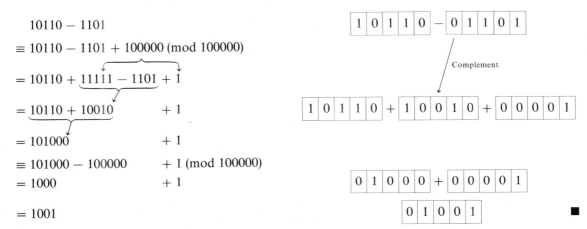

$10110 - 1101$

$\equiv 10110 - 1101 + 100000 \pmod{100000}$

$= 10110 + \underbrace{11111 - 1101 + 1}$

$= \underbrace{10110 + 10010} \qquad + 1$

$= 101000 \qquad\qquad + 1$

$\equiv 101000 - 100000 \qquad + 1 \pmod{100000}$

$= 1000 \qquad\qquad\quad + 1$

$= 1001$

In comparing the work on the right and left above, we notice that with the registers on the right side the addition of 100000 is ignored since the leftmost 1 cannot be stored in the five-digit registers; the subtraction of 100000 is unnecessary since the leftmost 1 of the sum

$$101000 = 10110 + 10010$$

is not carried by the five-digit register.

While the subtraction-by-complement method always yields the correct answer when smaller is subtracted from larger, the next example shows what happens when the technique is mechanically applied to the converse situation.

Example 2.6.7   **Subtraction by Complements: Smaller Minus Larger**

Compute

$$\begin{array}{r} 1011 \\ -11001 \\ \hline \end{array}$$

Recalling that

$$\text{Smaller} - \text{Larger} = -(\text{Larger} - \text{Smaller})$$

we have

$$\begin{array}{r} {\scriptstyle 0\ 1\ 10} \\ 1\,\overcancel{1}\,\overcancel{0}\,\overcancel{0}\,1 \\ -\ 1\ 0\ 1\ 1 \\ \hline 1\ 1\ 1\ 0 \end{array}$$

So the answer to the question is -1110.

However, employing the complement technique:

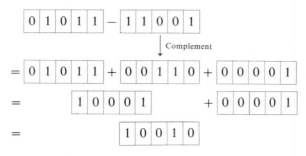

Observe that $10010 \neq -1110$. However, $10010 \equiv -1110 \pmod{100000}$ since

$$10010 - (-1110) = 10010 + 1110$$
$$= 100000$$

is a multiple of 100000.

Thus, in this situation the complement method using five-digit registers gives a (positive) result that is congruent mod 10^5 to the correct (negative) answer. ∎

EXERCISES 2.6

Unless indicated to the contrary, all integers here are given in decimal notation.

1. In each of the following, determine whether a is a multiple of m:

a. $a = 8, m = 1$ b. $a = 8, m = 2$
c. $a = 8, m = 3$ d. $a = -84, m = 7$
e. $a = -1402, m = 13$ f. $a = 0, m = 25$

2. In each of the following, determine whether a is a multiple of m:

a. $a = 9, m = 1$ b. $a = 9, m = 2$
c. $a = 9, m = 3$ d. $a = -107, m = 7$
e. $a = 0, m = 1$ f. $a = -4797, m = 41$

3. Determine whether each of the following is true or false:

a. $13 \equiv 7 \pmod 5$
b. $13 \equiv 8 \pmod 5$
c. $98 \equiv 71 \pmod 2$
d. $-7 \equiv 2 \pmod 3$
e. $-12 \equiv -32 \pmod{11}$
f. $0 \equiv -469 \pmod{20}$

4. Determine whether each of the following is true or false:

a. $12 \equiv 6 \pmod 3$
b. $31 \equiv 11 \pmod 4$
c. $50 \equiv -1 \pmod 6$
d. $-17 \equiv 2 \pmod 9$
e. $14 \equiv 0 \pmod{14}$
f. $-114 \equiv -162 \pmod{16}$

5. List the integers mod 5 and construct an addition table for them.

6. List the integers mod 6 and construct an addition table for them.

7. List the integers mod 12 (using 12 instead of 0) and construct an addition table for them.

8. Use clock arithmetic to compute the following:

a. $5 + 2$ b. $10 + 12$
c. $7 + 8$ d. $11 - 5$
e. $9 - 12$ f. $6 - 8$

Binary notation is used in the next six exercises. In exercises 9 and 10, determine whether the statements are true or false.

9. a. $11 \equiv 0 \pmod{10}$
    b. $101 \equiv 1 \pmod{10}$
    c. $110 \equiv 10 \pmod{10}$

d. $101 \equiv -1 \pmod{10}$
e. $-1 \equiv -100 \pmod{10}$
f. $1001 \equiv 101 \pmod{100}$

10. a. $110 \equiv 0 \pmod{10}$ **b.** $100 \equiv 0 \pmod{100}$
c. $100 \equiv 10 \pmod{100}$ **d.** $110 \equiv -10 \pmod{100}$
e. $101 \equiv 100 \pmod{10}$ **f.** $-110 \equiv -11 \pmod{10}$

In the next two exercises, use the two techniques illustrated in Example 2.6.6 to compute the desired differences.

11. $110101 - 10011$ **12.** $1100 - 1001$

Apply the methods of Example 2.6.7 to compute the next two differences.

13. $10011 - 110101$ **14.** $1001 - 1100$

In the next three exercises, use the definition of congruence to verify the given statement.

15. If $a \equiv b \pmod{m}$, then for any integer n

$a + n \equiv b + n \pmod{m}$

$a - n \equiv b - n \pmod{m}$

$a*n \equiv b*n \pmod{m}$

16. If $a \equiv b \pmod{m}$ and $c \equiv d \pmod{m}$, then

$a + c \equiv b + d \pmod{m}$

$a*c \equiv b*d \pmod{m}$

17. a. For any integer a, $a \equiv a \pmod{m}$.
b. If $a \equiv b \pmod{m}$ then $b \equiv a \pmod{m}$.
c. If $a \equiv b \pmod{m}$ and $b \equiv c \pmod{m}$ then $a \equiv c \pmod{m}$.

18. a. Construct a multiplication table for the integers mod 4.

b. Find all solutions (mod 4) to the congruence

$2*x \equiv 2 \pmod{4}$

19. a. Construct a multiplication table for the integers mod 5.
b. Find all solutions (mod 5) to the congruence

$2*x \equiv 2 \pmod{5}$

c. Is there a solution to the congruence

$x^2 \equiv -1 \pmod{5}$

20. Find all solutions (mod 15) to the congruence

$3*x \equiv 6 \pmod{15}$

21. If $\bar{x}$ is a solution to

$x^2 \equiv b \pmod{m}$

show that $m - \bar{x}$ is also a solution.

22. Use expanded-form notation to show that the following two statements about decimal integer k are equivalent:

a. k is a multiple of 9.
b. The sum of the digits of k is a multiple of 9.

23. The Division Algorithm was stated in exercise 43 at the end of Section 2.3. Use it to show that the following two statements are the same:

a. $a \equiv b \pmod{m}$.
b. a and b have the same remainders when divided by m.[†]

[†] The language Pascal contains a modulus operator MOD defined as follows:

a MOD b equals the integer remainder of a/b.

2.7

Units of Measurement

Anyone who has done the slightest amount of reading in computer science realizes that the field is filled with its own jargon, terminology, abbreviations, and acronyms. Some of these terms are units for measuring certain physical attributes of a computer system. The purpose of this section is to introduce several common units of measurement and to illustrate their use in numeric problems.

We begin with a list of measurement terms and their meanings:

Bit: A binary digit (either 0 or 1)

Byte: 8 bits (for example, 10010101)

BPI: Bytes per inch (for example, as stored on magnetic tape)

K: When used *alone* to designate memory storage capacity, this symbol equals 1024 storage units

KB: Kilobyte (that is, 1000 bytes)

KOPS: Thousands of operations per second

MB: Megabyte or "meg" (that is, one million bytes)

MTBF: Mean (or average) time between failures

Microsecond: One one-millionth (that is, 1/1,000,000) of a second

Nanosecond: One one-billionth (that is, 1/1,000,000,000) of a second

Example 2.7.1 **Memory Capacity**

A microcomputer is advertised as having 48K bytes of RAM (random access memory) and 5K bytes of ROM (read-only memory). Since the letter K is used alone, here it equals 1024, and this system contains

$$48 * 1024 = 49152 \text{ bytes of RAM}$$

and

$$5 * 1024 = 5120 \text{ bytes of ROM} \qquad \blacksquare$$

In contexts where K is adjoined to another symbol, like KB or KOPS, it simply means "thousand."

Example 2.7.2 **Conversion: Bit, Byte, KB, and MB**

Each row of the following table gives four equivalent forms of the same measurement, using the four different units of measurement given in the column headings. One should read each row from left to right, keeping in mind the following:

1. To convert bits to bytes, divide by 8.

2. To convert bytes to KB, divide by 1000.

3. To convert bytes to MB, divide by 1,000,000.

Bits	Bytes	KB	MB
2048	$\dfrac{2048}{8} = 256$	$\dfrac{256}{1000} = .256$	$\dfrac{256}{1,000,000} = .000256$
8000	$\dfrac{8000}{8} = 1000$	$\dfrac{1000}{1000} = 1$	$\dfrac{1000}{1,000,000} = .001$
20000	$\dfrac{20000}{8} = 2500$	$\dfrac{2500}{1000} = 2.5$	$\dfrac{2500}{1,000,000} = .0025$
16,000,000	$\dfrac{16,000,000}{8} = 2,000,000$	$\dfrac{2,000,000}{1000} = 2000$	$\dfrac{2,000,000}{1,000,000} = 2$

$\blacksquare$

Example 2.7.3   **MTBF**

A certain computer system operated the following numbers of hours between downtimes, that is, between failures in at least one component of the system:

2240,   3623,   3185

One computes MTBF in the same fashion as any other average:

$$\text{MTBF} = \frac{\text{Total number of operating hours}}{\text{Total number of failures}}$$

$$= \frac{2240 + 3623 + 3185}{3} = 3016 \text{ hours}$$ ∎

Example 2.7.4   **Processor Speed**

A particular computer processor can perform one addition (in binary notation, of course!) in .4 microsecond. That is,

1 addition = .4 microsecond

$$1 \text{ addition} = \frac{.4}{1,000,000} \text{ second}$$

1 addition = .0000004 second

To determine how many additions can be performed in one second, divide both sides of this last equation by .0000004, getting

$$\frac{1}{.0000004} \text{ additions} = 1 \text{ second}$$

2,500,000 additions = 1 second

So this processor can compute 2.5 million additions per second. ∎

   In the next two examples, it is helpful to note that phrases such as "inches per second" can be expressed in the form of a mathematical fraction,

$$\frac{\text{in.}}{\text{sec.}}$$

and measuring units may be canceled when such fractions are multiplied.

Example 2.7.5   **Transfer Rate**

Suppose we know that a given tape drive can move tape at a rate of 150 inches per second, and the tape density is 1800 BPI (bytes per inch). In this case the drive can transfer or receive

$$1800 \frac{\text{bytes}}{\text{in.}} * 150 \frac{\text{in.}}{\text{sec.}} = 270,000 \frac{\text{bytes}}{\text{sec.}}$$

Hence, the transfer rate of this device is 270 KB per second. ∎

Example **2.7.6**   **Disk Capacity**

One type of magnetic disk pack contains ten hard disks, each side of which is used for storage. Each disk surface is divided into 200 concentric circular tracks and eight pie-shaped sectors, and each track sector can hold 500 bytes (see Figure 23). Then the total storage capacity of this disk pack equals

$$10 \text{ disks} * 2 \frac{\text{surfaces}}{\text{disk}} * 200 \frac{\text{tracks}}{\text{surface}} * 8 \frac{\text{sectors}}{\text{tracks}} * 500 \frac{\text{bytes}}{\text{sector}}$$

$$= 16000000 \text{ bytes}$$

That is, this pack can hold 16 MB.   ∎

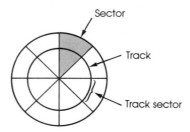

Figure 23

E X E R C I S E S **2.7**

1. Convert each of the following memory storage capacities into a number of bytes:

a. 6 K   **b.** 32 K   **c.** 48 K   **d.** 256 K

2. Convert each of the following memory storage capacities into a number of bytes:

a. 2 K   **b.** 16 K   **c.** 64 K   **d.** 1000 K

3. Complete the table below so that the measurements in each row are equivalent.

Bits	Bytes	KB	MB
5000	___	___	___
___	5000	___	___
___	___	500	___
___	___	___	50

4. Complete the table below so that the measurements in each row are equivalent.

Bits	Bytes	KB	MB
8192	___	___	___
___	4400	___	___
___	___	10	___
___	___	___	7

5. A component of a computer system has an MTBF rating of 2200 hours. Assuming that this component operates continuously, how many breakdowns of this component would you expect in a year?

6. A microcomputer printer had breakdowns after operating the following numbers of hours:

780   473   512   635   567

Compute the MTBF for this device.

7. The processor in a certain mainframe computer can perform one addition in 250 nanoseconds, one subtraction in 1 microsecond, one multiplication in 3 microseconds, and one division in 7 microseconds. Determine how many additions this processor can compute in one second. Subtractions? Multiplications? Divisions?

8. If a processor can compute 8 million additions per second, how long does it take to perform just one addition?

9. Complete the table below so that the measurements in each row are equivalent.

Seconds	Microseconds	Nanoseconds
.002	___	___
___	.6	___
___	___	42

10. Complete the table below so that the measurements in each row are equivalent.

Seconds	Microseconds	Nanoseconds
.00075	_____	_____
_____	.15	_____
_____	_____	510

11. If you want a tape drive with density rating 1620 BPI to transfer 160,000 characters (bytes) per second, how many inches of tape must the drive move per second?

12. How many characters (bytes) can be stored on a reel of magnetic tape 2400 feet long with density 1600 BPI?

13. Each surface of a particular type of magnetic disk is divided into 200 tracks and 20 sectors. If each track sector holds 500 bytes, how many such disks will be needed if one desires a total storage capacity of 80 MB?

14. One side of a floppy diskette can hold 128 KB. If the diskette is divided into 16 sectors, how many bytes can each sector hold?

15. A particular processor is rated at 110 KOPS. How long will it take this processor to perform one billion operations?

16. Why do you think that K in storage units (see Example 2.7.1) equals 1024 instead of 1000?

CHAPTER 3

Sets, Combinatorics, and Probability

The theory of sets is a fundamental and unifying topic within the many branches of the field of mathematics. And understanding the rudiments of sets and set operations is helpful for anyone interested in computer science. The following excerpts were taken from the glossary of an introductory text on computers and data processing.[†]

Database A set of data which is a part or the whole of another set of data, and consisting of at least one file, that is sufficient for a given purpose or for a given data processing system. As an example, the data collected for each individual belonging to VISA or Master Card credit card charging system.

Field In a record, a specified area used for a particular category of data, for example, a group of card columns in which a wage rate is recorded.

File A set of related records treated as a unit. For instant, in stock control, a file could consist of a set of invoices. The records are kept in some order by invoice number, customer name, or vendor number. The types are master, transaction, indexed, sequential, hierarchical, and inverted.

Record A collection of related data or words treated as a unit. For example, in stock control each invoice could constitute one record. Compare *file*.

In this chapter, the first four sections introduce material on sets. Section 3.3 deals with an algebraic structure of a general nature that has application to sets and, as we shall see in Chapter 4, an important use in the design of computer hardware. The last four sections use set theory as a foundation to develop skills in solving certain types of counting and probability problems.

[†] From *Data Processing: Computers in Action*, 2nd ed., by Perry Edwards and Bruce Broadwell, c 1982 by Wadsworth, Inc. Reprinted by permission of Wadsworth Publishing Company, Belmont, California 94002.

3.1

Sets and Subsets

Definition

> A *set* is a collection of objects, each of which is called an ***element*** or ***member*** of the set.
>
> **Notation**
>
> > Capital italic letters are used to represent sets.
> > Lowercase italic letters or numbers are used to represent elements.
> > Braces { } are used to identify the elements of a set.
> > The symbol $\in$ means "is an element of."
> > The symbol $\notin$ means "is not an element of."

Example 3.1.1   **Sets**

$A = \{a, b, p, m\}$ is a set, and

$\quad a \in A \quad m \in A \quad c \notin A$

$B = \{0, 7, -5.2, \frac{1}{2}, 3\}$ is a set, and

$\quad 0 \in B \quad 1 \notin B \quad \frac{1}{3} \notin B$ ∎

Definition

> Set A is a ***subset*** of set B if each element in A is also an element in B.
>
> **Notation:** $A \subseteq B$ means "A is a subset of B."
> $\qquad\qquad A \nsubseteq B$ means "A is *not* a subset of B."

Example 3.1.2   **Subsets**

Let

$$A = \{1, 2, 3\} \quad B = \{0, 1, 2, 3, 4, 5, 6\} \quad C = \{2, 3, 4\}$$

Then

$\quad A \subseteq A \qquad B \subseteq B \qquad C \subseteq C$

$\quad A \subseteq B \qquad C \subseteq B$

$\quad A \nsubseteq C \qquad B \nsubseteq C \qquad C \nsubseteq A$ ∎

Example 3.1.3   **Subsets**

Certain sets occur with such frequency that they have standard representations. For instance, the sets of integers, rational numbers, and real numbers are often denoted by

$\quad \mathbb{Z} \quad \mathbb{Q} \quad \mathbb{R}$

respectively, a convention that will be observed in this text. Therefore,

$$\mathbb{Z} \subseteq \mathbb{Q} \subseteq \mathbb{R}$$

but

$$\mathbb{R} \nsubseteq \mathbb{Q} \nsubseteq \mathbb{Z} \qquad \blacksquare$$

Definition

Set A is *equal* to set B if

$$A \subseteq B \quad \text{and} \quad B \subseteq A$$

Notation: $A = B$ means that "A is equal to B."
$A \neq B$ means that "A is not equal to B."

This definition says that two sets are equal if they have the same elements. If the elements are listed within braces, the order in which the elements are written is not important.

Example 3.1.4

Equality

$$\{a, e, f\} = \{a, f, e\} = \{e, a, f\}$$
$$\{1, 2, 3\} \neq \{1, 2, 4\}$$
$$\mathbb{Z} \neq \mathbb{Q} \neq \mathbb{R} \qquad \blacksquare$$

From the two previous examples, we have

$$\mathbb{Z} \subseteq \mathbb{Q} \subseteq \mathbb{R}$$

but

$$\mathbb{Z} \neq \mathbb{Q} \neq \mathbb{R}$$

In this type of situation, we may write

$$\mathbb{Z} \subsetneq \mathbb{Q} \subsetneq \mathbb{R}$$

or simply

$$\mathbb{Z} \subset \mathbb{Q} \subset \mathbb{R}$$

and say that $\mathbb{Z}$ is a *proper* subset of $\mathbb{Q}$, and $\mathbb{Q}$ is a *proper* subset of $\mathbb{R}$.

Notation

The symbol $\subset$ means $\subseteq$ but $\neq$.

In a particular application involving sets, the sets under consideration are usually drawn from a single underlying set.

Definitions

> The ***universal*** set in a particular example is that underlying set of which the other sets under discussion are subsets.
>
> The ***empty*** set is the set with no elements.
>
> **Notation:** U represents the universal set.
> ϕ represents the empty set.

It follows from the definitions that for any set A, $\phi \subseteq A$. The empty set ϕ plays a role within set theory that is analogous to the role of zero in our familiar number systems.

Example 3.1.5

Universal and Empty Sets

Let U be the alphabet; then

$$\phi \subset \{a\} \subset U$$
$$\phi \subset \{a, b, c\} \subset U$$
$$\phi \subset \{a, b, \ldots, z\} = U$$

Let $U = \mathbb{Z}$, then

$$\phi \subset \{-2, 5\} \subset U$$
$$\phi \subset \{0, 1, 2, \ldots, 10\} \subset U$$
$$\phi \subset \{75\} \subset U$$ ■

Venn diagrams, named after English logician John Venn (1834–1923), are geometric devices that are helpful in representing sets and their relationships. Such diagrams use a large rectangle to represent the universal set and circles or ovals within the rectangle to represent subsets of that universal set (see Figure 1.) The elements of a set are those points on and within the circle or oval.

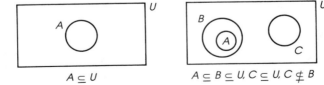

Figure 1 $A \subseteq U$ $A \subseteq B \subseteq U, C \subseteq U, C \nsubseteq B$

Example 3.1.6

Venn Diagrams

If U is the set of all living Americans
  F is the set of all living Americans over 50
  S is the set of all living Americans over 70
  T is the set of all living Americans under 30

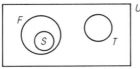

Figure 2

then we could use a Venn diagram to represent these sets, as shown in Figure 2. Here it is understood that the set F includes everything inside the region labeled F, even region S. ∎

Up to this point, we have specified elements within a set by listing them between braces. When the set has a large number of elements, this method is tedious and perhaps unclear; for instance,

$$A = \{a, b, \dots, z\}$$

is supposed to represent the alphabet. When the set has infinitely many elements, this listing method may be impossible to use with precision; for instance, how does one list the set of all fractions whose denominators are powers of 2? In these instances, it is preferable to use *set-builder notation*.

Notation

The symbol

$$\{x \mid x \text{ has property } P\}$$

means the set of *all* elements having a specified property P.

Example 3.1.7   **Set-Builder Notation**

$\{k \mid k \in \mathbb{Z} \text{ and } 1 \le k \le 2000\}$ represents all integers from 1 to 2000

$\{x \mid x = k/3 \text{ where } k \in \mathbb{Z}\}$ represents all fractions with denominator equal to 3

$\{w \mid w \text{ is a word beginning with "}b\text{"}\}$ represents all words beginning with the letter b

$\{p/q \mid p, q \in \mathbb{Z} \text{ and } q \ne 0\}$ represents $\mathbb{Q}$, the set of rational numbers

$\{p/2^n \mid n, p \in \mathbb{Z} \text{ and } n > 0\}$ represents all fractions whose denominators are powers of 2

$\{2k \mid k \in \mathbb{Z}\}$ represents all even integers

$\{2k + 1 \mid k \in \mathbb{Z}\}$ represents all odd integers ∎

EXERCISES 3.1

1. List the elements in each set.

a. $\{n, a, m, e\}$
b. $\{2, 3, \dots, 13\}$
c. $\{z \mid z \in \mathbb{Z} \text{ and } 1 < z < 14\}$
d. $\{t \mid t \text{ is a letter in the word "algorithm"}\}$
e. $\{x \mid x \in \mathbb{R} \text{ and } x^2 = 9\}$
f. $\{z \mid z \in \mathbb{Z} \text{ and } z \ge 10\}$
g. $\{2/m \mid m \in \mathbb{Z} \text{ and } m \ne 0\}$

2. List the elements in each set.

a. $\{w, e, z, \text{the}, r\}$
b. $\{-3, -2, -1, 0, 1, 2, 3\}$
c. $\{z \mid z \in \mathbb{Z} \text{ and } -3 \le z \le 3\}$
d. $\{s \mid s \text{ is a syllable in the word "flowchart"}\}$
e. $\{y \mid y \in \mathbb{R}, y > 0, \text{ and } y^3 = -1\}$
f. $\{-5z \mid z \in \mathbb{Z} \text{ and } z > 0\}$
g. $\{m/2 \mid m \in \mathbb{Z}\}$

3. Use the listing method, if possible, to describe the given sets.

 a. Set of positive integers less than 8
 b. Set of nonnegative integers less than 8
 c. Set of vowels in the alphabet
 d. Set of months whose names contain the letter n
 e. Set of women presidents of the United States
 f. Set of odd positive integers
 g. Set of real numbers between 0 and 1

4. Use the listing method, if possible, to describe the given sets.

 a. Set of letters in alphabet preceding h
 b. Set of positive integers not exceeding 7
 c. Set of consonants in the alphabet
 d. Set of months with fewer than 28 days
 e. Set of states in the United States not bordering another state
 f. Set of multiples of 4
 g. Set of fractions between 0 and 1

5. Do exercise 3 using set-builder notation.

6. Do exercise 4 using set-builder notation.

7. In each of the following, determine whether listing or set-builder notation would yield the better expression of the given set and express the set in that form:

 a. Last names of U.S. presidents entering office since 1960
 b. All odd integers
 c. All numbers of the form 10^n, n an integer
 d. All binary digits

8. In each of the following, determine whether listing or set-builder notation would yield the better expression of the given set and express the set in that form:

 a. All leap years in the 1980s
 b. All positive integers
 c. All fractions
 d. The integers modulo 4

9. Determine whether each of the following is true or false:

 a. $5 \in \{x \mid x^2 = 25\}$ **b.** $\phi \in \{1, 2, 3, 4\}$
 c. $\{a, b, t\} = \{t, a, b\}$ **d.** $0 \in \phi$
 e. $\phi \subset \phi$ **f.** $\phi \subset \mathbb{Z}$
 g. $\{-5, -1\} \subseteq \mathbb{Z}$

10. Determine whether each of the following is true or false:

 a. $20 \subseteq \{5, 10, 15, \ldots\}$
 b. $\phi \subseteq \{1, 2, 3, 4\}$

 c. $\{1, 2, 3,\} = \{z \mid z \in \mathbb{Z} \text{ and } 0 < z < 4\}$
 d. $\phi \in \{0\}$
 e. $\phi \subseteq \phi$
 f. $\mathbb{Z} \subset \phi$
 g. $\{\frac{1}{2}, 1\} \subseteq \mathbb{Z}$

11. Draw a Venn diagram that includes each of the following sets. Be sure to include and identify the universal set.

 F = all employees who work at least 40 hours

 G = all employees who work at least 50 hours

 H = all employees who work at least 60 hours

12. Draw a Venn diagram that includes each of the following sets. Be sure to include and identify the universal set.

 H = all adults who completed high school

 C = all adults who completed college

 N = all adults who did *not* complete elementary school

In the next two exercises, the following data set is used:

Name	Hourly pay rate	Number of hours worked
Bashful	$4.25	44
Bonzo	3.25	37
Doc	4.75	39
Dopey	3.55	47
Grumpy	3.95	43
Happy	5.00	49
Hondo	5.75	40
Sleepy	3.35	45
Sneezy	4.00	41
Snoopy	5.50	35

where

 U(*universal set*) = *set of all names*

 $R_{>x}$ = *names of all those whose rate exceeds* x

 $R_{<x}$ = *names of all those whose rate is less than* x

 $H_{>x}$ = *names of all those whose hours exceed* x

 $H_{<x}$ = *names of all those whose hours are less than* x

13. List the elements of the sets

 $R_{>4}$ $R_{>5}$ $R_{>6}$ $H_{>45}$ $H_{<50}$ $H_{>30}$

Which of these are subsets of another?

14. List the elements of the sets

 $R_{<5}$ $R_{>5}$ $R_{>3}$ $H_{<40}$ $H_{>40}$ $H_{<35}$

Which of these are subsets of another?

When working with real numbers $\mathbb{R}$*, the following interval notation is often used, where* $a, b \in \mathbb{R}$:

$$(a, b) = \{x \mid x \in \mathbb{R} \text{ and } a < x < b\}$$
$$[a, b] = \{x \mid x \in \mathbb{R} \text{ and } a \leq x \leq b\}$$
$$(a, b] = \{x \mid x \in \mathbb{R} \text{ and } a < x \leq b\}$$
$$[a, b) = \{x \mid x \in \mathbb{R} \text{ and } a \leq x < b\}$$

Here (a, b) *and* $[a, b]$ *are called an **open interval** and **closed interval**, respectively, and* $(a, b]$ *and* $[a, b)$ *are each half-open, half-closed intervals. Exercises 15 through 18 use this notation.*

15. Write the following intervals in set-builder form:

a. $(-1, 2)$ **b.** $[\frac{1}{2}, 5)$ **c.** $(0, \frac{12}{5}]$
d. $[-5, 6]$ **e.** $[2, 1]$

16. Write the following intervals in set-builder form:

a. $(-\frac{1}{2}, 4)$ **b.** $[-1, 7)$ **c.** $(-\frac{3}{2}, 0]$
d. $[-10, -2]$ **e.** $[2, 2)$

17. Express each of the following in interval notation:

a. $\{x \mid x \in \mathbb{R} \text{ and } 0 \leq x \leq 9\}$
b. $\{y \mid y \in \mathbb{R} \text{ and } -\sqrt{2} < y \leq 15\}$
c. $\{w \mid w \in \mathbb{R} \text{ and } -2 < w < -1\}$

18. Express each of the following in interval notation:

a. $\{x \mid x \in \mathbb{R} \text{ and } -\frac{3}{2} \leq x < -1\}$
b. $\{y \mid y \in \mathbb{R} \text{ and } 0 < y < 1\}$
c. $\{w \mid w \in \mathbb{R} \text{ and } -\frac{1}{2} \leq w \leq -\frac{1}{10}\}$

3.2

Set Operations

Recall from our work in Section 1.5 that a binary operation is an operation that has precisely two inputs and one output. The operations presented in this section are binary operations, but the two inputs and single output are sets rather than numbers.

Definition

The ***union of two sets A and B*** is the collection of all elements belonging to A or to B; that is,

$$\{x \mid x \in A \quad \text{or} \quad x \in B\}$$

Notation: $A \cup B$

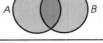

$A \cup B$ is shaded.

Note: In this definition, the word *or* is used in the inclusive sense of "*and/or*."

Definition

The ***intersection of two sets A and B*** is the collection of all elements belonging to both A and B; that is,

$$\{x \mid x \in A \quad \text{and} \quad x \in B\}$$

Notation: $A \cap B$

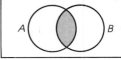

$A \cap B$ is shaded.

Example 3.2.1 **Union and Intersection**

Let

$$A = \{-1, 0, 1, 2, 3, 4\} \qquad B = \{1, 3, 7, 10\} \qquad C = \{0, 1, 3, 7\}$$
$$D = \{-5, 0, 5\} \qquad\qquad M = \{a, e, f\} \qquad N = \{g, m, n, p, x\}$$

Then

$$A \cup B = \{-1, 0, 1, 2, 3, 4, 7, 10\} \qquad A \cap B = \{1, 3\}$$
$$A \cup C = \{-1, 0, 1, 2, 3, 4, 7\} \qquad A \cap C = \{0, 1, 3\}$$
$$A \cup D = \{-5, -1, 0, 1, 2, 3, 4, 5\} \qquad A \cap D = \{0\}$$
$$B \cup D = \{-5, 0, 1, 3, 5, 7, 10\} \qquad B \cap D = \phi$$
$$A \cup \mathbb{Z} = \mathbb{Z} \qquad\qquad A \cap \mathbb{Z} = A$$
$$M \cup N = \{a, e, f, g, m, n, p, x\} \qquad M \cap N = \phi \qquad ■$$

Two sets whose intersection equals ϕ are called **disjoint**. Above the sets B and D are disjoint, as are M and N.

Definition

The **difference of two sets A and B** is the collection of the elements in A that are *not* in B; that is

$$\{x \mid x \in A \quad \text{and} \quad x \notin B\}.$$

Notation: $A - B$

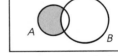

$A - B$ is shaded.

Definition

The **complement** of set A in universal set U is the set difference $U - A$.

Notation: $\bar{A}$

$\bar{A}$ is shaded.

Example 3.2.2 **Difference and Complement**

Let

$$A = \{0, 1, 3, 5\}$$
$$B = \{1, 2, 3, 4, 5, 6, 7, 8, 9, 10\}$$
$$E = \{2z \mid z \in \mathbb{Z}\} = \text{set of all even integers}$$
$$U = \mathbb{Z}$$

Then

$$A - B = \{0\}$$
$$B - A = \{2, 4, 6, 7, 8, 9, 10\}$$
$$U - E = \bar{E} = \{2z + 1 \mid z \in \mathbb{Z}\} = \text{set of all odd integers}$$
$$E - U = \phi$$
$$\bar{A} = \{z \mid z \in \mathbb{Z} \text{ and } z \neq 0, 1, 3, 5\} = \text{all integers except } 0, 1, 3, 5$$
$$\bar{B} = \{z \mid z \in \mathbb{Z} \text{ and } z \neq 1, 2, 3, 4, 5, 6, 7, 8, 9, 10\}$$
$$\quad = \text{all integers except } 1, 2, 3, 4, 5, 6, 7, 8, 9, 10$$

In the next example, let us *assume that the universal set U is finite;* that is, U has finitely many elements.

Example 3.2.3   **Flowcharts for Set Operations**

The four flowcharts in Figures 3 through 6 will produce listings of those elements in the union $A \cup B$, intersection $A \cap B$, difference $A - B$, and complement $\bar{A}$ respectively, where the universal set U is finite. ■

In the first section of this chapter, we defined set equality; namely, $A = B$ means $A \subseteq B$ and $B \subseteq A$. We shall conclude this section by illustrating a method of verifying set inclusions or set equalities. Each verification will be preceded by a Venn diagram that suggests the correctness of the statement.

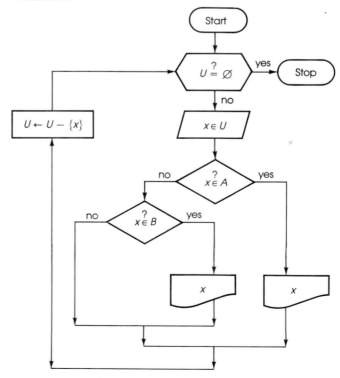

Figure 3 $A \cup B$

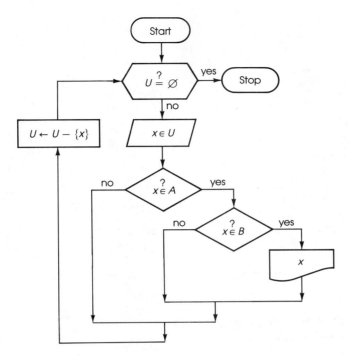

Figure 4 $A \cap B$

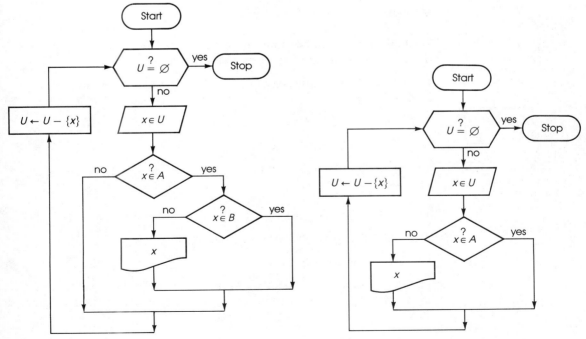

Figure 5 $A - B$

Figure 6 $\bar{A}$

Example 3.2.4

Proof of Set Inclusion

Verify that for any sets A and B,

$$(A \cap B) \subseteq (A \cup B)$$

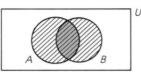

Figure 7

Motivation: Note that in Figure 7 the shaded region $A \cap B$ is inside the striped region $A \cup B$.

Verification: Let x represent an arbitrarily chosen element in set $A \cap B$. Then[†]

$$x \in A \cap B$$

$$\xrightarrow{\text{Def. } \cap} \quad x \in A \quad \text{and} \quad x \in B$$

$$\xrightarrow{\text{Language}} \quad x \in A \quad \text{and/or} \quad x \in B$$

$$\xrightarrow{\text{Def. } \cup} \quad x \in A \cup B$$

Since x is representative of those elements in $A \cap B$, then

$$(A \cap B) \subseteq (A \cup B)$$

∎

Example 3.2.5

Proof of Set Equality

For sets A and B, verify that

$$A - B = A \cap \bar{B}$$

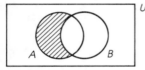

$A - B$ is striped.

$\bar{B}$ is horizontally striped.
A is vertically striped.
$A \cap \bar{B}$ is crosshatched.

Figure 8

Motivation: In Figure 8, the designated regions $A - B$ and $A \cap \bar{B}$ coincide.

Verification: Remember that the definition of set equality involves two inclusions, so there are two parts to this proof:

Let x be a representative element in $A - B$. Then

$$x \in A - B$$

$$\xrightarrow{\text{Def. } -} x \in A \quad \text{and} \quad x \notin B$$

$$\xrightarrow{\text{Def. } \bar{B}} x \in A \quad \text{and} \quad x \in \bar{B}$$

$$\xrightarrow{\text{Def. } \cap} x \in A \cap \bar{B}$$

So $\quad (A - B) \subseteq (A \cap \bar{B})$.

Let x be a representative element in $A \cap \bar{B}$. Then

$$x \in A \cap \bar{B}$$

$$\xrightarrow{\text{Def. } \cap} x \in A \quad \text{and} \quad x \in \bar{B}$$

$$\xrightarrow{\text{Def. } \bar{B}} x \in A \quad \text{and} \quad x \notin B$$

$$\xrightarrow{\text{Def. } -} x \in A - B$$

So $\quad (A \cap \bar{B}) \subseteq (A - B)$.

Because of these two set inclusions, we can say

$$(A - B) = (A \cap \bar{B})$$

∎

In the work above, the steps on the right are the reverse of those on the left. Sometimes that is possible and sometimes it is not, as the next example shows.

[†] Here the symbol → means that the statement to the right of the arrow follows directly from the statement preceding the arrow for the reason indicated above the arrow.

Example 3.2.6   **Proof of Set Equality**

For any sets A, B, and C, show that the distributive law

$$A \cap (B \cup C) = (A \cap B) \cup (A \cap C)$$

is valid.

Motivation: Since in Figure 9 region

$$A \cap (B \cup C)$$

on the top and region

$$(A \cap B) \cup (A \cap C)$$

on the bottom coincide, then the equality seems plausible.

Verification: Again, we have two inclusions to verify.

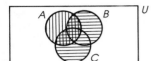

A is vertically striped.
$B \cup C$ is horizontally striped.
$A \cap (B \cup C)$ is crosshatched.

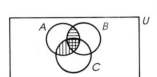

$A \cap B$ is horizontally striped.
$A \cap C$ is vertically striped.
$(A \cap B) \cup (A \cap C)$ is all striped area.

Figure 9

Let x be a representative element of $A \cap (B \cup C)$. Then

$$x \in A \cap (B \cup C)$$

$$\xrightarrow{\text{Def. } \cap} x \in A \quad \text{and} \quad x \in B \cup C$$

$$\xrightarrow{\text{Def. } \cup} x \in A \quad \text{and either} \quad x \in B$$
$$\text{or} \quad x \in C$$

Case 1: $x \in B$. Then

$$x \in A \quad \text{and} \quad x \in B$$

$$\xrightarrow{\text{Def. } \cap} x \in A \cap B$$

$$\xrightarrow{\text{Def. } \cup} x \in (A \cap B) \cup (A \cap C)$$

Case 2: $x \in C$. Then

$$x \in A \quad \text{and} \quad x \in C$$

$$\xrightarrow{\text{Def. } \cap} x \in A \cap C$$

$$\xrightarrow{\text{Def. } \cup} x \in (A \cap B) \cup (A \cap C)$$

So

$$A \cap (B \cup C) \subseteq (A \cap B) \cup (A \cap C)$$

Now let x be a representative element of $(A \cap B) \cup (A \cap C)$.

Case 1: $x \in A \cap B$. Then

$$x \in A \cap B$$

$$\xrightarrow{\text{Def. } \cap} x \in A \quad \text{and} \quad x \in B$$

$$\xrightarrow{\text{Def. } \cup} x \in A \quad \text{and} \quad x \in B \cup C$$

$$\xrightarrow{\text{Def. } \cap} x \in A \cap (B \cup C)$$

Case 2: $x \in A \cap C$. Then

$$x \in A \cap C$$

$$\xrightarrow{\text{Def. } \cap} x \in A \quad \text{and} \quad x \in C$$

$$\xrightarrow{\text{Def. } \cup} x \in A \quad \text{and} \quad x \in B \cup C$$

$$\xrightarrow{\text{Def. } \cap} x \in A \cap (B \cup C)$$

So

$$(A \cap B) \cup (A \cap C) \subseteq A \cap (B \cup C)$$

∎

In each of the three preceding examples, the Venn diagram alone does not constitute a proof, because it shows only one of several possible relationships among the given sets.

EXERCISES 3.2

1. If $U = \{-5, -4, \ldots, 9, 10\}$ and
$A = \{0, 1, 2, 3, 4, 5\}$ $B = \{-5, 0, 5, 10\}$
$C = \{1\}$

list the elements in each of the following sets:

a. $A \cup B$ **b.** $A \cup C$ **c.** $B \cup C$
d. $A \cap B$ **e.** $A \cap C$ **f.** $B \cap C$

g. $A - B$ **h.** $B - A$ **i.** $A - C$
j. $C - A$ **k.** $B - C$ **l.** $C - B$
m. $\bar{A}$ **n.** $\bar{B}$ **o.** $\bar{C}$

2. If $U = \{a, b, \ldots, m\}$ and

$M = \{d, e, g\}$ $N = \{a, m\}$ $P = \{b, d, f, h, j, l\}$

list the elements in each of the following sets:

a. $M \cup N$ **b.** $M \cup P$ **c.** $N \cup P$
d. $M \cap N$ **e.** $M \cap P$ **f.** $N \cap P$
g. $M - N$ **h.** $N - M$ **i.** $M - P$
j. $P - M$ **k.** $N - P$ **l.** $P - N$
m. $\bar{M}$ **n.** $\bar{N}$ **o.** $\bar{P}$

3. For each of the following, draw a flowchart similar to those in Example 3.2.3, assuming the universal set U is finite:

a. $A \cup \bar{B}$ **b.** $\overline{A \cap B}$
c. Shaded region in

4. For each of the following, draw a flowchart similar to those in Example 3.2.3, assuming the universal set U is finite:

a. $A \cap \bar{B}$ **b.** $\overline{A \cup B}$
c. Shaded region in

5. a. Redraw the flowchart for $A \cap B$ in Figure 4 to make the algorithm more efficient by reducing the number of passes through the loop-controlling hexagon symbol.
  b. Same as part (a), but for $A \cup B$ in Figure 3.

6. Redraw the flowchart for $A - B$ in Figure 5 so as to make the algorithm more efficient by reducing the number of passes through the loop-controlling hexagon symbol.

7. Why was it necessary to state that the universal set U in Example 3.2.3 is finite?

8. In Example 3.2.3, what is the purpose of the box

$$U \leftarrow U - \{x\}$$

in each flowchart? Would it be acceptable to delete this portion of the flowchart? Why?

9. a. Redraw the flowchart for $A \cap B$ in Figure 4 so that it will work for finite sets A and B while the universal set U contains infinitely many elements.

b. Same as part (a), but for $A \cup B$ in Figure 3.

10. Redraw the flowchart for $A - B$ in Figure 5 so that it will work for finite set A but the universal set U contains infinitely many elements.

11. Illustrate each of the following statements by means of a Venn diagram:

a. $A \subseteq (A \cup B)$
b. $A \cup \bar{B} = \overline{B - A}$
c. $\overline{A \cap B} = \bar{A} \cup \bar{B}$
d. $(A \cup B) \cup C = A \cup (B \cup C)$

12. Illustrate each of the following statements by means of a Venn diagram:

a. $(A \cap B) \subseteq A$
b. $\overline{A \cup B} = \bar{A} \cap \bar{B}$
c. $A \cup (B \cap C) = (A \cup B) \cap (A \cup C)$
d. $(A \cap B) \cap C = A \cap (B \cap C)$

13. Verify each of the statements in exercise 11 by using the definitions of set operations.

14. Verify each of the statements in exercise 12 by using the definitions of set operations.

Exercises 15 through 18 refer to the data set and notation used in exercises 13 and 14 in Exercises 3.1.

15. Describe the sets verbally and then list the elements.

a. $R_{>4} \cup R_{<5}$ **b.** $H_{<45} \cap H_{>34}$
c. $R_{>4} \cap H_{<40}$ **d.** $\bar{H}_{>40}$

16. Describe the sets verbally and then list the elements.

a. $R_{>4} \cap R_{<5}$ **b.** $H_{<40} \cap H_{>45}$
c. $R_{<4} \cup H_{<40}$ **d.** $\bar{R}_{>4}$

17. In each of the following, determine the correct set representation (operations are acceptable, if needed):

a. Those employees who make less than $5 per hour and who work more than 40 hours per week.
b. Those employees who make more than $5 or less than $4 per hour.
c. Those employees who work between 40 and 50 hours per week.
d. Those employees who work 40 hours or less per week.

18. In each of the following, determine the correct set representation (operations are acceptable, if needed):

a. Those employees who work less than 41 hours per week and who make more than $5 per hour.
b. Those employees who make between $4 and $5 per hour.

c. Those employees who work more than 45 or less than 40 hours per week

d. Those employees who make $4 per hour or more

Exercises 19 through 21 use the interval notation used in exercises 15 through 18 in Exercises 3.1.

19. Express each of the following in interval form:

a. $[-5, 1] \cap [0, 4]$ **b.** $[-5, 1] \cap (0, 4]$

c. $[-5, 1) \cap [0, 4]$ **d.** $[-5, 1) \cap (0, 4]$

e. $[-5, 1] \cup [0, 4]$ **f.** $(-5, 1] \cup [0, 4]$

g. $[-5, 1] \cup [0, 4)$ **h.** $(-5, 1] \cup [0, 4)$

i. $\overline{[-5, 1]} \cap (0, 4)$ **j.** $[-5, 1] \cap \overline{(0, 4)}$

20. Express each of the following in interval form:

a. $[-2, 0] \cap [-1, \frac{1}{2}]$ **b.** $[-2, 0] \cap (-1, \frac{1}{2}]$

c. $[-2, 0) \cap [-1, \frac{1}{2}]$ **d.** $[-2, 0) \cap (-1, \frac{1}{2}]$

e. $[-2, 0] \cup [-1, \frac{1}{2}]$ **f.** $(-2, 0] \cup [-1, \frac{1}{2}]$

g. $[-2, 0] \cup [-1, \frac{1}{2})$ **h.** $(-2, 0] \cup [-1, \frac{1}{2})$

i. $\overline{[-2, 0]} \cap (-1, \frac{1}{2}]$ **j.** $[-2, 0) \cap \overline{(-1, \frac{1}{2}]}$

21. Describe as best you can each of the following sets:

a. $\mathbb{Z} \cap [0, 1]$ **b.** $\mathbb{Z} \cap (0, 1)$

c. $\mathbb{Q} \cap [0, 1]$ **d.** $\mathbb{Q} \cap (0, 1)$

e. $\mathbb{R} \cap [0, 1]$ **f.** $\mathbb{R} \cap (0, 1)$

3.3

Boolean Algebra

For two main reasons, we shall now pause to generalize on some of the ideas presented in the chapter so far. First, by generalizing, we shall be able to gather more information about sets. But more importantly, such generalizing will help us with the logical design of the electronic circuits used in computers, a topic discussed in Chapter 4.

The algebraic structure defined below is named in honor of English logician George Boole (1815–1864), one of the key contributors to the field of mathematical logic.

Definition

Let B be a set on which is defined the binary operations $\vee$ and $\wedge$. Then *the system* $(B, \vee, \wedge)$ *is called a Boolean algebra* if it satisfies each of the following four properties.

1. The operations $\vee$ and $\wedge$ are *commutative*; that is,

$$a \vee b = b \vee a \quad \text{and} \quad a \wedge b = b \wedge a \qquad \text{for all } a, b \in B$$

2. Each operation is *distributive* over the other; that is,

$$a \vee (b \wedge c) = (a \vee b) \wedge (a \vee c)$$

and

$$a \wedge (b \vee c) = (a \wedge b) \vee (a \wedge c)$$

for all $a, b, c, \in B$.

3. There are elements 0 and 1 in B with the properties

$$a \vee 0 = a \quad \text{and} \quad a \wedge 1 = a \qquad \text{for all } a \in B$$

4. For each $a \in B$, there is an element $a' \in B$, called the *complement of a*, with the properties

$$a \vee a' = 1 \quad \text{and} \quad a \wedge a' = 0$$

Example 3.3.1   **Boolean Algebra**

Let I be any nonempty set and let $\mathscr{P}(I)$ be the set of all subsets of I; that is,

$$\mathscr{P}(I) = \{A \mid A \subseteq I\}$$

The set $\mathscr{P}(I)$ is called the **power set of** I, and the system $(\mathscr{P}(I), \cup, \cap)$ satisfies each of the following properties:

1S.[†] The operations $\cup$ and $\cap$ are *commutative*; that is,

$$A \cup B = B \cup A \quad \text{and} \quad A \cap B = B \cap A \qquad \text{for all } A, B \in \mathscr{P}(I)$$

2S. Each operation is *distributive* over the other; that is, for all A, B, C $\in \mathscr{P}(I)$,

$$A \cup (B \cap C) = (A \cup B) \cap (A \cup C)$$
$$A \cap (B \cup C) = (A \cap B) \cup (A \cap C)$$

3S. $\phi, I \in \mathscr{P}(I)$ and have the properties

$$A \cup \phi = A \quad \text{and} \quad A \cap I = A \qquad \text{for all } A \in \mathscr{P}(I).$$

4S. For each $A \in \mathscr{P}(I)$, the *complement* $\bar{A} = I - A \in \mathscr{P}(I)$ has the properties

$$A \cup \bar{A} = I \quad \text{and} \quad A \cap \bar{A} = \phi$$

The second distributive property given in (2S) was verified in Example 3.2.6 by the method introduced in that section. The others can similarly be verified by using the definitions of the various set operations. As a result, *the system*

$$(\mathscr{P}(I), \cup, \cap)$$

is an illustration of a Boolean algebra.

Again, let us return to the more general setting.

Facts

If system $(B, \vee, \wedge)$ is a Boolean algebra, then the following properties hold:

5. The operations $\vee$ and $\wedge$ are **associative**; that is,

$$a \vee (b \vee c) = (a \vee b) \vee c$$

and

$$a \wedge (b \wedge c) = (a \wedge b) \wedge c$$

for all $a, b, c \in B$.

---

† The suffix **S** stand for "sets."

6. Each $a \in B$ is ***idempotent*** with respect to each of the operations; that is,

$$a \vee a = a \quad \text{and} \quad a \wedge a = a$$

for all $a \in B$.

7. For each $a \in B$,

$$a \vee 1 = 1 \quad \text{and} \quad a \wedge 0 = 0$$

8. For each $a \in B$, complement a' is unique and $(a')' = a$.

9. $1' = 0$ and $0' = 1$.

10. ***De Morgan's laws*** hold; that is,

$$(a \vee b)' = a' \wedge b' \qquad \text{and} \qquad (a \wedge b)' = a' \vee b'$$

for all $a, b \in B$.

Proofs showing how properties 5 through 10 can be logically deduced from properties 1 through 4 are found in many books on set theory or abstract algebra,[†] so we shall accept these results as valid without repeating that work here.

Example 3.3.2

Additional Set Properties

Since the preceding facts apply to all Boolean algebras, we can apply them in particular to the Boolean algebra of sets presented in the preceding example.

For the system $(\mathcal{P}(I), \cup, \cap)$,

5S. The operations $\cup$ and $\cap$ are *associative*; that is,

$$A \cup (B \cup C) = (A \cup B) \cup C$$

and

$$A \cap (B \cap C) = (A \cap B) \cap C \qquad \text{for all } A, B, C \in \mathcal{P}(I)$$

6S. Each $A \in \mathcal{P}(I)$ is *idempotent* with respect to each of the operations; that is,

$$A \cup A = A \quad \text{and} \quad A \cap A = A \quad \text{for all } A \in \mathcal{P}(I)$$

7S. For each $A \in \mathcal{P}(I)$,

$$A \cup I = I \quad \text{and} \quad A \cap \phi = \phi$$

8S. For each $A \in \mathcal{P}(I)$, complement $\bar{A}$ is unique and

$$\bar{\bar{A}} = A$$

[†] David M. Burton, *Introduction to Modern Abstract Algebra* (Reading, Mass: Addison-Wesley, 1967) is one such book.

9S. $\bar{I} = \phi$ and $\bar{\phi} = I$.

10S. *De Morgan's laws* hold; that is,

$$\overline{A \cup B} = \bar{A} \cap \bar{B} \qquad \text{and} \qquad \overline{A \cap B} = \bar{A} \cup \bar{B} \qquad \text{for all } A, B, \in \mathscr{P}(I) \qquad \blacksquare$$

The next five examples show how set properties 1S through 10S can be used to simplify expressions involving several set operations. Properties used will be cited to the right side of the expression.

Example 3.3.3 **Simplification**

$$
\begin{aligned}
& A \cap (A \cup B) & \\
&= (A \cap A) \cup (A \cap B) & 2S \\
&= \quad A \quad \cup (A \cap B) & 6S \\
&= (A \cap I) \cup (A \cap B) & 3S \\
&= A \cap (I \cup B) & 2S \\
&= A \cap (B \cup I) & 1S \\
&= A \cap I & 7S \\
&= \quad A & 3S
\end{aligned}
$$

So $A \cap (A \cup B) = A$. $\qquad \blacksquare$

Example 3.3.4 **Simplification**

$$
\begin{aligned}
& (A \cap B) \cup (A \cap \bar{B}) \cup (\bar{A} \cap B) \cup (\bar{A} \cap \bar{B}) & \\
&= (A \cap (B \cup \bar{B})) \cup (\bar{A} \cap (B \cup \bar{B})) & 2S \\
&= (A \cap I) \cup (\bar{A} \cap I) & 4S \\
&= \quad A \cup \bar{A} & 3S \\
&= \quad I & 4S
\end{aligned}
$$

Hence, $(A \cap B) \cup (A \cap \bar{B}) \cup (\bar{A} \cap B) \cup (\bar{A} \cap \bar{B}) = I$. $\qquad \blacksquare$

Example 3.3.5 **Right Distributive Properties**

$$
\begin{aligned}
& (A \cup B) \cap C & \\
&= C \cap (A \cup B) & 1S \\
&= (C \cap A) \cup (C \cap B) & 2S \\
&= (A \cap C) \cup (B \cap C) & 1S
\end{aligned}
$$

As a result, we have

$$(A \cup B) \cap C = (A \cap C) \cup (B \cap C)$$

In a similar fashion, we can derive

$$(A \cap B) \cup C = (A \cup C) \cap (B \cup C)$$

These are called *right distributive properties* to distinguish them from the left distributive properties stated in 2S. $\qquad \blacksquare$

In the next example, we shall make use of the identity

$$A - B = A \cap \bar{B}$$

from Example 3.2.5.

Example 3.3.6   Set Identity

$$(A - B) - C$$

$= (A \cap \bar{B}) - C$	Example 3.2.5
$= (A \cap \bar{B}) \cap \bar{C}$	Example 3.2.5
$= A \cap (\bar{B} \cap \bar{C})$	5S
$= A \cap (\overline{B \cup C})$	10S
$= A - (B \cup C)$	Example 3.2.5

Thus, $(A - B) - C = A - (B \cup C)$.   ■

Example 3.3.7   Simplification

$$(A - B) - A$$

$= A - (B \cup A)$	Example 3.3.6
$= A - (A \cup B)$	1S
$= (A - A) - B$	Example 3.3.6
$= \phi - B$	Definition of subtraction
$= \phi$	Definition of subtraction

So, $(A - B) - A = \phi$.   ■

Awareness of the properties of the Boolean algebra of sets can sometimes help in constructing more efficient algorithms, as the next example illustrates.

Example 3.3.8   Algorithm Improvement

A certain town of 50000 residents is known to contain 35000 people aged 65 or over and 6000 people with physical handicaps. We shall construct an algorithm that will take in the name NAM, address ADD, age AGE, and handicap HAN (coded 1 for yes and 0 for no) of each person in the town and print out a list of names and addresses of those handicapped persons aged 65 or over.

Letting

S = set of all persons aged 65 or over

H = set of all persons with a physical handicap

we seek the names and addresses of all those in the set $S \cap H$. The flowchart in Figure 10 illustrates one algorithm that accomplishes the task.

Note that with the flowchart in Figure 10, there will be 50000 age questions and 35000 yes answers to the age question, so that the handicap

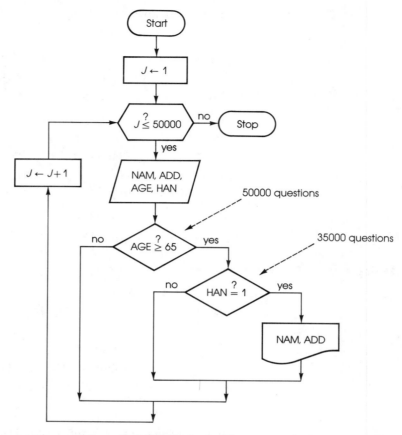

Figure 10

question will be asked 35000 times. Thus, a total of 85000 questions must be answered before the entire population has been processed.

However, if one remembers the set property

$$S \cap H = H \cap S$$

then one could modify the flowchart as shown in Figure 11. Here, the handicap question will be asked 50000 times with only 6000 yes answers; so the age question will be asked only 6000 times. Hence, a total of only 56000 questions will be asked using this algorithm—quite a saving in effort and time (possibly expensive computer time!) as compared with the first algorithm.   ∎

One final note here. Several of the Boolean algebra properties listed above can be extended to apply to additional sets. For instance, the commutative property 1S written for two sets can be extended to three or more sets; for example,

$$A \cap B \cap C = A \cap C \cap B = B \cap A \cap C = B \cap C \cap A = C \cap A \cap B$$
$$= C \cap B \cap A$$

where parentheses can be placed anywhere.

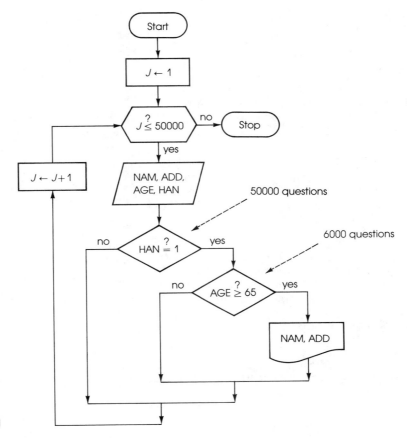

Figure 11

EXERCISES **3.3**

1. Verify that properties 1S, 2S, 4S, 5S, 6S, 8S, and 10S of the text hold for sets

$I = \{0, 1, \ldots, 12\}$ $A = \{3, 4, 5, 6, 7\}$

$B = \{5, 6, 7, 8, 9\}$ $C = \{0, 3, 6, 9, 12\}$

2. Verify that properties 1S, 2S, 4S, 5S, 6S, 8S, and 10S of the text hold for sets

$I = \{b, d, e, k, m, p, s, t, w\}$ $A \doteq \{b, m, p, s, w\}$

$B = \{d, e, m, t\}$ $C = \{b, d, k, p, s, t, w\}$

In exercises 3 through 8, use the Boolean algebra properties 1S through 10S and specify a reason at each step.

3. Simplify the following:

a. $A \cup (A \cap B)$
b. $(A \cup B) \cap (\bar{A} \cup B) \cap (A \cup \bar{B})$
c. $(A \cap B) \cup (\bar{A} \cap B)$

4. Simplify the following:

a. $(A \cap I) \cup (A \cap B)$
b. $(A \cup B) \cap (A \cup \bar{B})$
c. $(A \cap B) \cup (\bar{A} \cap B) \cup (A \cap \bar{B})$

5. Verify the right distributive law:

$(A \cap B) \cup C = (A \cup C) \cap (B \cup C)$

6. Verify the following generalized distributive laws:

$A \cap (B \cup C \cup D) = (A \cap B) \cup (A \cap C) \cup (A \cap D)$

$A \cup (B \cap C \cap D) = (A \cup B) \cap (A \cup C) \cap (A \cup D)$

7. Use the identity $A - B = A \cap \bar{B}$ to verify the following:

a. $A - (B \cap C) = (A - B) \cup (A - C)$

b. $A \cap (B - C) = (A \cap B) - (A \cap C)$
$\qquad\qquad = (A \cap B) - C$

8. Use the identity $A - B = A \cap \bar{B}$ to verify the following:

a. $A - (B - C) = (A - B) \cup (A \cap C)$

b. $A \cup (B - C) = (A \cup B) - (C - A)$

9. Using the information given in Example 3.3.8, construct a flowchart that will take the same input and print out a list of names and addresses of those persons aged 65 or over but *not* handicapped. Which Boolean property helps to improve the efficiency of your algorithm?

10. Using the same information given in Example 3.3.8, construct a flowchart that will take the same input and print out a list of names and addresses of those persons aged 65 or over, or handicapped. Which Boolean property helps to improve the efficiency of your algorithm?

Exercises 11 through 14 also relate to Example 3.3.8 of the text, but with the additional information that 11000 residents have an annual income of less than $10000. In each of these exercises, you must construct a flowchart that, in addition to name NAM, address ADD, age AGE, and handicap HAN, will also take in the annual income INC of each person in town. The flowchart must produce the names and addresses of those groups specified in the exercise. In each case, indicate which Boolean property helps to simplify, or make more efficient, the algorithm.

11. Those who are 65 (or over) and handicapped, or 65 (or over) and with incomes under $10000

12. Those who are 65 (or over) or handicapped, and 65 (or over) or with incomes under $10000

13. Those who are 65 (or over) or handicapped or with incomes under $10000

14. Those who are 65 (or over) and handicapped and with incomes under $10000

15. a. Illustrate De Morgan's laws by drawing Venn diagrams.

b. Verify De Morgan's laws by using the definition of set equality and set operations.

(These two methods were illustrated in Section 3.2.)

16. Can you guess what De Morgan's laws might look like if extended to three sets A, B, and C? Can you verify them?

3.4

Partitions, Cartesian Products, and Relations

Just as the operations of union, intersection, and difference allow us to form new sets from given ones, so other important methods can be used to construct sets. The first presented here is that of partitioning, that is, breaking a given set into nonoverlapping pieces. After presenting the notions of Cartesian products and relations, the last result of the section connects a particular type of relation (namely, an equivalence relation) back to the idea of partition. In the context of a time-sharing computer system, the term *partition* means that the core memory is divided into distinct portions, which are allotted to users on the system at a given time.

Definition

> A *partition of a set S* is a collection of subsets of S with the property that each element in S is an element of precisely one subset.

Example 3.4.1

Partitions

a. If $S = \{0, 1, \ldots, 10\}$ then

$$P_1 = \{\{0, 2, 4, 6, 8, 10\}, \{1, 3, 5, 7, 9\}\}$$
$$P_2 = \{\{0, 3, 6\}, \{2, 4\}, \{1\}, \{5, 7, 8, 9, 10\}\}$$
$$P_3 = \{\{1\}, \{2, 3\}, \{0, 4, 5, \ldots, 10\}\}$$

are three different partitions of S.

b. Let $B = \{000, 001, 010, 100, 011, 101, 110, 111\}$, the set of all three-digit binary strings. Then

$$P_1 = \{\{000, 011, 101, 110\}, \{001, 010, 100, 111\}\}$$
$$P_2 = \{\{000\}, \{001, 010, 100\}, \{011, 101, 110\}, \{111\}\}$$

are two different partitions of B.   ■

Example 3.4.2

Partition Formed by Two Sets

The sets $A, B \subseteq U$ create a partition of $U = \{P_1, P_2, P_3, P_4\}$ where

$$P_1 = A - B \qquad P_2 = B - A$$
$$P_3 = A \cap B \qquad P_4 = U - (A \cup B)$$

as shown in Figure 12. Note that $P_3 = \phi$ if A and B are disjoint.   ■

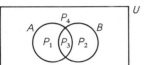

Figure 12

Example 3.4.3

Partitions: Familiar Number Systems

a. $\{\{2k \mid k \in \mathbb{Z}\}, \{2k + 1 \mid k \in \mathbb{Z}\}\}$

is a partition of integers $\mathbb{Z}$ into subsets of even and odd integers, respectively.

b. $\{\mathbb{Z}, \{m/n \mid m, n \in \mathbb{Z}, n \neq 0, m$ not multiple of $n\}\}$

is a partition of rational numbers $\mathbb{Q}$ into subsets of integers and nonintegers, respectively.

c. $\{\mathbb{Q}, \mathbb{R} - \mathbb{Q}\}$ is a partition of real numbers $\mathbb{R}$ into rationals and irrationals, respectively.   ■

Example 3.4.4

Partition: Integers mod 4

Recall the set of integers modulo 4

$$\{0, 1, 2, 3\}$$

discussed in Example 2.6.2. If a is any one of these four integers, let

$$[a] = \{x \in \mathbb{Z} \mid x \equiv a \ (\text{mod } 4)\}$$

Then using the definition of congruence mod 4, it is easy to check that

$$[0] = \{\dots, -16, -12, -8, -4, 0, 4, 8, 12, 16, 20, \dots\}$$
$$[1] = \{\dots, -15, -11, -7, -3, 1, 5, 9, 13, 17, 21, \dots\}$$
$$[2] = \{\dots, -14, -10, -6, -2, 2, 6, 10, 14, 18, 22, \dots\}$$
$$[3] = \{\dots, -13, -9, -5, -1, 3, 7, 11, 15, 19, 23, \dots\}$$

Therefore, we see that

$$\{[0], [1], [2], [3]\}$$

is a partition of $\mathbb{Z}$.   ■

Example 3.4.5

Nonpartitions

a. The collection of sets $\{\mathbb{Z}, \mathbb{Q}, \mathbb{R} - \mathbb{Q}\}$ is *not* a partition of real numbers $\mathbb{R}$ since

$$\mathbb{Z} \cap \mathbb{Q} = \mathbb{Z} \neq \phi$$

b. The power set (that is, set of all subsets) of any set S with more than one element is not a partition of S, because each element of S will belong to more than one subset of S. For instance, if $S = \{a, b\}$ then

$$P(S) = \{\phi, \{a\}, \{b\}, \{a, b\}\}$$ ∎

Definition

> The **Cartesian product of two sets A and B** is the set
>
> $$\{(a, b) \mid a \in A, b \in B\}$$
>
> where an element (a, b) is called an **ordered pair** with first coordinate a and second coordinate b.
>
> **Notation:** $A \times B$
>
> **Extension:** This definition can be generalized to any number n of sets
>
> $$A_1 \times A_2 \times \cdots \times A_n$$
>
> and the elements $(a_1, a_2, \ldots, a_n)$ are called **ordered n-tuples**.

Example 3.4.6

Cartesian Product

a. If $A = \{1, 2\}$ and $B = \{a, b, c\}$ then

$$A \times B = \{(1, a), (1, b), (1, c), (2, a), (2, b), (2, c)\}$$

b. If $Z_3 = \{0, 1, 2\}$ then

$$Z_3 \times Z_3 = \{(0, 0), (0, 1), (0, 2), (1, 0), (1, 1), (1, 2), (2, 0), (2, 1), (2, 2)\}$$

c. If $Z_2 = \{0, 1\}$ then

$$Z_2 \times Z_2 \times Z_2 = \{(0, 0, 0), (0, 0, 1), (0, 1, 0), (1, 0, 0), (0, 1, 1), (1, 0, 1),$$
$$(1, 1, 0), (1, 1, 1)\}$$

d. $\mathbb{Z} \times (\mathbb{Z} - \{0\}) = \{(m, n) \mid m, n \in \mathbb{Z}, n \neq 0\}$

is an alternate way of representing the set of rational numbers. ∎

Example 3.4.7

Elements of Cartesian Products

Using the notation of the preceding example, then the triple, or 3-tuple,

$$(-2, 0, 962) \in \mathbb{Z} \times \mathbb{Z} \times \mathbb{Z}$$

and the 5-tuple

$$(1, 0, 1, 1, 0) \in Z_2 \times Z_2 \times Z_2 \times Z_2 \times Z_2$$ ∎

Definitions

> *A relation from A to B* is a subset of $A \times B$. A relation from A to A (that is, a subset of $A \times A$) is called a *relation on A.*
>
> **Notation:** If R is a relation from A to B, then we write
>
> $a \, R \, b$ in place of $(a, b) \in R$
>
> $a \, \not{R} \, b$ in place of $(a, b) \notin R$

Example 3.4.8

Relations

Referring to the parts of Example 3.4.6 above, we have

a. $R = \{(1, c), (2, a), (2, c)\}$ is a relation from A to B, where $1 \, R \, c$, $2 \, R \, a$, $2 \, R \, c$; but $2 \, \not{R} \, b$, $1 \, \not{R} \, a$, $1 \, \not{R} \, b$.
b. $R = \{(0, 0), (0, 1), (1, 1), (2, 1)\}$ is a relation on Z_3, where $0 \, R \, 0$, $0 \, R \, 1$, $1 \, R \, 1$, $2 \, R \, 1$; but $0 \, \not{R} \, 2$, $1 \, \not{R} \, 2$. ■

Sometimes it is helpful to represent a relation on a set pictorially by means of a *directed graph.*[†]

Example 3.4.9

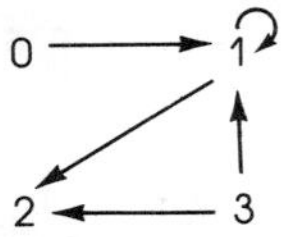

Relation: Directed Graph

Let $Z_4 = \{0, 1, 2, 3\}$. Then the relation on Z_4

$$\{(0, 1), (1, 1), (3, 2), (1, 2), (3, 1)\}$$

can be drawn as a directed graph (see Figure 13), where each arrow $a \to b$ represents the ordered pair (a, b). ■

Figure 13

Definition

> A relation R on set A is called an *equivalence relation on A* if it satisfies each of the following properties:
>
> **1.** R is *reflexive*; that is, for each $a \in A$,
>
> $a \, R \, a$
>
> **2.** R is *symmetric*; that is, for all $a, b \in A$,
>
> if $a \, R \, b$, then $b \, R \, a$
>
> **3.** R is *transitive*; that is, for all $a, b, c \in A$,
>
> if $a \, R \, b$ and $b \, R \, c$, then $a \, R \, c$
>
> **Notation:** Usually the symbol $\sim$ is used in place of R.

Example 3.4.10

Nonequivalence Relation

The relation R in Example 3.4.8b is neither reflexive ($2 \, \not{R} \, 2$) nor symmetric ($2 \, R \, 1$, yet $1 \, \not{R} \, 2$), so it is not an equivalence relation on Z_3. ■

[†] Graphs and trees are formally defined in Section 7.1.

Example 3.4.11

Equivalence Relation

The relation R on $Z_3 = \{0, 1, 2\}$ defined by

$$R = \{(0, 0), (1, 1), (2, 2), (1, 2), (2, 1)\}$$

is an equivalence relation on Z_3 because:

1. R is reflexive:

$$0 \, R \, 0, \quad 1 \, R \, 1, \quad 2 \, R \, 2$$

2. R is symmetric:

$$1 \, R \, 2 \quad \text{and} \quad 2 \, R \, 1$$

3. R is transitive:

$$1 \, R \, 2 \quad 2 \, R \, 1 \quad \text{and} \quad 1 \, R \, 1$$
$$2 \, R \, 1 \quad 1 \, R \, 2 \quad \text{and} \quad 2 \, R \, 2$$

∎

Example 3.4.12

Equivalence Relation: Binary Strings

Let

$$B = \{000, 001, 010, 100, 011, 101, 110, 111\}$$

be the set of all three-digit binary strings, and for each $s \in B$, let $\sum s$ represent the sum of the bits of s. Define R as follows:

For $s_1, s_2 \in B$,

$$s_1 \, R \, s_2 \quad \text{means} \quad \sum s_1 = \sum s_2$$

For example, $001 \, R \, 010$ and $110 \, R \, 101$, but $010 \, \not{R} \, 011$. Then

1. R is reflexive: For any $s \in B$,

$$\sum s = \sum s, \qquad \text{so } s \, R \, s$$

2. R is symmetric: For $s_1, s_2 \in B$, if $s_1 \, R \, s_2$, then $\sum s_1 = \sum s_2$, so

$$\sum s_2 = \sum s_1 \quad \text{or} \quad s_2 \, R \, s_1$$

3. R is transitive: For $s_1, s_2, s_3 \in B$, if $s_1 \, R \, s_2$ and $s_2 \, R \, s_3$, then $\sum s_1 = \sum s_2$ and $\sum s_2 = \sum s_3$, so

$$\sum s_1 = \sum s_2 = \sum s_3 \quad \text{or} \quad s_1 \, R \, s_3$$

The above work verifies that R is an equivalence relation on B. But, continuing with this example, for each $s \in B$, let

$$[s] = \{t \in B \,|\, t \, R \, s\}$$

called the **equivalence class of element s**. In this case,

$$[000] = \{t \in B \,|\, t \, R \, 000\} = \{t \in B \,|\, \sum t = 0\} = \{000\}$$
$$[001] = \{t \in B \,|\, t \, R \, 001\} = \{t \in B \,|\, \sum t = 1\} = \{001, 010, 100\}$$

$$[011] = \{t \in B \,|\, t \, R \, 011\} = \{t \in B \,|\, \textstyle\sum t = 2\} = \{011, 101, 110\}$$
$$[111] = \{t \in B \,|\, t \, R \, 111\} = \{t \in B \,|\, \textstyle\sum t = 3\} = \{111\}$$

Using this same process, one can verify that

$$[001] = [010] = [100]$$
$$[011] = [101] = [110]$$

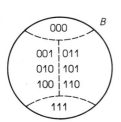

Figure 14

We have performed these computations to indicate that the four equivalence classes of R

$$[000] = \{000\} \qquad\qquad [001] = \{001, 010, 100\}$$
$$[011] = \{011, 101, 110\} \qquad [111] = \{111\}$$

form a partition of set B, as shown in Figure 14. ∎

What happened in the preceding example is always the case!

Fact: If $\sim$ is an equivalence relation on set S, then the equivalence classes of $\sim$ form a partition of S. Conversely, a partition of set S can be used to define an equivalence relation on S whose equivalence classes are the given partition.

The proof of this fact can be found in most books[†] on elementary number theory, so we shall omit it here. Exercises 17 and 18 at the end of this section relate to portions of the proof.

Example 3.4.13

Equivalence Relation: Integers mod 4

In Example 3.4.4 we saw that

$$\{[0], [1], [2], [3]\}$$

formed a partition of $\mathbb{Z}$. So by the preceding fact, these subsets of $\mathbb{Z}$ are equivalence classes of some equivalence relation on $\mathbb{Z}$. As you might guess, the relation R defined by: For $a, b \in \mathbb{Z}$,

$$a \, R \, b \quad \text{means} \quad a \equiv b \,(\text{mod } 4)$$

can be shown (see exercise 14 below) to be an equivalence relation on $\mathbb{Z}$ with equivalence classes

$$\{[0], [1], [2], [3]\}$$

(See Figure 15.)

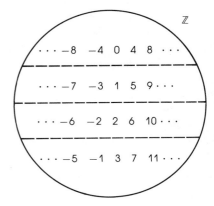

Figure 15 ∎

† David M. Burton, *Introduction to Modern Abstract Algebra* (Reading, Mass.: Addison-Wesley, 1967) pp. 10–11.

EXERCISES **3.4**

1. Given $S = \{-5, -4, \ldots, 4, 5\}$, determine if each of the following is a partition of S:

a. $\{\{-5\}, \{-4\}, \{-3\}, \{-2\}, \{-1\}, \{0\}, \{1\}, \{2\}, \{3\}, \{4\}, \{5\}\}$

b. $\{\{-5\}, \{-4\}, \{-3\}, \{-2\}, \{-1\}, \{1\}, \{2\}, \{3\}, \{4\}, \{5\}\}$

c. $\{\{-5, -4, -3\}, \{-3, -2, -1\}, \{-1, 0, 1\}, \{1, 2, 3\}, \{3, 4, 5\}\}$

d. $\{\{-5, -4, -2, -1\}, \{0\}, \{1, 2, 3, 4, 5\}\}$

e. $\{\{5, -3\}, \{-2, 0, 4, 1\}, \{3, 2, -4\}, \{-1\}, \{-5\}\}$

f. $\{\{1, -4, 3\}, \{0, 2, 5, 1\}, \{4, -2, -3, -4\}\}$

2. Given $B = \{1100, 0110, 0011, 1010, 1001, 0101\}$, determine if each of the following is a partition of B:

a. $\{\{1100, 0110, 0011, 1010, 1001, 0101\}\}$

b. $\{\{1100, 1001\}, \{0110, 0101\}, \{0011, 1010\}\}$

c. $\{\{1100, 0110, 0011\}, \{1010, 0101\}\}$

d. $\{\{1100, 1001\}, \{1001, 0101\}, \{0011, 1010\}\}$

e. $\{\{1100, 1001\}, \{0110, 0101\}, \{0011, 1010\}, \{0110\}\}$

f. $\{\{1100, 0110, 0011\}, \{1010\}, \{1001, 0101\}\}$

3. Are the following partitions of $\mathbb{Z}$?

a. $\{\{k \in \mathbb{Z} \mid k > 0\}, \{k \in \mathbb{Z} \mid k < 0\}\}$

b. $\{\{k \in \mathbb{Z} \mid k \geq 2\}, \{k \in \mathbb{Z} \mid k \leq 2\}\}$

c. $\{\{k \in \mathbb{Z} \mid k > -5\}, \{k \in \mathbb{Z} \mid k \leq -5\}\}$

4. Are the following partitions of $\mathbb{R}$?

a. $\{\mathbb{R}, \phi\}$

b. $\{\mathbb{Z}, \mathbb{R} - \mathbb{Z}\}$

c. $\{\mathbb{Z}, \mathbb{Q}, \mathbb{R} - \mathbb{Q}\}$

5. a. Draw a Venn diagram with universal set U and subsets A, B, and C, each of which intersects the other.

b. List the elements in the partition of U formed by the diagram in (a).

6. Use the notion of congruence mod 7 (see Examples 2.6.2 and 3.4.4) to list a partition of $\mathbb{Z}$ containing seven elements.

In the next two exercises, let $A = \{5\}$, $B = \{a, b, c, d\}$, $Z_2 = \{0, 1\}$, *and* $Z_3 = \{0, 1, 2\}$.

7. a. List the elements in $A \times B$, $B \times B$, $B \times Z_2$, and $A \times A \times A$.

b. Describe the elements of $A \times \mathbb{Z}$.

8. a. List the elements in $A \times A$, $A \times Z_3$, $B \times Z_3$, $Z_2 \times Z_2 \times Z_2$.

b. Describe the elements of $Z_2 \times \mathbb{R}$.

9. For each of the following relations on $Z_7 = \{0, 1, 2, 3, 4, 5, 6\}$, draw the associated directed graph:

a. $\{(0, 0), (1, 3), (5, 3), (3, 5), (2, 5), (4, 5)\}$

b. $\{(0, 1), (1, 2), (2, 3), (3, 4), (4, 5), (5, 6), (6, 0)\}$

c. $\{(0, 2), (1, 3), (2, 4), (3, 5), (4, 6), (5, 0), (3, 3)\}$

10. For each of the directed graphs in Figure 16, find the associated relation on Z_5.

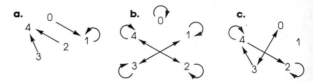

Figure 16

11. Which of the following relations on $Z_3 = \{0, 1, 2\}$ are equivalence relations on Z_3?

a. $R = \{(0, 0), (1, 1), (1, 2), (2,1)\}$

b. $R = \{(0, 0), (1, 1), (2, 2), (1, 2)\}$

c. $R = \{(0, 0), (1, 1), (2, 2), (0, 1), (1, 2)\}$

d. $R = \{(0, 0), (1, 1), (2, 2), (0, 1), (1, 0), (1, 2), (2, 1), (0, 2), (2, 0)\}$

e. $R = \{(0, 0), (1, 1), (2, 2), (1, 2), (2, 1)\}$

12. For those relations in exercise 11 that are equivalence relations, find the equivalence classes.

13. Draw a directed graph for each of the relations in exercise 11. Then translate the properties of an equivalence relation into statements about a directed graph.

14. Let R be a relation on $\mathbb{Z}$ defined as follows:

$a\, R\, b$ means $a \equiv b \pmod{4}$

Use the definition of congruence mod 4 to show that this relation is

a. reflexive
(that is, $a \equiv a \pmod{4}$ for each $a \in \mathbb{Z}$),

b. symmetric
(that is, if $a \equiv b \pmod{4}$, then $b \equiv a \pmod{4}$),

c. transitive
(that is, if $a \equiv b \pmod{4}$ and $b \equiv c \pmod{4}$, then $a \equiv c \pmod{4}$).

15. Use the definition of set equality to prove:

a. $A \times (B \cup C) = (A \times B) \cup (A \times C)$

b. $A \times (B \cap C) = (A \times B) \cap (A \times C)$

c. $A \times (B - C) = (A \times B) - (A \times C)$

16. Referring to Example 3.4.2, use the definition of set equality to prove that

$$(A - B) \cup (B - A) \cup (A \cap B) \cup [U - (A \cup B)] = U$$

17. For any given equivalence relation on set S, prove the following:

    a. Any two equivalence classes are either identical or disjoint.

b. The collection of equivalence classes forms a partition of S.

18. Let I be a set and $\{S_i | i \in I\}$ a partition of set S. Define relation R on S by: For $a, b \in S$

    $a \, R \, b$ means $a, b \in S_i$ for some $i \in I$

    a. Show that R is an equivalence relation on S.

    b. Show that the equivalence classes of R are S_i, $i \in I$.

3.5

Counting Principles

This section and the next present several tools that are frequently useful in solving elementary problems in counting. Recall from Example 3.4.2 that two sets can be used to partition the universal set into four subsets.

Example 3.5.1

Counting

In a certain population of 50000, it is known that 6000 have physical handicaps, 11000 have a yearly income below $10000, and 4500 fall in both of these categories. Thus, a Venn diagram can be used to determine the numbers of people in various categories. Letting

$$H = \text{set of all persons with physical handicaps}$$
$$I = \text{set of all persons with income below \$10000}$$
$$\#H = \text{number of people in } H = 6000$$
$$\#I = \text{number of people in } I = 11000$$

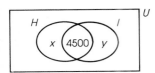

Figure 17

we have (see Figure 17)

$4500 + x = \#H$	$4500 + y = \#I$
$4500 + x = 6000$	$4500 + y = 11000$
$x = 1500$	$y = 6500$

Figure 18

These results are shown in Figure 18. In general, if we let $\#T$ mean the number of elements in finite set T, then

$\#(H \cap I) = 4500$ people have both handicap and income below $10000

$\#(H \cup I) = 1500 + 4500 + 6500 = 12500$ have either handicap or income below $10000

$\#(H - I) = 1500$ have handicap and income at least $10000

$\#(I - H) = 6500$ have income below $10000 and no physical handicap

    ■

We should make a few comments here about the preceding example. First, the given fact that $\#U = 50000$ was unused, irrelevant information.

That is quite typical of "real-world" problems: Often some of the given data for a particular problem are unrelated to the solution. Second, it is easy to verify the correctness of the equation

$$\#(H \cup I) = \#H + \#I - \#(H \cap I)$$

This relationship holds in general, but will be stated here without proof.

Fact

> For finite sets A and B,
>
> $$\#(A \cup B) = \#A + \#B - \#(A \cap B)$$
>
> **Extension:** For finite sets A, B, and C,
>
> $$\#(A \cup B \cup C) = \#A + \#B + \#C - \#(A \cap B)$$
> $$- \#(B \cap C) - \#(A \cap C) + \#(A \cap B \cap C)$$

Thus, to determine a particular number in one of the above equations, all other numbers in that equation must be known.

Example 3.5.2 **Counting**

In a certain population of 50000, it is known that

6000 have physical handicaps,
11000 have yearly incomes below $10000,
35000 are age 65 or over,
3200 are in all three categories,
4500 are handicapped with incomes below $10000,
7200 are age 65 or over with income below $10000, and
39800 are in one of the three categories.

We can determine how many people are handicapped and age 65 or over as follows: Let H, I, $\#H$, and $\#I$ be as in the previous example, and let

S = set of all persons of age 65 or over
$\#S$ = number of people in S

Then we can translate the given data into set notation as:

$$\#H = 6000$$
$$\#I = 11000$$
$$\#S = 35000$$
$$\#(H \cap I \cap S) = 3200$$
$$\#(H \cap I) = 4500$$
$$\#(I \cap S) = 7200$$
$$\#(H \cup I \cup S) = 39800$$

Applying the fact stated above, we have

$$\#(H \cup I \cup S) = \#H + \#I + \#S - \#(H \cap I) - \#(I \cap S)$$
$$- \#(H \cap S) + \#(H \cap I \cap S)$$
$$39800 = 6000 + 11000 + 35000 - 4500 - 7200$$
$$- \#(H \cap S) + 3200$$
$$39800 = 40300 - \#(H \cap S) + 3200$$
$$39800 = 43500 - \#(H \cap S)$$

So $\#(H \cap S) = 3700$; that is, 3700 people are handicapped and 65 or over.

◼

How the above work can be applied in a computer setting is illustrated in the next example.

Example 3.5.3   **Algorithm Efficiency**

Using the data of the preceding example, construct an algorithm that will take in the name NAM, address ADD, age AGE, income INC, and handicap HAN (coded 1 for yes and 0 for no) of each person, and print out a list of the names and addresses of all those handicapped persons aged 65 or over with incomes less than $10000. (The reader might want to take a moment to review Example 3.3.8 before proceeding.)

If we use the notation from above, we need to identify the persons in the set

$$H \cap I \cap S$$

Recalling that

$$H \cap I \cap S = H \cap S \cap I = S \cap H \cap I = S \cap I \cap H$$
$$= I \cap S \cap H = I \cap H \cap S$$

we can ask the three questions

HAN = 1?   INC < 10000?   AGE ≥ 65?

in any order we choose. From our discussion in Example 3.3.8, we know that the fewer questions asked, the less computer time is required and the more efficient the algorithm. Since set H has fewer people than I or S, then the question

HAN = 1?

which has only 6000 yes answers, should be asked first. Which of the other two questions should be asked next? Since

$$\#(H \cap I) = 4500 \quad \text{and} \quad \#(H \cap S) = 3700$$

then

AGE ≥ 65?

which has 3700 yes answers, should be asked next. This leaves

INC < 10000?

to be asked last. The flowchart for this algorithm is in Figure 19. ■

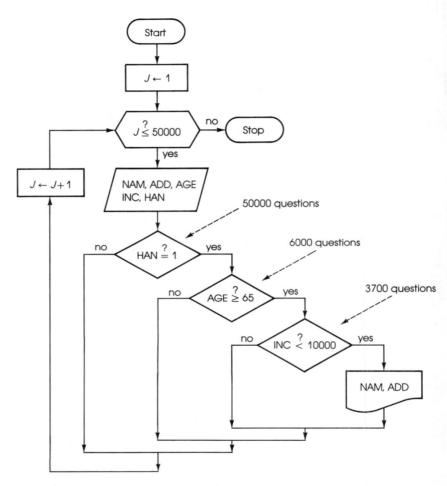

Figure 19

The following counting tool, although simply stated, is nevertheless very useful.

Fundamental Counting Principle

If one event can be accomplished *m* ways and another can be accomplished *n* ways, then both can be accomplished

$$m * n$$

ways.

Example 3.5.4   **Fundamental Counting Principle**

In some versions of the BASIC programming language, a numeric variable can be represented by either a letter of the alphabet or a letter followed by a decimal digit. There are 26 alphabet letters, and for the pairs consisting of a letter followed by a digit there are by the Fundamental Counting Principle

$$26 * 10 = 260$$

possible combinations. (For example, A7, M2, and T8 are three possibilities.) So such versions of BASIC contain

$$260 + 26 = 286$$

symbols for numeric variables.   ∎

Example 3.5.5   **Fundamental Counting Principle**

Let sets A and B have five and four elements, respectively. Elements in $A \times B$ are ordered pairs (a, b). Since there are five possible ways to fill the first coordinate and four ways to fill the second, there are $5 \times 4 = 20$ elements in $A \times B$. Similarly, there are $5 \times 5 = 25$ elements in $A \times A$ and $4 \times 4 = 16$ elements in $B \times B$.   ∎

The Fundamental Counting Principle (FCP) can be extended to any number of events, as the next few examples illustrate.

Example 3.5.6   **Extended FCP: Binary Strings**

a. In a five-bit string

there are exactly two ways of filling each of the five positions, namely, by placing a 0 or 1 in that position. Thus, there are

$$2 * 2 * 2 * 2 * 2 = 2^5 = 32$$

such five-bit strings.

b. By similar reasoning, there are

$$2^8 = 256$$

binary strings of eight digits—that is, 256 different bytes. Thus, the commonly used eight-bit binary codes ASCII and EBCDIC (see Section 2.5) can code up to 256 different characters.   ∎

Example 3.5.7   **Extended FCP: Floating Point Numbers**

A certain register contains eight positions to store a floating-point number:

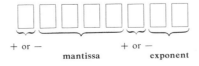

+ or −   + or −
    mantissa   exponent

The number of different floating-point numbers that this register can hold using binary digits is

$$2 * 2 * 2 * 2 * 2 * 2 * 2 * 2 = 2^8 \qquad = 256$$

using octal digits is

$$2 * 8 * 8 * 8 * 8 * 2 * 8 * 8 = 2^2 * 8^6 = 1,048,576$$

using decimal digits is

$$2 * 10 * 10 * 10 * 10 * 2 * 10 * 10 = 2^2 * 10^6 = 4,000,000$$

using hexadecimal digits is

$$2 * 16 * 16 * 16 * 16 * 2 * 16 * 16 = 2^2 * 16^6 = 67,108,864 \qquad \blacksquare$$

Example 3.5.8   **Extended FCP: Subsets of a Set**

Let S be a set with six elements

Since any subset is formed by saying yes or no to each of the six positions, then there are

$$2 * 2 * 2 * 2 * 2 * 2 = 2^6 = 64$$

different subsets of S. This work can be generalized to show that

a set of n elements has 2^n different subsets. $\blacksquare$

We conclude this section with illustrations of a geometric counting tool, *tree graphs*.

Example 3.5.9   **Tree Graph**

A professional basketball playoff series consists of (at most) three games between teams A and B; the winner is the first to win two games. The tree diagram in Figure 20 provides a convenient way to list all possible outcomes.

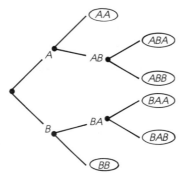

Figure 20

Here, a dot represents a game, and the letter A or B the winner of that game. The combination AB, for example, means that A won the first game and B won the second. The six possible outcomes are circled. ∎

Example 3.5.10   **Tree Graph**

Suppose we are to list all possible five-bit strings whose digits add to 2. Of course, we could list all possible five-bit strings and choose those whose digits add to 2. But the tree diagram in Figure 21 is a systematic and shorter way. The six circled strings with fewer than five bits must be augmented with zeroes on the right to yield the ten desired strings:

$$11000 \quad 01100 \quad 10100 \quad 00110 \quad 01010$$
$$10010 \quad 00011 \quad 00101 \quad 01001 \quad 10001$$

∎

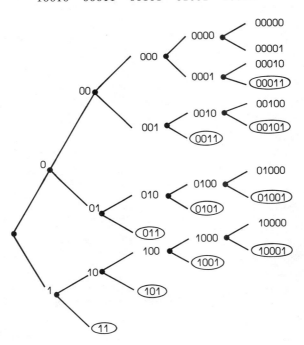

Figure 21

EXERCISES 3.5

In the first four exercises, find the number of elements in the indicated set.

	#A	#B	#(A ∩ B)	#(A ∪ B)
1.	200	125	72	?
2.	172	321	?	433
3.	94	?	25	202
4.	?	192	77	314

5. In a certain city with 460000 eligible voters, 52% voted in the 1976 presidential election, 56% voted in the 1980 presidential election, and 45% voted in both. How many voters went to the polls in one or both of these elections?

6. Of 2042 deaths recorded over one year at an urban hospital, 550 were cancer related, 732 were due to heavy smoking, and 980 were in one or both of these categories. How many of these deaths involved both cancer and heavy smoking?

7. A survey of 8000 people yielded the following data on the modes of public transportation used during 1982:

2200 used a bus,
1500 used a train,
800 used a plane,
200 used a bus and a plane,
350 used a plane and a train,
450 used a bus and a train, and
150 used a bus, a train, and a plane.

How many of the 8000 respondents used at least one of the three modes during the year?

8. A poll concerning breakfast cereals *A*, *B*, and *C* in which 2000 children responded yielded the following results:

720 ate *A*,
680 ate *B*,
950 ate *C*,
370 ate *A* and *C*,
420 ate *A* and *B*,
320 ate *B* and *C*, and
220 ate *A* and *B* and *C*.

How many of the children polled ate none of the three cereals?

In the next four exercises, the algorithms constructed should be efficient in the sense of Example 3.5.3.

9. Construct a flowchart for exercise 5 that will read in the names of all eligible voters and print out only the names of those who voted in both elections.

10. Construct a flowchart for exercise 6 that will take in the names of the deceased and print out only the names of those whose deaths involved cancer but not heavy smoking.

11. Draw a flowchart for exercise 7 that will take in the names of those surveyed and print out only the names of those who used a bus and train but not a plane.

12. Draw a flowchart for exercise 8 that will take in the names of those children polled and produce only the names of those children who ate none of the cereals.

13. If sets *A* and *B* have 7 and 3 elements, respectively, then how many elements are in the following:

a. $A \times B$ **b.** $B \times A$
c. $A \times A$ **d.** $B \times B \times B$

How many subsets does *A* have?

14. How many different outcomes are possible if

a. one die is thrown?
b. two dice are thrown?
c. three dice are thrown?

15. If a computer system ID consists of a single letter followed by three decimal digits, how many different IDs are possible?

16. If each license plate for a particular state consists of two letters (except the letter *O*) followed by four decimal digits, how many different license plates are possible?

17. Consider a word to be any sequence of letters formed from the eight letters *a*, *b*, *e*, *m*, *n*, *r*, *s*, and *t*, where repetitions are allowed. Find the number of possible:

a. three-letter words
b. three-letter words beginning with *t*
c. three-letter words with *e* appearing exactly once

18. How many different strings of six characters can be formed using the following:

a. binary digits   **b.** octal digits
c. decimal digits   **d.** hexadecimal digits

19. If it takes 5 seconds to write one eight-bit binary string, how long would it take to list all such strings?

20. In how many ways can five prizes be distributed to five children?

21. Construct a tree graph to find all possible three-bit strings whose digits add to 2.

22. Construct a tree graph to find all possible four-bit strings whose digits add to 3.

23. Use a tree graph to list all possible outcomes in a five-game baseball playoff, the winner being the first to win three games.

24. Use a tree graph to determine all possible outcomes in a two-person tennis championship, the winner being the first to win two sets in a row or a total of three sets.

25. Draw a Venn diagram that relates to Example 3.5.2 and place the appropriate positive integer in each portion of the partition.

26. Use the given formula for $\#(A \cup B)$ to derive the formula for $\#(A \cup B \cup C)$.

27. Verify that the Fundamental Counting Principle gave the correct results in Example 3.5.5 by listing the elements in $A \times A$, $A \times B$, and $B \times B$.

28. For $n = 4$, verify the result given in the last line of Example 3.5.8 by listing all the subsets of $\{a, b, c, d\}$.

3.6

Permutations and Combinations

The Fundamental Counting Principle can also be applied to a situation slightly different from those described in the preceding section.

Example 3.6.1   **Fundamental Counting Principle**

The number of distinct octal numbers

a. with three digits

8 possible choices for each position

is $8 * 8 * 8 = 512$

b. with three digits, *no two of which are the same,*

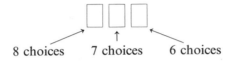

8 choices   7 choices   6 choices

is $8 * 7 * 6 = 336$ ∎

The last product in this example relates to the notion of *factorial*.

Definition

For a given positive integer n, the product

$$1 * 2 * 3 * 4 * \cdots * n$$

is called **n factorial**.

Notation: $n!$

Also, it is useful to define $0! = 1$.

Example 3.6.2   **Factorial**

$$1! = 1$$
$$2! = 1 * 2 = 2$$
$$3! = 1 * 2 * 3 = 6$$
$$5! = 1 * 2 * 3 * 4 * 5 = 120$$
$$8! = 1 * 2 * 3 * 4 * 5 * 6 * 7 * 8 = 40320$$

Example 3.6.3 Factorial

We can now rewrite the result in Example 3.6.1(b). The number of three-digit octal numbers with no repeating digits is

$$8 * 7 * 6 = \frac{8 * 7 * 6 * 5 * 4 * 3 * 2 * 1}{5 * 4 * 3 * 2 * 1} = \frac{8!}{5!} = \frac{8!}{(8 - 3)!} \qquad \blacksquare$$

In other words, what we have found here is the number of arrangements of the eight octal digits $0, 1, \ldots, 7$ taken three at a time. More generally,

Definitions

a. A *permutation of n objects* is an arrangement of n objects in a certain order.

b. A *permutation of n objects taken r at a time* is an arrangement in a certain order of r of the n objects.

Example 3.6.4 Permutations

a. *abcde*, *dbaec*, and *caedb* are some of the permutations of the five objects a, b, c, d, and e.

b. *lpmn*, *lmnp*, *mpko*, *mlkp*, and *klmn* are some of the permutations of the six objects k, l, m, n, o, and p taken four at a time. $\qquad \blacksquare$

Facts

a. $n!$ is the number of permutations of n objects.

b. $_nP_r = \dfrac{n!}{(n - r)!}$ is the number of permutations of n objects taken r at a time.

Example 3.6.5 Permutations

a. The number of permutations of five objects is $5! = 120$.
In Example 3.6.4(a), we listed just 3 of the 120 possibilities.

b. The number of permutations of six objects taken four at a time is

$$_6P_4 = \frac{6!}{(6 - 4)!} = \frac{6!}{2!} = \frac{6 * 5 * 4 * 3 * 2 * 1}{2 * 1}$$

$$= 6 * 5 * 4 * 3 = 360$$

So the previous example has listed only 5 of the 360 possibilities! $\qquad \blacksquare$

Example 3.6.6 Flowchart for a Factorial

The computation of a factorial involves repeated multiplications, so its flowchart, shown in Figure 22, involves a loop. For a reason that will become evident in the next example, we use F rather than n for the given nonnegative

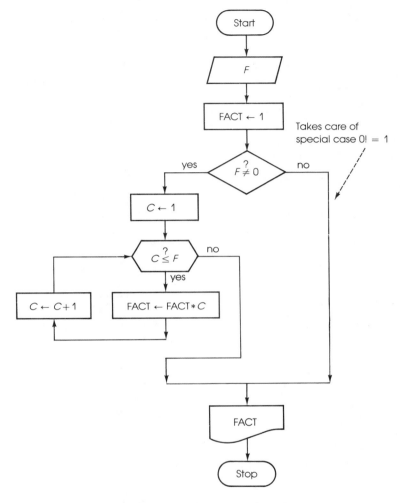

Figure 22

integer. In this flowchart, positive integer F is input and FACT, with value $F!$, is output. ∎

Example 3.6.7

Subalgorithms: Flowchart for $_nP_r$

The formula

$$_nP_r = \frac{n!}{(n-r)!}$$

involves the computation of two factorials. In constructing a flowchart to compute the value of $_nP_r$, it would be possible to include *two* slight variations of the flowchart in Figure 22, but that is unnecessary. Using a subalgorithm, or "subflowchart," that is accessed twice by the main algorithm is a shorter, more efficient method, as shown in Figure 23.

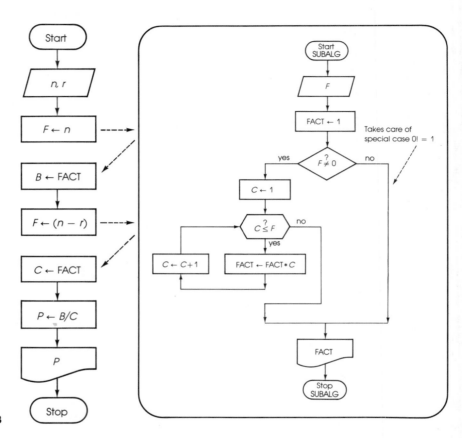

Figure 23

This flowchart takes positive integers n and r as input, and it outputs P, whose value is $_nP_r$. The dotted arrows indicate that after a subalgorithm is accessed and then completed at a particular step, we return to the next step in the main algorithm. ∎

Definition

> A *combination of n objects taken r at a time* is a selection without regard to order of r of the n objects.

Example 3.6.8 **Combinations**

Although *lpm*, *lmp*, *mlp*, *mpl*, *plm*, and *pml* are six permutations of the three letters *l*, *m*, and *p*, nevertheless they provide only *one* combination of the six objects k, l, m, n, o, and p taken three at a time. The same can be said for knp, kpn, nkp, npk, pnk, and pkn. ■

Fact

$$_nC_r = \frac{n!}{r!(n-r)!}$$

is the number of combinations of n objects taken r at a time.

Since $0! = 1$, it follows that $_nC_n = {_nC_0} = 1$.

Example 3.6.9

Combinations

The number of combinations of six objects taken

a. three at a time is

$$_6C_3 = \frac{6!}{3!(6-3)!} = \frac{6!}{3!3!} = \frac{6*5*4*3*2*1}{3*2*1*3*2*1} = 20$$

b. four at a time is

$$_6C_4 = \frac{6!}{4!(6-4)!} = \frac{6!}{4!2!} = \frac{6*5*4*3*2*1}{4*3*2*1*2*1} = 15$$ ■

Example 3.6.10

Combinations

On an exam, a student must answer any 10 of 12 given questions. The number of ways that the student can complete the exam is

$$_{12}C_{10} = \frac{12!}{10!(12-10)!} = \frac{12!}{10!2!} = \frac{12*11}{2} = 66$$ ■

Example 3.6.11

Combinations

Example 3.5.10 used a tree to list all possible five-bit strings whose digits add to 2. Realizing that any such string

must have the 1 bit in exactly two of the five positions and the 0 bit elsewhere, we know that the number of such strings is

$$_5C_2 = \frac{5!}{2!(5-2)!} = \frac{5!}{2!3!} = \frac{5*4}{2} = 10$$ ■

In algebra one calls the sum

$$x + y$$

of two variables x and y a **binomial** and learns how to multiply such expressions. In particular,

$$(x + y)^2 = x^2 + 2xy + y^2$$
$$(x + y)^3 = x^3 + 3x^2y + 3xy^2 + y^3$$
$$(x + y)^4 = x^4 + 4x^3y + 6x^2y^2 + 4xy^3 + y^4$$
$$(x + y)^5 = x^5 + 5x^4y + 10x^3y^2 + 10x^2y^3 + 5xy^4 + y^5$$

The coefficients on the right sides of these equations are called **binomial coefficients**. The concluding example of this section illustrates a connection between these coefficients and the idea of combination, a relationship that will be helpful in our work with probability in Section 3.8.

Example 3.6.12

Binomial Coefficients

a. $(x + y)^1 = x + y$
$$= 1 * x + 1 * y$$
$$= {}_1C_1 * x + {}_1C_0 * y$$

b. $(x + y)^2 = (x + y)(x + y)$
$$= x^2 + \underbrace{xy + yx} + y^2$$
$$= x^2 + {}_2C_1 * xy + y^2$$
$$= {}_2C_2 * x^2 + {}_2C_1 * xy + {}_2C_0 * y^2$$

c. $(x + y)^3 = (x + y)(x + y)(x + y)$
$$= (x + y)(x^2 + xy + yx + y^2)$$
$$= x^3 + \underbrace{x^2y + xyx + yx^2} + \underbrace{y^2x + yxy + xy^2} + y^3$$
$$= x^3 + \qquad {}_3C_2 * x^2y \qquad + \qquad {}_3C_1 * xy^2 \qquad + y^3$$
$$= {}_3C_3 * x^3 + {}_3C_2 * x^2y + {}_3C_1 * xy^2 + {}_3C_0 * y^3$$

Generalizing from these results, it is possible to show that the coefficients in the expansion of

$$(x + y)^n \qquad n \text{ a positive integer}$$

are

$${}_nC_n, {}_nC_{n-1}, {}_nC_{n-2}, \ldots, {}_nC_2, {}_nC_1, {}_nC_0 \qquad \blacksquare$$

EXERCISES 3.6

1. Compute each of the following:

    a. $4!$ **b.** $7!$ **c.** ${}_5P_2$

    d. ${}_7P_4$ **e.** ${}_5C_4$ **f.** ${}_8C_4$

2. Compute each of the following:

    a. $6!$ **b.** $9!$ **c.** ${}_5C_2$

    d. ${}_8P_6$ **e.** ${}_7C_3$ **f.** ${}_8C_6$

3. a. Use the formula to compute ${}_4P_2$. Verify this answer by listing all permutations of $\{A, B, C, D\}$ taken two at a time.

    b. Use the formula to compute ${}_5C_2$. Verify this answer by listing all combinations of $\{A, B, C, D, E\}$ taken two at a time.

4. a. Use the formula to compute $_5P_3$. Verify this answer by listing all permutations of $\{A, B, C, D, E\}$ taken three at a time.
b. Use the formula to compute $_5C_3$. Verify this answer by listing all combinations of $\{A, B, C, D, E\}$ taken three at a time.

5. How many distinct three-digit hexadecimal numbers are there with the following:

   a. no repeating digits
   b. exactly one repeating digit
   c. at most one repeating digit

6. How many distinct four-digit octal numbers are there with the following:

   a. no repeating digits
   b. exactly one repeating digit
   c. at most one repeating digit

7. How many different ways can seven books be arranged on a shelf?

8. How many different ways can a family of six stand in line?

9. How many possible travel itineraries are there for a political candidate who wants to visit three of six cities?

10. Given 25 players to choose from, how many baseball lineups (that is, batting orders) of 9 players can be presented to the umpire?

11. How many starting teams of 5 players can be chosen from a basketball squad of 12?

12. How many possible 5-card hands can be dealt from a 52-card poker deck?

13. How many eight-bit strings are there whose digits add to 5?

14. How many eight-bit strings are there with exactly four 1 bits?

15. For the initial value $F = 8$, do a trace of the flow-chart in Figure 22 at the point preceding

$$\text{FACT} \leftarrow \text{FACT} * C$$

for the variables F, C, and FACT.

16. Perform a trace as described in exercise 15, but for initial value $F = 6$.

17. Modify Figure 22 so that both of the special cases $0! = 1$ and $1! = 1$ are handled by the same branch of the flowchart.

18. Construct a flowchart for $_nC_r$ that uses the computation of factorial as a subalgorithm.

19. Use the Fundamental Counting Principle to verify that $n!$ equals the number of permutations of n objects.

20. Verify the formula for $_nP_r$ by generalizing the work in Examples 3.6.1 and 3.6.2.

21. a. Verify the expressions for $(x + y)^n$, $n = 2, 3, 4, 5$, preceding **Example 3.6.12**.
b. Verify for $n = 4, 5$ the generalization expressed at the end of Example 3.6.12.

22. The binomial coefficients in the expansion of $(x + y)^n$ for $n = 0, 1, 2, 3, 4, 5$ are

$n = 0$						1					
$n = 1$					1		1				
$n = 2$				1		2		1			
$n = 3$			1		3		3		1		
$n = 4$		1		4		6		4		1	
$n = 5$	1		5		10		10		5		1

This is a portion of what is called *Pascal's triangle*, named after the eminent French mathematician–philosopher Blaise Pascal (1623–1662).

   a. Can you discern a pattern for these coefficients?
   b. Write the binomial coefficients for $n = 6, 7, 8, 9$.

23. Verify that:

   a. $_nC_n = {}_nC_0 = 1$ **b.** $_nC_r = {}_nC_{n-r}$

3.7

Elementary Probability

Our world is filled with phenomena whose outcomes exhibit uncertainty. For instance, a typical laboratory experiment may have many possible outcomes. *Probability* is the branch of mathematics that studies such phenomena and their behavior *in the long run*. The set of all possible outcomes (or events) of a given experiment (or observation) is called the **sample space** for the experiment (or observation). Unless specified to the contrary, the sample spaces of this text will all be finite.

Definition

Let S be a sample space for an experiment that can be repeated under identical conditions, and let $A \subseteq S$. The ***probability of event*** A is the fraction

$$\frac{\# A}{\# S}$$

Notation: $\Pr(A)$

Note: Here we assume that there is no bias in favor of particular elements of S; that is, each of the possible outcomes is *equally likely*.

In applying this definition, it is helpful to think of event A as a subset of S with a particular property. Thus, the complement $\bar{A}$ of A is that subset of S *without* that particular property. Also, observe that the numerator of the fraction in the definition is always nonnegative and less than or equal to the denominator, so for any A,

$$0 \leq \Pr(A) \leq 1$$

Since $\# A + \# \bar{A} = \# S$, it follows that $\Pr(A) + \Pr(\bar{A}) = 1$. If $\Pr(A) = 0$, then outcome A is called ***impossible***; if $\Pr(A) = 1$, then A is ***certain***.

It is important to realize that the probability $\Pr(A)$ represents the relative frequency with which outcome A occurs as the experiment is performed a large number of times under identical conditions. So, $\Pr(A) = \frac{2}{7}$ should be taken to mean that if the experiment is repeated many times, say 7000, event A will occur about 2000 times and A will not occur (or $\bar{A}$ *will* occur) about 5000 times. In fact, a more general definition would identify $\Pr(A)$ as the relative, limiting frequency as the number of repetitions of the experiment increases without bound.

Example 3.7.1   **Probability**

a. Consider the experiment of one coin toss. Writing T for tail and H for head, the sample space is simply

$$S = \{H, T\}$$

and

$$\Pr(H) = \frac{\#\{H\}}{\# S} = \frac{1}{2} \qquad \Pr(T) = \frac{\#\{T\}}{\# S} = \frac{1}{2}$$

(To simplify notation here, we write $\Pr(a)$ rather than $\Pr(\{a\})$ for the one-element subset $\{a\} \subseteq S$.) In many coin tosses, say 1000, you can expect about 500 heads and about 500 tails.

b. Consider the experiment of one roll of a die. Then

$$S = \{1, 2, 3, 4, 5, 6\}$$

and

$$\text{Pr(even outcome)} = \text{Pr}(\{2, 4, 6\}) = \frac{\#\{2, 4, 6\}}{\#S} = \frac{3}{6} = \frac{1}{2}$$

$$\text{Pr(odd outcome)} = \text{Pr}(\{1, 3, 5\}) = \frac{\#\{1, 3, 5\}}{\#S} = \frac{3}{6} = \frac{1}{2}$$

$$\text{Pr(multiple of 3 outcome)} = \text{Pr}(\{3, 6\}) = \frac{\#\{3, 6\}}{\#S} = \frac{2}{6} = \frac{1}{3}$$

$$\text{Pr(rolling 5)} = \text{Pr}(5) = \frac{\#\{5\}}{\#S} = \frac{1}{6}$$

$$\text{Pr(not 4 or 6)} = \text{Pr}(\{1, 2, 3, 5\}) = \frac{\#\{1, 2, 3, 5\}}{\#S} = \frac{4}{6} = \frac{2}{3}$$

Thus, in 6000 rolls of a die, one would expect about 3000 even outcomes, about 3000 odd outcomes, about 2000 outcomes of 3 or 6, about 1000 outcomes of 5, and about 4000 outcomes of 1, 2, 3, or 5.

c. Let the experiment be one random selection from the collection S of all four-bit strings, and let

$$A = \{s \in S \mid s \text{ has exactly two bits}\}$$

For example, $0110 \in A$ but $0100 \notin A$. Then

$$\text{Pr}(A) = \frac{\#A}{\#S} = \frac{{}_4C_2}{2*2*2*2} = \frac{\frac{4!}{2!(4-2)!}}{16} = \frac{6}{16} = \frac{3}{8}$$

So, if 8000 such selections were made, approximately 3000 would have exactly two 1 bits. ∎

In the next example, we use the set operations $\cup$ and $\cap$ to mean "or" and "and," respectively, which is consistent with their use earlier in this chapter.

Example 3.7.2

Mutually Exclusive Events

If the experiment is one draw from a 52-card poker deck, then the sample space S is the deck of cards. Let Q and T represent the subsets of queens and tens, respectively. Then

$$\text{Pr(queen)} = \text{Pr}(Q) = \frac{\#Q}{\#S} = \frac{4}{52} = \frac{1}{13}$$

$$\text{Pr(ten)} = \text{Pr}(T) = \frac{\#T}{\#S} = \frac{4}{52} = \frac{1}{13}$$

$$\text{Pr(queen or ten)} = \text{Pr}(Q \cup T) = \frac{\#(Q \cup T)}{\#S} = \frac{8}{52} = \frac{2}{13}$$

$$\text{Pr(queen and ten)} = \text{Pr}(Q \cap T) = \frac{\#(Q \cap T)}{\#S} = \frac{0}{52} = 0 \qquad ∎$$

We should make two comments here. First, since $\Pr(Q \cap T) = 0$, then $Q \cap T$ is an impossible event. This means that Q and T cannot occur simultaneously on one draw, so we call them *mutually exclusive* events. Second,

$$\Pr(Q \cup T) = \Pr(Q) + \Pr(T)$$

These comments can be generalized.

Definition

> Events A and B are called ***mutually exclusive*** if $\Pr(A \cap B) = 0$.

Fact

> **Addition Rule**
>
> If A and B are mutually exclusive events, then
>
> $$\Pr(A \cup B) = \Pr(A) + \Pr(B)$$

As we'll see in the next example, this fact can be extended to a finite union of events that are pairwise mutually exclusive.

Example 3.7.3   **Mutually Exclusive Events**

Table 1 gives information about a particular freshman college class; students were allowed to choose one major only—no dual majors permitted.

Table 1

Sex \ Major	A (Arts)	B (Business)	E (Engineering)	S (Science)	Total
M (Male)	83	226	118	93	520
F (Female)	82	241	102	169	594
Total	165	467	220	262	1114

Suppose that one student is randomly chosen from this class of 1114 freshmen. Then the sample space SS is the set of students, and

$$\Pr(\text{male in arts}) = \Pr(M \cap A) = \frac{\#(M \cap A)}{\# SS} = \frac{83}{1114} = .0745$$

$$\Pr(\text{female}) = \Pr(F) = \frac{\# F}{\# SS} = \frac{594}{1114} = .5332$$

$$\Pr(\text{science}) = \Pr(S) = \frac{\# S}{\# SS} = \frac{262}{1114} = .2352$$

$$\Pr(\text{female in business}) = \Pr(F \cap B) = \frac{\#(F \cap B)}{\# SS} = \frac{241}{1114} = .2163$$

Table 2

Sex \ Major	A	B	E	S	Total
M	.0745	.2029	.1059	.0835	.4668
F	.0736	.2163	.0916	.1517	.5332
Total	.1481	.4192	.1975	.2352	1.0000

Table 2 includes these four results as well as the other probabilities that the chosen student will fall into one of various categories. Each probability can be found by dividing the given table entry by 1114, the total number of students. Notice that the outcomes A, B, E, and S are pairwise mutually exclusive, and hence

$$P(SS) = \Pr(A \cup B \cup E \cup S)$$
$$= \Pr(A) + \Pr(B) + \Pr(E) + \Pr(S)$$
$$= .1481 + .4192 + .1975 + .2352$$
$$= 1$$

Similarly,

$$\Pr(M) = \Pr(M \cap A) + \Pr(M \cap B) + \Pr(M \cap E) + \Pr(M \cap S)$$
$$= .0745 + .2029 + .1059 + .0835$$
$$= .4668$$

and

$$\Pr(F) = \Pr(F \cap A) + \Pr(F \cap B) + \Pr(F \cap E) + \Pr(F \cap S)$$
$$= .0736 + .2163 + .0916 + .1517$$
$$= .5332$$

Since the probabilities relating to major (namely, .1481, .4192, .1975, and .2352) and sex (namely, .4668 and .5332) appear in the outside margins of Table 2, they are called *marginal probabilities* for this experiment. ∎

Example 3.7.4 **Nonexclusive Events**

Again, let our sample space be a 52-card poker deck, and let J and D be the subsets of jacks and diamonds, respectively. If one draw is made from the deck, then

$$\Pr(\text{jack}) = \Pr(J) = \frac{\#J}{\#S} = \frac{4}{52} = \frac{1}{13}$$

$$\Pr(\text{diamond}) = \Pr(D) = \frac{\#D}{\#S} = \frac{13}{52} = \frac{1}{4}$$

$$\Pr(\text{jack or diamond}) = \Pr(J \cup D) = \frac{\#(J \cup D)}{\#S} = \frac{16}{52} = \frac{4}{13}$$

$$\Pr(\text{jack of diamonds}) = \Pr(J \cap D) = \frac{\#(J \cap D)}{\#S} = \frac{1}{52}$$

From these results, we observe the relationship

$$Pr(J \cup D) = Pr(J) + Pr(D) - Pr(J \cap D)$$

$$\frac{4}{13} = \frac{1}{13} + \frac{1}{4} - \frac{1}{52}$$

This observation can be extended.　　　　　　　　　　　■

Fact

> **Generalized Addition Rule**
>
> For any events A and B,
>
> $$Pr(A \cup B) = Pr(A) + Pr(B) - Pr(A \cap B)$$

EXERCISES **3.7**

1. For each of the following, specify the sample space and then find the probability.

  a. Getting a tail on one coin toss
  b. Not getting a tail on one coin toss
  c. Getting both a head and a tail on one coin toss
  d. Getting 6 on one roll of a die
  e. Getting a 1, 2, 3, 4, 5, or 6 on one roll of a die
  f. Getting a 10 on one roll of two dice
  g. Not getting a 4 or 5 on one roll of a die
  h. Getting a 7 in one draw from a poker deck
  i. Not getting a 7 in one draw from a poker deck
  j. Getting a face card (jack, queen, or king) in one draw from a poker deck
  k. Not getting a spade in one draw from a poker deck
  l. Getting a red 9 in one draw from a poker deck

2. For each of the following, specify the sample space and then find the probability.

  a. Getting a head on one coin toss
  b. Not getting a head on one coin toss
  c. Getting either a head or tail on one coin toss
  d. Getting a 1 or 2 on one roll of a die
  e. Getting a 9 on one roll of a die
  f. Getting a 9 on one roll of two dice

  g. Not getting a 3 on one roll of a die
  h. Getting a club in one draw from a poker deck
  i. Not getting a club in one draw from a poker deck
  j. Getting a number card in one draw from a poker deck
  k. Not getting a red face card in one draw from a poker deck
  l. Getting the 5 of spades in one draw from a poker deck

3. Let S be the collection of all eight-bit strings. If one random draw is made from S, find the probability of picking the following:

  a. A string with exactly three 1 bits
  b. A string whose bits add to four
  c. A string with either six or seven 1 bits

4. Let S be the collection of all six-bit strings. If one random draw is made from S, find the probability of picking the following:

  a. A string with exactly one 0 bit
  b. A string whose bits add to two
  c. A string with more than three 1 bits

5. In a certain city, voters must register as Democrat, Republican, or Independent. Table 3 gives the numbers

Table 3

Party \ Income ($1000s)	L (Low: below 15)	M (Middle: 15 to 40)	H (High: above 40)	Total
D (Democratic)	8143	20890	4137	33170
I (Independent)	1444	5302	3095	9841
R (Republican)	4785	9049	15667	29501
Total	14372	35241	22899	72512

of registered voters according to party affiliation and income. Using this table, determine the probability that a registered voter selected at random will be:

  a. Republican
  b. high income
  c. Democrat and middle income
  d. Independent and low income

Also, convert Table 3 into a table of probabilities, like Table 2.

  6. A particular town is divided into two school districts. The numbers of residents in certain age ranges in each of the districts are tabulated in Table 4. Using this table, determine the probability that a person selected at random will be:

  a. A resident of district R
  b. A youth
  c. An adult from district T
  d. A preschooler from district R

Also, convert Table 4 into a table of probabilities similar to Table 2.

In each of the next four exercises, compute the four probabilities specified and then use them to verify the equation.

$$\Pr(A \cup B) = \Pr(A) + \Pr(B) - \Pr(A \cap B)$$

Then determine if A and B are mutually exclusive.

  7. Using Table 3, compute the following:

  a. $\Pr(L)$ **b.** $\Pr(M)$
  c. $\Pr(L \cup M)$ **d.** $\Pr(L \cap M)$

  8. Using Table 4, compute

  a. $\Pr(K)$ **b.** $\Pr(A)$
  c. $\Pr(K \cup A)$ **d.** $\Pr(K \cap A)$

  9. Using Table 3, compute

  a. $\Pr(R)$ **b.** $\Pr(H)$
  c. $\Pr(R \cup H)$ **d.** $\Pr(R \cap H)$

10. Using Table 4, compute

  a. $\Pr(T)$ **b.** $\Pr(Y)$
  c. $\Pr(T \cup Y)$ **d.** $\Pr(T \cap Y)$

11. Verify the statement $\Pr(A) + \Pr(\bar{A}) = 1$ for the following:

  a. $A = R$ of Table 3   **b.** $A = K$ of Table 4

12. If the probability of getting a head in one coin flip is $\frac{1}{2}$, why don't you get one head every two times you flip a coin?

*The next four exercises use the notion of **odds**. The **odds are a to b in favor of event** A if*

$$\frac{\Pr(A)}{\Pr(\bar{A})} = \frac{a}{b}$$

For instance, if $\Pr(A) = \frac{1}{3}$ and $\Pr(\bar{A}) = \frac{2}{3}$, then $\Pr(A)/\Pr(\bar{A}) = \frac{1}{2}$ and the odds are 1 to 2 in favor of event A.

13. What are the odds in favor of rolling a 2 or 3 on one roll of a die?

14. If the odds are 3 to 1 against a team winning a game, what is the probability that the team will win?

15. What are the odds against a randomly chosen eight-bit string having exactly four 1 bits?

16. What are the odds in favor of randomly choosing the null string 000000 from the collection of all six-bit strings?

*The last two exercises involve the idea of **expected payoff** of a game, which is defined as the probability of winning times the monetary payoff for a win. This number represents the average dollar payoff when the game is played many times.*

17. A game consists of a player rolling one die. The operator will pay the player $3 should he or she roll a 3, and nothing otherwise.

  a. What is the expected payoff of the game?
  b. Would you play the game if the dealer charged $.25 per game to play? $1 to play?

18. A game consists of a player picking one card from a 52-card poker deck. The operator pays the player $1 if a face card (jack, queen, or king) is drawn, and nothing otherwise.

  a. What is the expected payoff of the game?
  b. Would you play the game if the dealer charged $.25 per game to play? $1 to play?

Table 4

District \ Age	K (Preschool: under 6)	Y (Youth: 6–21)	A (Adult: over 21)	Total
R	325	562	864	1751
T	298	649	909	1856
Total	623	1211	1773	3607

3.8

Repeated Trials

When an experiment that has exactly two outcomes, such as flipping a coin, is repeated, these repetitions are called ***Bernoulli trials***, after the Swiss mathematician Jakob Bernoulli (1654–1705).

Example 3.8.1

Bernoulli Trials

Suppose a coin is tossed three times. Using H for head and T for tail, the sample space S is

$$\{HHH, HHT, HTH, HTT, THH, THT, TTH, TTT\}$$

and

$$\Pr(HHH) \qquad = \frac{\#\{HHH\}}{\#S} \qquad = \frac{1}{8}$$

$$\Pr(TTH) \qquad = \frac{\#\{TTH\}}{\#S} \qquad = \frac{1}{8}$$

$$\Pr(HHH \text{ or } TTT) \quad = \frac{\#\{HHH, TTT\}}{\#S} \qquad = \frac{2}{8} = \frac{1}{4}$$

$$\Pr(\text{exactly two heads}) = \frac{\#\{HHT, HTH, THH\}}{\#S} = \frac{3}{8}$$

Now let

H_1 be the outcome H on the first toss,
T_2 be the outcome T on the second toss, and
H_3 be the outcome H on the third toss.

Then we have

$$\Pr(H_1) = \frac{\#H_1}{\#S} = \frac{\#\{HHH, HHT, HTH, HTT\}}{\#S} = \frac{4}{8} = \frac{1}{2}$$

$$\Pr(T_2) = \frac{\#T_2}{\#S} = \frac{\#\{HTH, HTT, TTH, TTT\}}{\#S} = \frac{4}{8} = \frac{1}{2}$$

$$\Pr(H_3) = \frac{\#H_3}{\#S} = \frac{\#\{HHH, HTH, THH, TTH\}}{\#S} = \frac{4}{8} = \frac{1}{2}$$

and

$$\Pr(H_1 \cap T_2 \cap H_3) = \frac{\#(H_1 \cap T_2 \cap H_3)}{\#S} = \frac{\#\{HTH\}}{\#S} = \frac{1}{8}$$

So, in this situation we have the relationship

$$Pr(H_1 \cap T_2 \cap H_3) = Pr(H_1) * Pr(T_2) * Pr(H_3)$$ ■

This last equation leads to the definition of ***independent events***.

Definition

Outcomes $A_1, A_2, \ldots, A_n$ are called ***independent*** if

$$Pr(A_1 \cap A_2 \cap \cdots \cap A_n) = Pr(A_1) * Pr(A_2) * \cdots * Pr(A_n)$$

In some situations, like the preceding example, the notation of independence means that the occurrence or nonoccurrence of one outcome in no way affects the occurrence or nonoccurrence of other outcomes. However, in other cases, like the next example, independence is not easily characterized or understood.

Example 3.8.2   **Independence**

a. Consider the births of two children, where the sex of each child is noted using M for male and F for female. Let A be the outcome of at most one male, and B the outcome of at least one of each sex. Then the sample space is

$$S = \{FF, FM, MF, MM\}$$

and

$$Pr(A) \quad = \frac{\#A}{\#S} \quad = \frac{\#\{FF, FM, MF\}}{\#S} = \frac{3}{4}$$

$$Pr(B) \quad = \frac{\#B}{\#S} \quad = \frac{\#\{FM, MF\}}{\#S} \quad = \frac{2}{4} = \frac{1}{2}$$

$$Pr(A \cap B) = \frac{\#(A \cap B)}{\#S} = \frac{\#\{FM, MF\}}{\#S} \quad = \frac{2}{4} = \frac{1}{2}$$

But

$$Pr(A) * Pr(B) = \frac{3}{4} * \frac{1}{2} = \frac{3}{8}$$

so

$$Pr(A \cap B) \neq Pr(A) * Pr(B)$$

Thus A and B are *not* independent events.

b. Now let's consider the sex of three children. The sample space here is

$$S = \{FFF, FFM, FMF, FMM, MFF, MFM, MMF, MMM\}$$

Taking outcomes A and B to be the same as described in part (a), we have

$$\Pr(A) \quad = \frac{\#A}{\#S} \quad = \frac{\#\{FFF, FFM, FMF, MFF\}}{\#S}$$

$$= \frac{4}{8} = \frac{1}{2}$$

$$\Pr(B) \quad = \frac{\#B}{\#S} \quad = \frac{\#\{FFM, FMF, FMM, MFF, MFM, MMF\}}{\#S}$$

$$= \frac{6}{8} = \frac{3}{4}$$

$$\Pr(A \cap B) = \frac{\#(A \cap B)}{\#S} = \frac{\#\{FFM, FMF, MFF\}}{\#S}$$

$$= \frac{3}{8}$$

However, in this case

$$\Pr(A) * \Pr(B) = \frac{1}{2} * \frac{3}{4} = \frac{3}{8} = \Pr(A \cap B)$$

so that A and B *are* independent. ∎

The binomial coefficients mentioned at the end of Section 3.6 arise naturally in computing certain probabilities, as the next few examples demonstrate.

Example 3.8.3   **Binomial Distribution**

Let's return to the setting of Example 3.8.1, where a coin is tossed three times. These three coin flips are Bernoulli trials, and the outcomes are independent. If x equals the number of heads appearing, then

$$\Pr(x = 0) = \frac{\#\{TTT\}}{\#S} \qquad = \frac{1}{8} = .125$$

$$\Pr(x = 1) = \frac{\#\{HTT, THT, TTH\}}{\#S} = \frac{3}{8} = .375$$

$$\Pr(x = 2) = \frac{\#\{HHT, HTH, THH\}}{\#S} = \frac{3}{8} = .375$$

$$\Pr(x = 3) = \frac{\#\{HHH\}}{\#S} \qquad = \frac{1}{8} = .125$$

These results are summarized in the **binomial probability distribution** of Figure 24; on the right side of that figure is a graphic representation of the distribution called a **line chart**. Note that the sum of the probabilities equals 1. The distribution is called *binomial* because the probabilities can be expressed

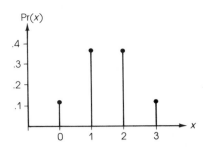

$x =$ # heads	$\Pr(x)$
0	.125
1	.375
2	.375
3	.125
	1.000

Figure 24

in the form

$$\Pr(x = 0) = .125 = 1 * (.5)^3 \qquad\quad = {}_3C_0 * \Pr(T)^3$$
$$\Pr(x = 1) = .375 = 3 * (.5)^2 * (.5) = {}_3C_1 * \Pr(T)^2 * \Pr(H)$$
$$\Pr(x = 2) = .375 = 3 * (.5) \ * (.5)^2 = {}_3C_2 * \Pr(T) \ * \Pr(H)^2$$
$$\Pr(x = 3) = .125 = 1 * (.5)^3 \qquad\quad = {}_3C_3 * \Pr(H)^3$$

The four terms on the right sides of these equations are the terms in the expansion of

$$[\Pr(H) + \Pr(T)]^3 \qquad\qquad\blacksquare$$

One area of computer science in which the notion of probability is useful is that of *transmission error*. Recall from our discussion in Chapter 2 that at operating machine level, all characters, both numeric and otherwise, are coded in binary notation. Since all machines have failure rates, an important question arises: When a binary string is transmitted from one device to another within the computer system (for example, from core memory to disk storage), what is the probability that the binary string received is the same as the one sent?

Example 3.8.4   **Transmission Error**

A particular computer system uses a four-bit BCD code for each character. The probability of error in the transmission of binary strings from core memory to the printer is found to be

.1 per bit

(in reality, transmission error probability is much smaller than this), and the accuracy of transmission of one bit in no way affects the accuracy of transmission of others. Since

$$\Pr(\text{incorrect bit}) + \Pr(\text{correct bit}) = 1$$

the probability of correct transmission must be

.9 per bit

Consider the reception of one four-bit string by the printer. There are $2^4 = 16$ possible strings that can be received by the printer; these constitute our sample space S. For instance, if

$BBBB$

represents the string sent by the memory, and if we let F represent an incorrect or faulty bit in the string received by the printer, then

$BFFB \in S$

is that received string with incorrect second and third bits. Since we are given that error transmission is bitwise independent, then

$$\Pr(BBBB) = \Pr(B) * \Pr(B) * \Pr(B) * \Pr(B) = .9 * .9 * .9 * .9 = .6561$$
$$\Pr(BFFB) = \Pr(B) * \Pr(F) * \Pr(F) * \Pr(B) = .9 * .1 * .1 * .9 = .0081$$
$$\Pr(BFBB) = \Pr(B) * \Pr(F) * \Pr(B) * \Pr(B) = .9 * .1 * .9 * .9 = .0729$$
$$\Pr(FBFF) = \Pr(F) * \Pr(B) * \Pr(F) * \Pr(F) = .1 * .9 * .1 * .1 = .0009$$
$$\Pr(FFFF) = \Pr(F) * \Pr(F) * \Pr(F) * \Pr(F) = .1 * .1 * .1 * .1 = .0001$$

Now let x equal the number of incorrect bits in the received string. For $x = 0$, $BBBB$ is the only possibility, so

$$\Pr(x = 0) = \Pr(BBBB) = .6561$$

For $x = 1$, there are four (pairwise mutually exclusive) possibilities:

$$\Pr(FBBB) = \Pr(F) * \Pr(B) * \Pr(B) * \Pr(B) = .1 * .9 * .9 * .9 = .0729$$
$$\Pr(BFBB) = \Pr(B) * \Pr(F) * \Pr(B) * \Pr(B) = .9 * .1 * .9 * .9 = .0729$$
$$\Pr(BBFB) = \Pr(B) * \Pr(B) * \Pr(F) * \Pr(B) = .9 * .9 * .1 * .9 = .0729$$
$$\Pr(BBBF) = \Pr(B) * \Pr(B) * \Pr(B) * \Pr(F) = .9 * .9 * .9 * .1 = .0729$$

Therefore,

$$\begin{aligned}
\Pr(x = 1) &= \Pr(\{FBBB, BFBB, BBFB, BBBF\}) \\
&= \Pr(\{FBBB\} \cup \{BFBB\} \cup \{BBFB\} \cup \{BBBF\}) \\
&= \Pr(FBBB) + \Pr(BFBB) + \Pr(BBFB) + \Pr(BBBF) \\
&= \quad .0729 \quad + \quad .0729 \quad + \quad .0729 \quad + \quad .0729 \\
&= .2916
\end{aligned}$$

Note that we can rewrite these two results in the form

$$\begin{aligned}
\Pr(x = 0) &= .6561 \\
&= 1 * .6561 \\
&= {}_4C_0 * (.9)^4 * (.1)^0
\end{aligned}$$

and

$$\begin{aligned}
\Pr(x = 1) &= .2916 \\
&= 4 * .0729 \\
&= {}_4C_1 * (.9)^3 * (.1)^1
\end{aligned}$$

One can go through similar calculations for the other values of x, obtaining

$$Pr(x = 2) = {}_4C_2 * (.9)^2 * (.1)^2$$

$$= \frac{4!}{2!(4 - 2)!} * .81 * .01$$

$$= 6 * .0081$$

$$= .0486$$

$$Pr(x = 3) = {}_4C_3 * (.9)^1 * (.1)^3$$

$$= \frac{4!}{3!(4 - 3)!} * .9 * .001$$

$$= 4 * .0009$$

$$= .0036$$

$$Pr(x = 4) = {}_4C_4 * (.9)^0 * (.1)^4$$

$$= \frac{4!}{4!(4 - 4)!} * 1 * .0001$$

$$= 1 * .0001$$

$$= .0001$$

Observe that the five expressions above involving ${}_4C_0$, ${}_4C_1$, ${}_4C_2$, ${}_4C_3$, and ${}_4C_4$ are precisely the terms in the binomial expansion of $(.1 + .9)^4$. A summary of this binomial distribution is given in Figure 25. ∎

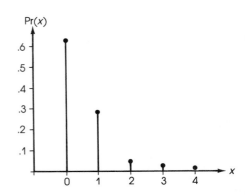

x = # incorrect bits	Pr(x)
0	.6561
1	.2916
2	.0486
3	.0036
4	.0001
	1.0000

Figure 25

In order to help detect and correct transmission errors of this type, an extra bit called a *parity bit* is usually placed at the left of binary strings to be transmitted. In an *even parity system*, for each string the parity bit is chosen as 0 or 1, whichever makes the sum of the digits even. A checking device at the receiving end then adds the digits to determine if the sum is even. If not, an error message is generated to alert the computer operator. In an *odd parity system*, the parity bit is chosen so as to make the sum odd.

Example **3.8.5**   **Even Parity System**

In the preceding example, suppose a parity bit is added to each four-bit string to be transmitted so as to make it an even parity system. For instance,

$$0110 \xrightarrow{\text{plus parity bit}} 00110$$

$$0111 \xrightarrow{\text{plus parity bit}} 10111$$

Now five-bit strings are involved, so consider the reception of one five-bit string by the printer. Again, let x equal the number of faulty bits in the five-bit string received by the printer. Our sample space now consists of 32 five-bit strings

$$S = \{BBBBB, BBBBF, BBBFB, \cdots, BFFFF, FFFFF\}$$

Then, using the method of the preceding example,

$$
\begin{aligned}
\Pr(x = 0) &= {}_5C_0 * (.9)^5 * (.1)^0 \\
&= 1 * .59049 * 1 \\
&= .59049
\end{aligned}
$$

$$
\begin{aligned}
\Pr(x = 1) &= {}_5C_1 * (.9)^4 * (.1)^1 \\
&= 5 * .6561 * .1 \\
&= .32805
\end{aligned}
$$

$$
\begin{aligned}
\Pr(x = 2) &= {}_5C_2 * (.9)^3 * (.1)^2 \\
&= 10 * .729 * .01 \\
&= .0729
\end{aligned}
$$

$$
\begin{aligned}
\Pr(x = 3) &= {}_5C_3 * (.9)^2 * (.1)^3 \\
&= 10 * .81 * .001 \\
&= .0081
\end{aligned}
$$

$$
\begin{aligned}
\Pr(x = 4) &= {}_5C_4 * (.9)^1 * (.1)^4 \\
&= 5 * .9 * .0001 \\
&= .00045
\end{aligned}
$$

$$
\begin{aligned}
\Pr(x = 5) &= {}_5C_5 * (.9)^0 * (.1)^5 \\
&= 1 * 1 * .00001 \\
&= .00001
\end{aligned}
$$

These six results are the terms in the binomial expansion $(.1 + .9)^5$. Figure 26 gives a summary of this binomial distribution.

Note that if the parity check device at the receiving end works properly, then *all* received five-bit strings with an odd number of errors will be detected; those with an even number of errors will not be detected. So, for instance, the probability of a received string having one *undetected* error is 0. ∎

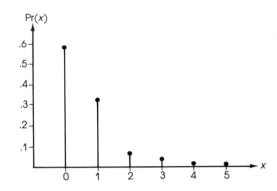

$x =$ # incorrect bits	$Pr(x)$
0	.59049
1	.32805
2	.0729
3	.0081
4	.00045
5	.00001
	1.00000

Figure 26

We should make one final remark here about the two preceding examples. The attachment of a parity bit to the left of the four-bit BCD code has the effect of *lowering* the probability of transmission error in the BCD code portion. In the case of no parity bit (Example 3.8.4), we have

$$Pr(\text{some error}) = Pr(x = 1) + Pr(x = 2) + Pr(x = 3) + Pr(x = 4)$$
$$= .2916 + .0486 + .0036 + .0001$$
$$= .3439$$

However, in the five-bit case (a parity bit adjoined to the four-bit BCD code), only a string with an even number of errors will not be detected; so

$$Pr(\text{undetected error in BCD code}) = Pr(x = 2) + Pr(x = 4)$$
$$= .0729 + .00045$$
$$= .07335$$

EXERCISES **3.8**

1. Suppose a coin is tossed four times.

   a. List the sample space and compute the following:

   $Pr(TTTT)$ $Pr(THTH)$

   $Pr(TTTT \text{ or } HHHH)$ $Pr(\text{exactly two tails})$

   b. Let X_i represent outcome X on toss i. Compute the following:

   $Pr(T_1)$ $Pr(H_2)$

   $Pr(H_3)$ $Pr(T_4)$

   $Pr(T_1 \cap H_2)$ $Pr(T_1 \cap H_2 \cap H_3)$

   $Pr(T_1 \cap H_2 \cap H_3 \cap T_4)$ $Pr(THHT)$

   Are the events T_1, H_2, H_3, and T_4 independent according to the definition of the text?

2. Suppose a coin is tossed five times.

   a. List the sample space and compute the following:

   $Pr(HHHHH)$

   $Pr(\text{first toss is head})$

   $Pr(THTHT \text{ or } HTHTH)$

   $Pr(\text{exactly three heads})$

   b. Let X_i represent outcome X on toss i. Compute the following:

   $Pr(H_1)$ $Pr(T_2)$

   $Pr(T_3)$ $Pr(H_4)$

   $Pr(T_5)$ $Pr(H_1 \cap T_2)$

   $Pr(H_1 \cap T_2 \cap T_3)$

   $Pr(H_1 \cap T_2 \cap T_3 \cap H_4)$

   $Pr(H_1 \cap T_2 \cap T_3 \cap H_4 \cap T_5)$

   $Pr(HTTHT)$

   Are the events H_1, T_2, T_3, H_4, and T_5 independent?

3. Suppose a five-bit string is randomly generated.

   a. List the sample space and compute the following:

   Pr(00000)   Pr(first bit is 0)

   Pr(10101 or 01010)   Pr(exactly three 0 bits)

   b. Let X_i represent the outcome of X on the ith bit. Compute the following:

   $Pr(0_1)$ $Pr(1_2)$

   $Pr(1_3)$ $Pr(0_4)$

   $Pr(1_5)$ $Pr(0_1 \cap 1_2)$

   $Pr(0_1 \cap 1_2 \cap 1_3)$

   $Pr(0_1 \cap 1_2 \cap 1_3 \cap 0_4)$

   $Pr(0_1 \cap 1_2 \cap 1_3 \cap 0_4 \cap 1_5)$

   Pr(01101)

   Are the events 0_1, 1_2, 1_3, 0_4, and 1_5 independent?

4. Suppose a four-bit string is randomly generated.

   a. List the sample space and compute the following:

   Pr(1111)   Pr(1010)

   Pr(1111 or 0000)   Pr(exactly two 1 bits)

   b. Let X_i represent the outcome of X on the ith bit. Compute the following:

   $Pr(1_1)$ $Pr(0_2)$

   $Pr(0_3)$ $Pr(1_4)$

   $Pr(1_1 \cap 0_2)$ $Pr(1_1 \cap 0_2 \cap 0_3)$

   $Pr(1_1 \cap 0_2 \cap 0_3 \cap 1_4)$ Pr(1001)

   Are the events 1_1, 0_2, 0_3, and 1_4 independent?

5. Let a coin be tossed three times.

   a. Write the sample space.

   b. If A is the event of tail on the first toss and B the event of exactly two tails in a row, then compute Pr(A), Pr(B), and Pr($A \cap B$).

   c. Are events A and B independent?

6. Let a coin be tossed three times.

   a. Write the sample space.

   b. If A is the event of tail on the second toss and B the event of exactly two tails in a row, then compute Pr(A), Pr(B), and Pr($A \cap B$).

   c. Are events A and B independent?

7. Consider the births of three children.

   a. Write the sample space of all possible combinations for sex of the children.

   b. If A is the event of the second child being female and B the event of exactly two females in a row, then compute Pr(A), Pr(B), and Pr($A \cap B$).

   c. Are events A and B independent?

8. Consider the births of three children.

   a. Write the sample space of all possible sexes of the children.

   b. If A is the event of the first child being female and B the event of exactly two females in a row, then compute Pr(A), Pr(B), and Pr($A \cap B$).

   c. Are events A and B independent?

9. Suppose we use a random process to generate a three-bit string.

   a. List the elements of the sample space of all possible three-bit strings.

   b. If A is the event of at most one 1 bit and B the event of at least one 0 bit and one 1 bit, then compute Pr(A), Pr(B), and Pr($A \cap B$).

   c. Are events A and B independent?

10. Suppose we use a random process to generate a two-bit string.

   a. List the elements of the sample space of all possible two-bit strings.

   b. If A is the event of at most one 0 bit, and B the event of one 0 bit and one 1 bit, then compute Pr(A), Pr(B), and Pr($A \cap B$).

   c. Are the events A and B independent?

In exercises 11 through 18, an event and a random variable x are given. In each case construct the binomial probability distribution for x and the associated line chart.

11. A student completes a five-question true–false exam by guessing. Thus, the probability of the student answering any one of the five questions correctly is $\frac{1}{2}$. Let x equal the number of questions answered correctly out of five.

12. A coin is tossed four times. Let x equal the number of heads appearing in the four tosses.

13. A student has some knowledge of the subject matter of a five-question true–false test, and she estimates the probability of answering any one of the five questions correctly as .6. Let x equal the number of questions answered correctly out of five.

14. A coin is tossed four times, but the coin is weighted so that on each toss the probability of getting a head is .7. Let x equal the number of heads appearing in the four tosses.

15. An eight-bit string is randomly generated. Let x equal the number of 1 bits appearing in this string.

16. A four-bit string is randomly generated. Let x equal the number of 1 bits appearing in this string.

17. A computer system transmits eight-bit BCD strings from memory to hard disk storage, and it is known that the probability of error in this received string is .2 per bit. Let x equal the number of incorrect bits in the received eight-bit string.

18. A computer system transmits six-bit BCD strings from memory to hard disk storage, and it is known that the probability of error in this received string is .1 per bit. Let x equal the number of incorrect bits in the received six-bit string.

19. In an even parity system, what single bit would be affixed to the left of the following binary strings:

   a. 10111011   **b.** 10000101
   c. 00000000   **d.** 01101100

20. In an odd parity system, what single bit would be affixed to the left of the following binary strings:

   a. 10011000   **b.** 01000000
   c. 10000001   **d.** 10111101

21. Suppose the computer system of exercise 17 is converted to an odd parity system. Let x equal the number of incorrect bits in the received nine-bit string, and compute the binomial probability distribution for x.

22. Suppose the computer system of exercise 18 is converted to an odd parity system. Let x equal the number of incorrect bits in the received seven-bit string, and compute the binomial probability distribution for x.

23. a. In exercise 17, compute the probability of the received eight-bit string having at least one bit in error.

   b. In exercise 21, compute the probability of the received nine-bit string having at least one undetected error in the eight-bit BCD code portion.

24. a. In exercise 18, compute the probability of the received six-bit string having at least one bit in error.

   b. In exercise 22, compute the probability of the received seven-bit string having at least one undetected error in the six-bit BCD code portion.

25. Compute the probability of a detected error in the BCD code portion of the transmitted five-bit string, given the information provided in the last paragraph of this section.

26. Would the results of exercises 23(b), 24(b), or 25 change if the systems were even parity rather than odd parity?

CHAPTER 4

Computer-Related Logic

One of the preeminent beauties in the landscape of mathematics is the discovery of systems that seem unrelated on the surface yet underneath are governed by identical rules. When this occurs, whatever we learn in the one setting can be translated into equivalent results in the other.

At first glance, what does building sentences from propositions by using the connecting words *and, or,* and *not* have in common with building electronic circuits from switches by using series, parallel, and inverter connections? In context and external appearances, very little. But in algebraic properties, everything! It is precisely this relationship that we shall explore in the first three sections of this chapter. Then we shall apply some of this knowledge to the analysis and synthesis of computer-related electronic circuits (Sections 4.4 and 4.5), which will enable us to solve problems of the type shown on the next page.[†]

Finally, the material in Sections 4.6 and 4.7 on conditional and biconditional connectives and deductive arguments is intended for those who wish to continue the study of sentential logic introduced at the beginning of the chapter.

4.1

AND and OR Gates

While the term *proposition* is used in diverse ways in colloquial English, in mathematics it is used to mean any statement that is either true or false. For instance, the statements

$$2 + 3 = 5$$
$$6 - 4 = 1$$

We live on the planet Earth.
Del Ennis was National League rookie of the year in 1946.
The capital city of Albania is Kabul.

[†] Ronald J. Tocci, *Digital Systems: Principles and Applications,* revised and enlarged, © 1980, pp. 44–45. Reprinted by permission of Prentice-Hall, Inc., Englewood Cliffs, N.J.

Example: Simplify the logic circuit shown in Figure 2.22(a).

(a)

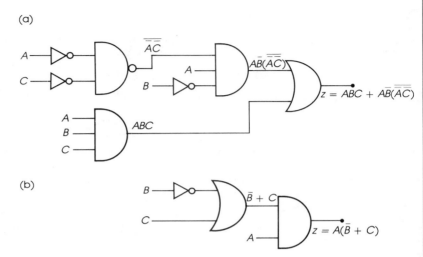

(b)

Figure 2.22

Solution: The first step is to determine the expression for the output. The result is
$$z = ABC + A\bar{B}.(\overline{\overline{A}\overline{C}})$$

Once the expression is determined, it is usually a good idea to break down all large inverter signs using De Morgan's theorems and then multiply out all terms.

$$z = ABC + A\bar{B}(\bar{\bar{A}} + \bar{\bar{C}})$$ [theorem (17)]
$$= ABC + A\bar{B}(A + C)$$
$$= ABC + A\bar{B}A + A\bar{B}C$$
$$= ABC + A\bar{B} + A\bar{B}C$$ [theorem (3)]

When the expression is in this form, we should look for common variables among the various terms with the intention of factoring. The first and third terms above have AC in common, which can be factored out:

$$z = AC(B + \bar{B}) + A\bar{B}$$

Since $B + \bar{B} = 1$, then

$$z = AC(1) + A\bar{B}$$
$$= AC + A\bar{B}$$

We can now factor out A, which results in

$$z = A(C + \bar{B})$$

This result can be simplified no further. Its circuit implementation is shown in Figure 2.22(b). It is obvious that the circuit in (b) is a great deal simpler than the original circuit in (a).

all have a certain truth value (namely, true, false, true, true, and false, respectively), and so each is a proposition. However, no truth value can be assigned to any of the statements

$2 + y = 5$
Hello!
Who won the NBA title in 1983?
He must be the meanest person alive!

and thus none of these is a proposition.

In speech and writing, we utilize propositions individually to form *simple sentences* or use certain connecting words to combine several propositions into one *compound sentence*. Two such connectives are the words *and*, denoted by $\wedge$, and *or*, denoted by $\vee$. (Note that these are the same symbols used in the definition of Boolean algebra in Section 3.3.) If we represent two propositions by the letters p and q, then Table 1 contains the *truth tables for $\wedge$ and $\vee$*. These tables summarize the truth values (T for "true" and F for "false") that we usually attach to "and" and "or" statements in our everyday language. Namely, an "and" sentence is true only when both components are true, and an "or" sentence is false only when both components are false. Here we use *or* in the inclusive "and/or" sense. The exclusive use of *or* will be discussed in Section 4.2.

At times it is helpful to express the truth tables in Table 1 as

$$T \wedge T = T \qquad T \wedge F = F \qquad F \wedge T = F \qquad F \wedge F = F$$

and

$$T \vee T = T \qquad T \vee F = T \qquad F \vee T = T \qquad F \vee F = F$$

Table 1

p	q	$p \wedge q$
T	T	T
T	F	F
F	T	F
F	F	F

p	q	$p \vee q$
T	T	T
T	F	T
F	T	T
F	F	F

Example 4.1.1

$\wedge$ and $\vee$ Truth Tables

Let's represent the five propositions stated at the beginning of this section by the letters r, s, u, v, and w, respectively. Then the form

$s \vee u$

symbolizes the sentence

$6 - 4 = 1$ or   we live on the planet Earth

And

$s \vee u = F \vee T = T$

so this sentence is true. The symbolic form

$r \wedge w$

represents the sentence

$2 + 3 = 5$ and   the capital city of Albania is Kabul

Since

$r \wedge w = T \wedge F = F$

this sentence is false. In a similar fashion, we could interpret the forms

$$r \wedge u \qquad u \wedge v \qquad v \vee w \qquad r \wedge r$$

each of which is true; and

$$r \wedge s \qquad u \wedge w \qquad s \vee w \qquad w \vee w$$

each of which is false. ■

Of course, one can construct sentences and logical forms that use several *and* and *or* connectives.

Example 4.1.2   **Construction of a Truth Table**

Given a logical form

$$(p \vee q) \wedge r$$

then we can construct a truth table for the expression as follows (refer to Table 2 at each step):

Step 1: Label the leftmost columns with the different propositions comprising the given expression. In this case there are three component propositions, p, q, and r.

Step 2: Below these column headings, list on separate lines all possible ways of assigning the truth values T and F. Here there are $2^3 = 8$ such ways.[†] In general, if there are N propositions, then by the Fundamental Counting Principle there are 2^N different truth assignments, each on a separate line.

Step 3: To the right of these columns, draw as many additional columns as there are connectives in the given expression. In this example there are two connectives, so make two additional columns.

Step 4: Place headings on these columns so as to proceed one connective at a time from the component propositions to the given expression, which should appear in the rightmost column. In our example, there is only one middle column, $p \vee q$, which precedes the rightmost column $(p \vee q) \wedge r$.

Step 5: Using the connective truth tables and the appropriate pair of columns, fill in *each* middle column with the correct truth values. Here the single middle column $p \vee q$ is completed using the p and q columns and the "or" truth table.

Step 6: Using the appropriate connective truth table and the appropriate pair of columns, fill in the rightmost column with the correct truth values. For instance, to determine the second entry in the rightmost column of our example, one takes the T under $p \vee q$ and "ands" it with the F in the r column

$$T \wedge F = F$$

to get the proper F in the right column. ■

[†] For a given number of propositions (in this case three), it is a good habit to *always* use the same pattern in listing the possible truth value combinations.

Table 2

Step 1			Steps 3 and 4			p	q	r	$p \vee q$	$(p \vee q) \wedge r$
p	q	r	$p \vee q$	$(p \vee q) \wedge r$						
T	T	T				T	T	T	T	T
T	T	F				T	T	F	T	F
T	F	T				T	F	T	T	T
T	F	F				T	F	F	T	F
F	T	T				F	T	T	T	T
F	T	F				F	T	F	T	F
F	F	T				F	F	T	F	F
F	F	F				F	F	F	F	F

Step 2 (under p, q, r on the left); Step 5 (under $p \vee q$ on the right); Step 6 (under $(p \vee q) \wedge r$ on the right).

The result of this six-step process, shown at the right in Table 2, is called the **truth table for $(p \vee q) \wedge r$** since it tabulates the truth values of $(p \vee q) \wedge r$ for all possible truth value combinations assigned to the component propositions p, q, and r. For instance, the second row of the table tells us that when p and q are true and r is false, then the sentence $(p \vee q) \wedge r$ is false.

Example 4.1.3   **Truth Table for a Sentence**

Consider the sentence[†]

Either it rains today or you and I will go to the game.

If we let

p = It rains today
q = You will go to the game
r = I will go to the game

then this sentence can be written symbolically as

$p \vee (q \wedge r)$

It is important to notice that the sentence with compound subject

You and I will go to the game

is translated into the compound sentence

You will go to the game and I will go to the game

or symbolically,

$q \wedge r$

Applying the six-step method described above, the truth table for this expression is given in Table 3. ∎

[†] Whenever *either/or* is used in this text, it shall be used in the inclusive "and/or" sense.

Table 3

p	q	r	$q \wedge r$	$p \vee (q \wedge r)$
T	T	T	T	T
T	T	F	F	T
T	F	T	F	T
T	F	F	F	T
F	T	T	T	T
F	T	F	F	F
F	F	T	F	F
F	F	F	F	F

In the truth tables of the two preceding examples, notice that the same order was used in the three leftmost columns, but the two rightmost columns differ in the second and fourth positions. (The work in the middle columns can be ignored!) This shows that the placement of parentheses in some expressions, like $p \vee q \wedge r$, is necessary to avoid ambiguity.

This next example uses the same connective in two places and has more than one intermediate column in the truth table.

Example 4.1.4   **Truth Table for a Symbolic Sentence**

Given the sentence

$$(p \vee q) \wedge (p \vee r)$$

the six-step method can be used to construct the truth table, as shown in Table 4. ∎

It is curious to note that the rightmost columns of Tables 3 and 4 coincide, while, of course, the same ordering was used in the columns headed p, q, and r. The meaning of this will be discussed in Section 4.2.

Now let's temporarily leave our work with propositions and sentences to discuss a topic that appears unrelated but is actually quite similar. Those of us who are fortunate enough to live in a society where electricity is plentiful are accustomed to using on–off switches on lights and appliances, even computers.

Table 4

p	q	r	$p \vee q$	$p \vee r$	$(p \vee q) \wedge (p \vee r)$
T	T	T	T	T	T
T	T	F	T	T	T
T	F	T	T	T	T
T	F	F	T	T	T
F	T	T	T	T	T
F	T	F	T	F	F
F	F	T	F	T	F
F	F	F	F	F	F

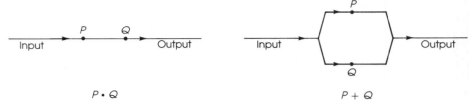

		$P \cdot Q$			$P + Q$
Figure 1		**Series Circuit**			**Parallel Circuit**

Table 5

P	Q	$P \cdot Q$	P	Q	$P + Q$
1	1	1	1	1	1
1	0	0	1	0	1
0	1	0	0	1	1
0	0	0	0	0	0

Using two such switches P and Q and sufficient electrical wiring, Figure 1 exhibits two elementary ways of building a circuit. The circuit on the left is called a *series circuit*, denoted by $P \cdot Q$, and that on the right is a *parallel circuit*, denoted by $P + Q$. The arrows indicate the direction of electrical flow, assuming a power supply is available. It is customary to symbolize the on or "current flowing" condition by the digit 1, and the off or "current not flowing" condition by the digit 0.

Table 5 gives the *output tables for* $\cdot$ *and* $+$. These tables summarize the output for the various on–off states of switches P and Q. Namely, a series circuit is on only if both switches are on, and a parallel circuit is off only if both switches are off.

Whenever it is helpful to do so, we may express the table entries in Table 5 in the form

$$1 \cdot 1 = 1 \qquad 1 \cdot 0 = 0 \qquad 0 \cdot 1 = 0 \qquad 0 \cdot 0 = 0$$

and

$$1 + 1 = 1 \qquad 1 + 0 = 1 \qquad 0 + 1 = 1 \qquad 0 + 0 = 0$$

One important note! The notation $\cdot$ and $+$ is used here because it is most frequently found in computer science texts, although it may not be the most appropriate notation. At any rate, when $\cdot$ and $+$ are employed in the context of circuits, you should avoid the temptation of referring to them as "times" and "plus," respectively, and instead think of them as "and" and "or." (Look at the circuit diagrams in Figure 1!)

Example 4.1.5

$\cdot$ and $+$ Output Tables

If switches R, S, and W are on, off, and on, respectively, then

$R \cdot S = 1 \cdot 0 = 0$, so series circuit $R \cdot S$ is off;
$R \cdot W = 1 \cdot 1 = 1$, so series circuit $R \cdot W$ is on;
$S + W = 0 + 1 = 1$, so parallel circuit $S + W$ is on; and
$S + S = 0 + 0 = 0$, so parallel circuit $S + S$ is off.

Inputs $\begin{array}{c} P \\ Q \end{array}$ —[AND gate] $\bullet \begin{array}{c} P \cdot Q \\ \text{Output} \end{array}$ Inputs $\begin{array}{c} P \\ Q \end{array}$ —[OR gate] $\bullet \begin{array}{c} P + Q \\ \text{Output} \end{array}$

Figure 2   **AND Gate**   **OR Gate**

In a similar fashion, we can show that

$$R + S \qquad R + W \qquad R \cdot R \qquad W \cdot W \qquad R + R \qquad W + W$$

are all on, and

$$S \cdot W \qquad S \cdot S$$

are off.   ∎

    Hopefully, by now you have discerned a certain congruity between the two superficially different topics presented in this section. For, if one replaces the symbols

$$p \qquad q \qquad \wedge \qquad T \qquad F$$

in the truth table for $p \wedge q$ (Table 1) by the symbols

$$P \qquad Q \qquad \cdot \qquad 1 \qquad 0$$

respectively, the result is the output table for $P \cdot Q$ (Table 5). Similarly, replacing

$$p \qquad q \qquad \vee \qquad T \qquad F$$

by

$$P \qquad Q \qquad + \qquad 1 \qquad 0$$

transforms the truth table for $p \vee q$ into the output table for $P + Q$. For this reason, the series and parallel circuits presented, which are the building blocks of complex electronic circuitry, are called **AND gates** and **OR gates**, respectively, and are drawn as shown in Figure 2. Here P and Q are thought of as inputs rather than switches; of course, the output tables for these circuits have already been given in Table 5.

Example 4.1.6   **Circuit/Output Table**

Given the form

$$(P + Q) \cdot R$$

we can draw the associated circuit by first drawing the OR gate within the parentheses

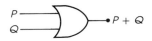

Table 6

P	Q	R	P + Q	(P + Q) · R		P	Q	R	P + Q	(P + Q) · R
1	1	1				1	1	1	1	1
1	1	0				1	1	0	1	0
1	0	1				1	0	1	1	1
1	0	0				1	0	0	1	0
0	1	1				0	1	1	1	1
0	1	0				0	1	0	1	0
0	0	1				0	0	1	0	0
0	0	0				0	0	0	0	0

Step 1 | Steps 3 and 4 ... Step 2 ... Step 5 ... Step 6

and then using this output $P + Q$ as one input to an AND gate.

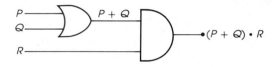

The output table for this circuit is constructed in steps analogous to those used in constructing truth tables for sentences (see Table 6).

Step 1: Label the first three columns with the inputs P, Q, and R.

Step 2: Beneath these headings, list the $2^3 = 8$ possible input states for P, Q, and R.

Step 3: Construct two additional columns, since there are two gates in the given circuit form.

Step 4: Place headings on these columns so as to proceed one gate at a time from the inputs to the given circuit, which should appear in the rightmost column. In this example, the one middle column is labeled with $P + Q$.

Step 5: Fill in each middle column with the correct output values,

Step 6: Fill in the rightmost column with the correct output values.

The result, shown at the right side of Table 6, is called the *output table for $(P + Q) \cdot R$*, since it lists the output values of circuit $(P + Q) \cdot R$ for all possible input values assigned to P, Q, and R. For instance, the second row of the table indicates that when P and Q are on and R is off, then circuit $(P + Q) \cdot R$ is off. ∎

Example 4.1.7 **Circuit/Output Table**

Consider the circuit

$$P + (Q \cdot R)$$

Table 7

P	Q	R	$Q \cdot R$	$P + (Q \cdot R)$
1	1	1	1	1
1	1	0	0	1
1	0	1	0	1
1	0	0	0	1
0	1	1	1	1
0	1	0	0	0
0	0	1	0	0
0	0	0	0	0

To diagram this circuit, we first draw the AND gate in parentheses

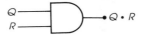

and then place this output $Q \cdot R$ as one input to an OR gate.

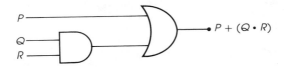

The output table, shown in Table 7, is constructed as described in the preceding example. ∎

Observe that the output tables in the two preceding examples are different, so that parentheses must be placed in the expression

$$P + Q \cdot R$$

in order to define a unique circuit.

A single letter used more than once in the symbolic form of a circuit represents the *same* input. This is the case in the next example.

Example 4.1.8   **Circuit/Output Table**

Given the circuit

$$(P + Q) \cdot (P + R)$$

then we first draw the two OR gates in parentheses

Table 8

P	Q	R	$P + Q$	$P + R$	$(P + Q) \cdot (P + R)$
1	1	1	1	1	1
1	1	0	1	1	1
1	0	1	1	1	1
1	0	0	1	1	1
0	1	1	1	1	1
0	1	0	1	0	0
0	0	1	0	1	0
0	0	0	0	0	0

but combine them using *just one* input line for P as follows:

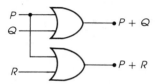

The desired circuit diagram is found by placing these two outputs, $P + Q$ and $P + R$, as inputs to an AND gate.

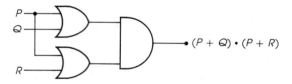

The output table for $(P + Q) \cdot (P + R)$ is given in Table 8. ∎

Again, it is curious to note that the two rightmost columns in Tables 7 and 8 coincide, and, as usual, the same ordering is employed in the input columns P, Q, and R.

In the preceding examples we started with symbolic forms and derived the associated circuit diagrams. Now let's consider the converse case. If you are given a circuit and asked to determine the associated symbolic form, start with those gates that are fed directly by lettered inputs and work from there.

Example 4.1.9

Symbolic Form from Circuit

In the circuit diagram

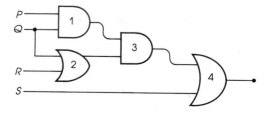

only gates 1 and 2 are fed directly by the lettered inputs.

Gate 1 is $P \cdot Q$ and   gate 2 is $Q + R$

Gate 3, an AND gate, is fed by these two outputs, so

gate 3 is $(P \cdot Q) \cdot (Q + R)$

Finally, gate 4, an OR gate, is fed by S and the gate 3 output, so

gate 4 is $[(P \cdot Q) \cdot (Q + R)] + S$

This last expression is the symbolic form of the given diagram.   ■

In this section, we have discussed the building of sentences from the connective words *and* and *or* and have related that to the construction of electronic circuits using AND and OR gates. We shall continue this idea in the next section by introducing additional connective words and electronic gates.

E X E R C I S E S 4.1

1. Determine whether each of the following is a proposition:

   a. An integer is either even or odd.
   b. $7 * 7 = 51$
   c. $3 * x = 21$
   d. $3 * x = 21$ has a solution.
   e. 1011 is a byte.
   f. The Earth is at least one million years old.
   g. He is the strongest.
   h. Please bring that package to me.

2. Determine whether each of the following is a proposition:

   a. Odd integers are divisible by 2.
   b. $8 * 8 = 64$
   c. $8 * 8$
   d. Mary!
   e. 2 is a bit.
   f. Was Dwight D. Eisenhower president of the United States?
   g. That's terrific!
   h. The year 1982 has passed.

3. Let p, q, and r be the respective propositions

George Orwell wrote *1984*.   (True)
William Shakespeare was a poet and dramatist.
  (True)
Carl Sandburg is still alive.   (False)

For each of the following, write the symbolic expression in prose form and determine its truth value:

   a. $p \wedge q$ **b.** $p \wedge r$ **c.** $q \wedge r$
   d. $p \vee q$ **e.** $p \vee r$ **f.** $q \vee r$

4. Let p, q, and r be the respective propositions

The Wright brothers flew their first plane at Kitty Hawk, North Carolina.   (True)

Alan Shepard was the first man on the moon.
  (False)
The gravity of the moon is greater than the gravity of earth.   (False)

For each of the following, write the symbolic expression in prose form and determine its truth value:

   a. $p \wedge q$ **b.** $p \wedge r$ **c.** $q \wedge r$
   d. $p \vee q$ **e.** $p \vee r$ **f.** $q \vee r$

5. In each of the following, use letters and the connectives $\wedge$ and $\vee$ to express the given sentence in symbolic form. Then construct a truth table for the form.

   a. It's raining and I have an umbrella.
   b. John or Mary is coming.
   c. She and I will study or we'll flunk.
   d. Tom and Sue will win or Pete and Sue will win.
   e. Tom and Sue will win or Pete and Joan will win.

6. In each of the following, use letters and the connectives $\wedge$ and $\vee$ to express the given sentence in symbolic form. Then construct a truth table for the form.

   a. Either that's green or I'm color-blind.
   b. He and I will go.
   c. She performed well and won the money or the trip.
   d. Jack or Jill will compete but Jack or Pat will qualify.
   e. Jack or Jill will compete but Mike or Pat will qualify.

7. Construct a truth table for each of the following symbolic expressions:

a. $p \wedge p$ **b.** $p \vee (q \wedge p)$
c. $(p \wedge q) \wedge r$ **d.** $p \vee (q \vee r)$
e. $(p \wedge r) \vee (q \vee r)$ **f.** $[p \wedge (q \wedge r)] \vee s$
g. $(p \wedge q) \wedge (r \vee s)$

8. Construct a truth table for each of the following symbolic expressions:

a. $p \vee p$ **b.** $(p \vee q) \wedge p$
c. $p \wedge (q \wedge r)$ **d.** $(p \vee q) \vee r$
e. $(p \vee r) \wedge (q \wedge r)$ **f.** $[p \vee (q \vee r)] \wedge s$
g. $(p \vee q) \vee (r \wedge s)$

9. Given switches R, S, W, and Y, which are on, off, on, and off, respectively, determine whether the following circuits are on or off:

a. $R \cdot S, R + S, R \cdot (S \cdot Y), (R + Y) \cdot W$
b. $S \cdot W, S + W, S \cdot (W \cdot Y), (S + R) \cdot W$
c. $W \cdot Y, W + Y, W \cdot (R \cdot R), (W + W) \cdot S$

10. Given switches P, Q, and R, which are off, off, and on, respectively, determine whether the following circuits are on or off:

a. $P \cdot P, P \cdot Q, P + P, P + R, P \cdot (Q + R)$
b. $Q \cdot Q, Q \cdot R, Q + Q, Q + R, Q \cdot (R + P)$
c. $R \cdot R, R \cdot P, R + R, R + P, R \cdot (P + Q)$

11. For each of the following expressions, draw the associated electronic circuit and construct the output table:

a. $P \cdot P$ **b.** $P + (Q \cdot P)$
c. $(P \cdot Q) \cdot R$ **d.** $P + (Q + R)$
e. $(P \cdot R) + (Q + R)$ **f.** $[P \cdot (Q \cdot R)] + S$
g. $(P \cdot Q) \cdot (R + S)$

12. For each of the following expressions, draw the associated electronic circuit and construct the output table:

a. $P + P$ **b.** $(P + Q) \cdot P$
c. $P \cdot (Q \cdot R)$ **d.** $(P + Q) + R$
e. $(P + R) \cdot (Q \cdot R)$ **f.** $[P + (Q + R)] \cdot S$
g. $(P + Q) + (R \cdot S)$

13. For each case in Figure 3, find a symbolic form for the given circuit.

a.

b.

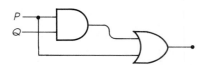

c.

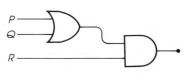

d.

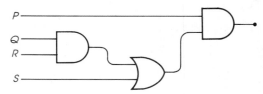

e.

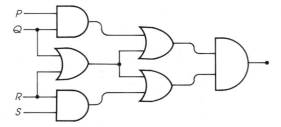

Figure 3

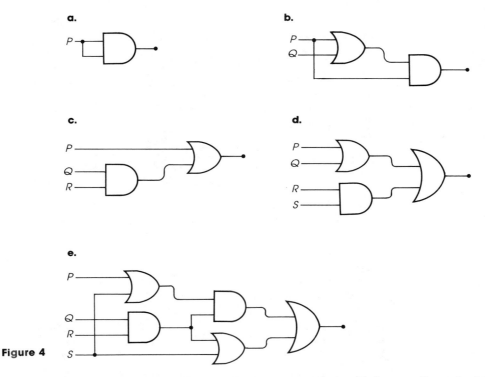

Figure 4

14. For each case in Figure 4, find a symbolic form for the given circuit.

15. How many rows are there in an output table with three inputs? Four? Five? Six? k?

16. Do you see any relationship between Examples 4.1.2 and 4.1.6? Examples 4.1.3 and 4.1.7? Examples 4.1.4 and 4.1.8?

4.2

NOT, NAND, and NOR Gates

In colloquial English, the word *not* is ordinarily used to modify a single proposition rather than to connect different propositions. If p represents a proposition, then the two propositions

$$p \qquad \text{not } p$$

Table 9

p	$\sim p$
T	F
F	T

have opposite truth values; this is tabulated in Table 9, where the symbol $\sim$ is used for the word *not*. (Alternate notations for $\sim p$ are $-p$, $\bar{p}$, and p'.) One could alternately express Table 9 in the form

$$\sim T = F \quad \text{and} \quad \sim F = T$$

In electronics, the analogue of the word *not* is the **NOT gate** or **inverter**. Symbolized by

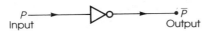

P— Input $\bar{P}$ Output

Table 10

P	$\bar{P}$
1	0
0	1

it reverses the on/off condition of the single input P, as shown in Table 10, which could also be written as

$$\bar{1} = 0 \quad \text{and} \quad \bar{0} = 1$$

The next few examples indicate that we work with *not* and NOT gates in much the same way as with *and*, *or*, AND gates, and OR gates.

Example 4.2.1

~ and − Tables

a. If we let p and q represent the propositions

American watercolorist Winslow Homer painted *Breezing Up*.   (True)
Artist Mary Cassatt was born in Paris.   (False)

respectively, then

$$\sim p$$

symbolizes the sentence

American watercolorist Winslow Homer did not paint *Breezing Up*.

which is false, since

$$\sim p = \sim T = F$$

Also,

$$p \wedge \sim q$$

represents the sentence

American watercolorist Winslow Homer painted *Breezing Up*, and artist Mary Cassatt was not born in Paris.

Since

$$p \wedge \sim q = T \wedge \sim F = T \wedge T = T$$

this sentence is true.

b. Let P and Q represent inputs that are off and on, respectively. Then

$$P \cdot \bar{Q}$$

symbolizes the circuit

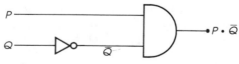

and

$$P \cdot \bar{Q} = 0 \cdot \bar{1} = 0 \cdot 0 = 0$$

so this circuit is off. Also

$$\bar{P} + Q$$

represents the circuit

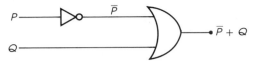

which is on, because

$$\bar{P} + Q = \bar{0} + 1 = 1 + 1 = 1$$ ∎

If we treat $\sim$ and $^-$ as connectives like $\wedge$, $\vee$, $\cdot$, and $+$, then the steps in constructing truth tables and output tables are like those of the preceding section.

Example 4.2.2

Truth Table/Output Table

a. Consider the sentence

Either he telephones tonight or you and I are not going to the game.

Letting

p = He telephones tonight

q = You are going to the game

r = I am going to the game

this sentence can be written in the form

$$p \vee (\sim q \wedge \sim r)$$

The truth table for this expression is given in Table 11.

b. Given the electric circuit

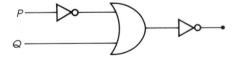

to find a symbolic form, we start with that gate which is directly fed by inputs P and Q, namely

$$\bar{P}$$

Table 11

p	q	r	$\sim q$	$\sim r$	$\sim q \wedge \sim r$	$p \vee (\sim q \wedge \sim r)$
T	T	T	F	F	F	T
T	T	F	F	T	F	T
T	F	T	T	F	F	T
T	F	F	T	T	T	T
F	T	T	F	F	F	F
F	T	F	F	T	F	F
F	F	T	T	F	F	F
F	F	F	T	T	T	T

Table 12

P	Q	$\bar{P}$	$\bar{P} + Q$	$\overline{\bar{P} + Q}$
1	1	0	1	0
1	0	0	0	1
0	1	1	1	0
0	0	1	1	0

and then proceed to the middle gate

$$\bar{P} + Q$$

which feeds the last, rightmost gate

$$\overline{\bar{P} + Q}$$

This circuit is represented by the form

$$\overline{\bar{P} + Q}$$

whose output table is shown in Table 12. ∎

<table>
<tr><td>

Example 4.2.3
</td><td>

Tautology/On Circuit

The tables for the sentence

$$p \vee \sim(p \wedge q)$$

and the equivalent circuit

$$P + (\overline{P \cdot Q})$$

are given in Table 13.

A sentence that is true for all possible truth values of its component propositions (that is, the rightmost column of its truth table contains all T's) is called a ***tautology***. An electric circuit whose output is on for all possible on/off combinations of its inputs (that is, the rightmost column of its output table contains all 1s) is called an ***on circuit***. ∎
</td></tr>
<tr><td>

Example 4.2.4
</td><td>

Contradiction/Off Circuit

Table 14 gives the tables for the sentence

$$\sim p \wedge (p \wedge q)$$
</td></tr>
</table>

Table 13

p	q	$p \wedge q$	$\sim(p \wedge q)$	$p \vee \sim(p \wedge q)$	P	Q	$P \cdot Q$	$\overline{P \cdot Q}$	$P + \overline{P \cdot Q}$
T	T	T	F	T	1	1	1	0	1
T	F	F	T	T	1	0	0	1	1
F	T	F	T	T	0	1	0	1	1
F	F	F	T	T	0	0	0	1	1

Table 14

p	q	$\sim p$	$p \wedge q$	$\sim p \wedge (p \wedge q)$	P	Q	$\bar{P}$	$P \cdot Q$	$\bar{P} \cdot (P \cdot Q)$
T	T	F	T	F	1	1	0	1	0
T	F	F	F	F	1	0	0	0	0
F	T	T	F	F	0	1	1	0	0
F	F	T	F	F	0	0	1	0	0

and its counterpart circuit

$$\bar{P} \cdot (P \cdot Q)$$

A sentence that is false for all possible truth values of its component propositions (that is, the rightmost column of its truth table contains all F's) is called a ***contradiction***. A circuit whose output is off for all possible on/off combinations of its inputs (that is, the rightmost column of its output table contains all 0's) is called an ***off circuit***. ∎

So far in this chapter, we have used the familiar equal sign in a limited way to evaluate the values of sentences or circuits with known input values; for instance, if $p = q = T$ then

$$p \wedge \sim q = T \wedge \sim T = T \wedge F = F$$

Now we want to employ this same equal sign in a more general, more formal way that will be applied throughout the remainder of this chapter.

Definitions

> **a.** *Two sentences s_1 and s_2 are equal*, written
>
> $$s_1 = s_2$$
>
> if each can be expressed in symbolic form using the same symbols and in such a way that the rightmost columns of their truth tables coincide.
>
> **b.** *Two circuits c_1 and c_2 are equal*, written
>
> $$c_1 = c_2$$
>
> if each can be expressed in symbolic form using the same symbols and in such a way that the rightmost columns of their output tables coincide.
>
> **Note:** Whenever one compares rightmost columns of tables, it is assumed that the values assigned the propositions (or inputs) are listed in the same order in each table.

In effect, equal sentences have the same truth values for *any* combination of truth values assigned to the component propositions. Likewise, equal circuits have the same output values for *any* combination of values assigned to the inputs. Referring to our work in the preceding section, as a result of

Table 15

p	q	r	$\sim q$	$\sim r$	$\sim q \vee \sim r$	$p \wedge (\sim q \vee \sim r)$	p	q	r	$q \wedge r$	$\sim(q \wedge r)$	$p \wedge \sim(q \wedge r)$
T	T	T	F	F	F	F	T	T	T	T	F	F
T	T	F	F	T	T	T	T	T	F	F	T	T
T	F	T	T	F	T	T	T	F	T	F	T	T
T	F	F	T	T	T	T	T	F	F	F	T	T
F	T	T	F	F	F	F	F	T	T	T	F	F
F	T	F	F	T	T	F	F	T	F	F	T	F
F	F	T	T	F	T	F	F	F	T	F	T	F
F	F	F	T	T	T	F	F	F	F	F	T	F

the truth tables in Examples 4.1.3 and 4.1.4, we can say

$$p \vee (q \wedge r) = (p \vee q) \wedge (p \vee r)$$

and from Examples 4.1.7 and 4.1.8, we have

$$P + (Q \cdot R) = (P + Q) \cdot (P + R)$$

Example 4.2.5 Equal Sentences

In Table 15 are the truth tables for the sentences

$$p \wedge (\sim q \vee \sim r) \quad \text{and} \quad p \wedge \sim(q \wedge r)$$

Since the rightmost columns of these two tables coincide, then

$$p \wedge (\sim q \vee \sim r) = p \wedge \sim(q \wedge r) \qquad \blacksquare$$

Example 4.2.6 Equal Circuits

The output tables for the circuits

$$\overline{P + Q} \quad \text{and} \quad \bar{P} \cdot \bar{Q}$$

are given in Table 16. Since the rightmost columns coincide,

$$\overline{P + Q} = \bar{P} \cdot \bar{Q} \qquad \blacksquare$$

As mentioned in the preceding section, we have been using the word *or* and the OR gate in the inclusive sense. That is, an "or" statement is considered to be true if one *or both* of the propositions is true, and the OR gate output is on if one *or both* of the inputs is on. However, sometimes there is need to use *or* in the exclusive sense. This means the "or" statement is true *only* when one proposition is true and the other is false, and the OR gate

Table 16

P	Q	$P + Q$	$\overline{P + Q}$	P	Q	$\bar{P}$	$\bar{Q}$	$\bar{P} \cdot \bar{Q}$
1	1	1	0	1	1	0	0	0
1	0	1	0	1	0	0	1	0
0	1	1	0	0	1	1	0	0
0	0	0	1	0	0	1	1	1

Table 17

p	q	$p \otimes q$	P	Q	$P \oplus Q$
T	T	F	1	1	0
T	F	T	1	0	1
F	T	T	0	1	1
F	F	F	0	0	0

output is on *only* when one input is on and the other off. If we use the symbols $\otimes$ and $\oplus$ for the ***exclusive "or"*** and ***exclusive OR gate***, respectively, we obtain the tables given in Table 17.

Also, the circuit symbol for an exclusive OR gate is

which is a slight alteration of the inclusive OR gate.

Example 4.2.7

Exclusive OR

a. The truth table for

$$(\sim p \wedge q) \vee (p \wedge \sim q)$$

is shown in Table 18. Since the right column coincides with that of $p \otimes q$, then

$$p \otimes q = (\sim p \wedge q) \vee (p \wedge \sim q)$$

b. The output table for $(\bar{P} \cdot Q) + (P \cdot \bar{Q})$ is given in Table 19. Since the right column coincides with that of $P \oplus Q$, then

$$P \oplus Q = (\bar{P} \cdot Q) + (P \cdot \bar{Q})$$ ∎

Table 18

p	q	$\sim p$	$\sim q$	$\sim p \wedge q$	$p \wedge \sim q$	$(\sim p \wedge q) \vee (p \wedge \sim q)$
T	T	F	F	F	F	F
T	F	F	T	F	T	T
F	T	T	F	T	F	T
F	F	T	T	F	F	F

Table 19

P	Q	$\bar{P}$	$\bar{Q}$	$\bar{P} \cdot Q$	$P \cdot \bar{Q}$	$\bar{P} \cdot Q + P \cdot \bar{Q}$
1	1	0	0	0	0	0
1	0	0	1	0	1	1
0	1	1	0	1	0	1
0	0	1	1	0	0	0

In electronics it is often necessary to follow one gate by the inverter, or NOT gate. For instance, the AND gate followed by the NOT gate would appear as

but for convenience this is abbreviated as

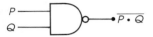

and called a *NAND gate*. Similarly, the inclusive and exclusive OR gates followed by NOT gates are abbreviated as

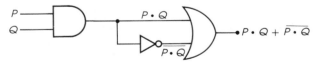

and are called the inclusive NOR gate, or just *NOR gate*, and the *exclusive NOR gate*, respectively. The output tables for these three gates are given in Table 20.

Example 4.2.8   **NAND Gate**

Consider the circuit $(P \cdot Q) + (\overline{P \cdot Q})$. It can be drawn *without* a NAND gate as

Table 20

NAND			NOR			Exclusive NOR		
P	Q	$\overline{P \cdot Q}$	P	Q	$\overline{P + Q}$	P	Q	$\overline{P \oplus Q}$
T	T	F	T	T	F	T	T	T
T	F	T	T	F	F	T	F	F
F	T	T	F	T	F	F	T	F
F	F	T	F	F	T	F	F	T

Table 21

P	Q	$P \cdot Q$	$\overline{P \cdot Q}$	$P \cdot Q + \overline{P \cdot Q}$
1	1	1	0	1
1	0	0	1	1
0	1	0	1	1
0	0	0	1	1

or *with* a NAND gate as

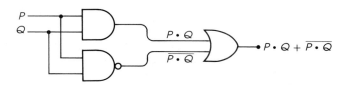

The output table in Table 21 indicates that this circuit is always on. ∎

Example 4.2.9

Universality of NAND Gates

The circuits

$$\overline{P \cdot P} \qquad \overline{(\overline{P \cdot Q}) \cdot (\overline{P \cdot Q})} \qquad \overline{(\overline{P \cdot P}) \cdot (\overline{Q \cdot Q})}$$

are shown in Figure 5 using only NAND gates. But the output tables for

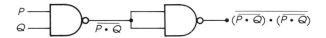

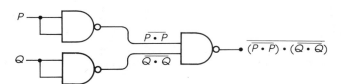

Figure 5

Table 22

P	$P \cdot P$	$\overline{P \cdot P}$
1	1	0
0	0	1

P	Q	$P \cdot Q$	$\overline{P \cdot Q}$	$(\overline{P \cdot Q}) \cdot (\overline{P \cdot Q})$	$\overline{(\overline{P \cdot Q}) \cdot (\overline{P \cdot Q})}$
1	1	1	0	0	1
1	0	0	1	1	0
0	1	0	1	1	0
0	0	0	1	1	0

P	Q	$P \cdot P$	$Q \cdot Q$	$\overline{P \cdot P}$	$\overline{Q \cdot Q}$	$(\overline{P \cdot P}) \cdot (\overline{Q \cdot Q})$	$\overline{(\overline{P \cdot P}) \cdot (\overline{Q \cdot Q})}$
1	1	1	1	0	0	0	1
1	0	1	0	0	1	0	1
0	1	0	1	1	0	0	1
0	0	0	0	1	1	1	0

these circuits in Table 22 show that

$$\overline{P \cdot P} = \bar{P}$$

$$\overline{(\overline{P \cdot Q}) \cdot (\overline{P \cdot Q})} = P \cdot Q$$

$$\overline{(\overline{P \cdot P}) \cdot (\overline{Q \cdot Q})} = P + Q$$

As a result, any electronic circuit built from AND gates, OR gates, and NOT gates can be reconstructed *entirely* from NAND gates. A similar statement can be made concerning the universality of NOR gates, the verification of which is the subject of exercise 23. ∎

EXERCISES **4.2**

1. If p, q, and r represent propositions with truth values false, true, and true, respectively, determine the truth value of each of the following sentences:

a. $\sim p$ **b.** $p \wedge \sim q$

c. $p \vee (q \wedge \sim r)$ **d.** $(p \wedge \sim q) \vee (q \vee \sim r)$

e. $p \wedge [\sim q \vee (q \vee \sim r)]$

2. If p, q, and r represent propositions with truth values true, false, and true, respectively, determine the truth value of each of the following sentences:

a. $\sim q$ **b.** $\sim p \vee q$

c. $(p \wedge \sim q) \vee r$ **d.** $(\sim p \vee q) \vee (p \wedge \sim r)$

e. $\sim p \vee [q \vee (p \wedge \sim r)]$

3. Let P, Q, and R be inputs that are on, off, and on, respectively. In each of the following circuits, determine whether the output is on or off:

a. $\bar{Q}$ **b.** $P + \bar{Q}$

c. $P \cdot (Q + \bar{R})$ **d.** $(\bar{P} \cdot Q) + (\bar{Q} + R)$

e. $\bar{P} \cdot [Q + (\bar{Q} + R)]$

4. Let P, Q, and R be inputs that are off, off, and on, respectively. In each of the following circuits, determine whether the output is on or off:

a. $\bar{P}$

b. $\bar{P} \cdot Q$

c. $(P + Q) \cdot \bar{R}$

d. $(P + \bar{Q}) + (\bar{P} \cdot R)$

e. $P + [\bar{Q} + (\bar{P} \cdot R)]$

5. Let p and q represent the respective propositions

Thomas Eakins was a twentieth-century American painter. (True)

Georges Seurat was a nineteenth-century French impressionist artist. (True)

For each of the following sentences, first determine the truth value and then express it in prose form:

a. $\sim p$

b. $p \wedge \sim q$

c. $p \vee \sim q$

d. $\sim p \vee \sim q$

6. Let p and q represent the respective propositions

Van Cliburn is an accomplished pianist. (True)

Arthur Rubenstein became famous as a cellist. (False)

For each of the following sentences, first determine the truth value and then express it in prose form:

a. $\sim q$

b. $\sim p \vee q$

c. $\sim p \wedge q$

d. $\sim p \wedge \sim q$

7. In each of the following, use letters and the connectives $\wedge$, $\vee$, and $\sim$ to express the given sentence in symbolic form. Then construct a truth table for the form.

a. It's raining but I don't have an umbrella.

b. Either he or I will not go.

c. She and I will study or the instructor will not pass us.

d. Tom and Sue will not win or Pete and Sue will not win.

e. Either Tom will win and Sue will not win, or Pete will win and Joan will not win.

8. In each of the following, use letters and the connectives $\wedge$, $\vee$, and $\sim$ to express the given sentence in symbolic form. Then construct a truth table for the form.

a. Either that's not green or I'm color-blind.

b. Neither Martha nor Amy is coming.

c. She did not perform well and lost both the money and the trip.

d. Jack or Jill will not compete, but Jack or Pat will not qualify.

e. Jack or Jill will not compete, but Mike or Pat will not qualify.

9. Write each of the circuits in Figure 6 in symbolic form and then construct an output table.

a.

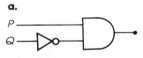

b.

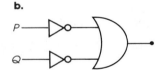

c.

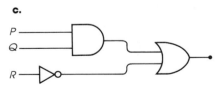

d.

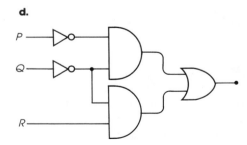

e.

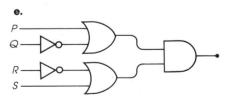

Figure 6

a.

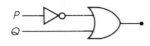

b.

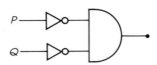

c.

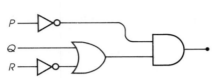

d.

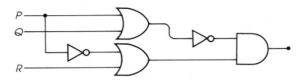

e.

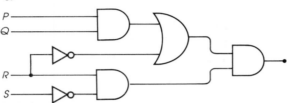

Figure 7

10. Write each of the circuits in Figure 7 in symbolic form and then construct an output table.

11. Construct a truth table for each of the following sentences:

 a. $\sim p \wedge q$
 b. $\sim(\sim p \vee q)$
 c. $p \wedge (q \vee \sim p)$
 d. $(p \wedge q) \vee \sim(p \wedge q)$
 e. $\sim(p \wedge q) \wedge r$
 f. $(p \wedge \sim q) \vee (r \wedge \sim q)$
 g. $(p \wedge \sim q) \vee (r \wedge \sim s)$

12. Construct a truth table for each of the following sentences:

 a. $p \vee \sim q$
 b. $\sim(p \wedge \sim q)$
 c. $p \vee (q \wedge \sim p)$
 d. $(p \vee q) \wedge \sim(p \vee q)$
 e. $p \wedge \sim(q \wedge r)$
 f. $(\sim p \vee q) \wedge (\sim p \vee r)$
 g. $(\sim p \vee q) \wedge (\sim s \vee r)$

13. Construct an output table for each of the following circuits:

 a. $P + \bar{Q}$
 b. $\overline{P \cdot \bar{Q}}$
 c. $P + (Q \cdot \bar{P})$
 d. $(P + Q) \cdot (\overline{P} + Q)$
 e. $P \cdot (\overline{Q \cdot R})$
 f. $(\bar{P} + Q) \cdot (\bar{P} + R)$
 g. $(\bar{P} + Q) \cdot (\bar{S} + R)$

14. Construct an output table for each of the following circuits:

 a. $\bar{P} \cdot Q$
 b. $\overline{\bar{P} + Q}$
 c. $P \cdot (Q + \bar{P})$
 d. $(P \cdot Q) + (\overline{P \cdot Q})$
 e. $(\overline{P \cdot Q}) \cdot R$
 f. $(P \cdot \bar{Q}) + (R \cdot \bar{Q})$
 g. $(P \cdot \bar{Q}) + (R \cdot \bar{S})$

15. Use the definition of sentence equality to verify each of the following:

 a. $p \vee q = q \vee p$
 b. $\sim(p \vee \sim q) = \sim p \wedge q$
 c. $(p \oslash q) \wedge p = p \wedge \sim q$
 d. $(p \wedge q) \wedge r = p \wedge (q \wedge r)$
 e. $p \wedge (\sim q \vee r) = (p \wedge \sim q) \vee (p \wedge r)$

16. Use the definition of sentence equality to verify each of the following:

 a. $p \wedge q = q \wedge p$
 b. $\sim(\sim p \wedge q) = p \vee \sim q$
 c. $q \wedge (p \oslash q) = \sim p \wedge q$
 d. $(p \vee q) \vee r = p \vee (q \vee r)$
 e. $p \vee (q \wedge \sim r) = (p \vee q) \wedge (p \vee \sim r)$

17. Use the definition of circuit equality to verify each of the following:

  a. $P \cdot Q = Q \cdot P$
  b. $\overline{P \cdot Q} = P + \bar{Q}$
  c. $Q \cdot (P \oplus Q) = \bar{P} \cdot Q$
  d. $(P + Q) + R = P + (Q + R)$
  e. $P + (Q \cdot \bar{R}) = (P + Q) \cdot (P + \bar{R})$
  f. $P + (\bar{P} \cdot Q) = P + Q$

18. Use the definition of circuit equality to verify each of the following:

  a. $P + Q = Q + P$
  b. $\overline{P + \bar{Q}} = \bar{P} \cdot Q$
  c. $(P \oplus Q) \cdot P = P \cdot \bar{Q}$
  d. $(P \cdot Q) \cdot R = P \cdot (Q \cdot R)$
  e. $P \cdot (\bar{Q} + R) = P \cdot \bar{Q} + P \cdot R$
  f. $(\bar{P} + Q) \cdot (P + Q) = Q$

19. In each case, draw the given circuit two ways: first with the appropriate NAND and NOR gates, and then without. Finally, construct the output table.

  a. $(P + Q) \cdot (\overline{P + Q})$ **b.** $(P \cdot Q) \oplus (\overline{P \cdot Q})$
  c. $P + \bar{Q} \cdot R$ **d.** $(\overline{P \oplus Q}) \cdot P$

20. In each case, draw the given circuit two ways: first

with the appropriate NAND and NOR gates, and then without. Finally, construct the output table.

  a. $(P \oplus Q) \cdot (\overline{P \oplus Q})$ **b.** $(P + Q) + (\overline{P \cdot Q})$
  c. $(\overline{P + Q}) \cdot Q$ **d.** $P \cdot (\overline{Q + R})$

21. Using the results of Example 4.2.9, redraw the circuit of exercise 9(c) so that it consists entirely of NAND gates.

22. Using the results of Example 4.2.9, redraw the circuit of exercise 10(c) so that it consists entirely of NAND gates.

23. Verify the following circuit equalities:

$$\overline{P + P} = \bar{P}$$

$$\overline{(\overline{P + Q}) + (\overline{P + Q})} = P + Q$$

$$\overline{(\overline{P + P}) + (\overline{Q + Q})} = P \cdot Q$$

Therefore, any electronic circuit built entirely from AND gates, OR gates, and NOT gates can be reconstructed entirely from NOR gates.

24. Use the result of exercise 23 to redraw the circuit of exercise 10(c) so that it consists entirely of NOR gates.

25. Use the result of exercise 23 to redraw the circuit of exercise 9(c) so that it consists entirely of NOR gates.

4.3

Boolean Algebras of Sentences and Circuits

In the first two sections of this chapter, we've illuminated a fundamental relationship between the construction of sentences from connecting words and the construction of electronic circuits from gates. Here we intend to formalize that relationship by employing the unifying notion of Boolean algebra (introduced in Section 3.3, where we discussed the Boolean algebra of sets). In so doing, we shall gather additional tools that will assist us in the analysis and synthesis of circuits.

But first let's clarify some of the terms and symbols we've been using. As mentioned before, a *proposition* is any statement that can be determined as being true or false. By the word *sentence*, we mean either a single proposition or the result of combining propositions by finitely many applications of the connectives $\wedge$, $\vee$, and $\sim$. Since such sentences also have truth value, we now use the lowercase letters *p*, *q*, *r*, and so on, to represent *either* propositions or sentences.

Therefore, given a nonempty set of propositions, let $\mathscr{S}$ be the collection of *all* sentences formed from these propositions. Using the method of verifying sentence equalities demonstrated in the preceding section, the statements in the following box can be proved. The suffix P stands for proposition.

Sentence Facts

	Name of Property
1P. For any $p, q \in \mathscr{S}$	Commutative
$\qquad p \vee q = q \vee p \qquad$ and $\qquad p \wedge q = q \wedge p$	
2P. For any $p, q, r \in \mathscr{S}$	Distributive
$\qquad p \vee (q \wedge r) = (p \vee q) \wedge (p \vee r)$	
and	
$\qquad p \wedge (q \vee r) = (p \wedge q) \vee (p \wedge r)$	
3P. For any false proposition $F \in \mathscr{S}$ and true proposition $T \in \mathscr{S}$	Identity
$\qquad p \vee F = p \qquad$ and $\qquad p \wedge T = p$	
for all $p \in \mathscr{S}$.	
4P. For each $p \in \mathscr{S}$, then $\sim p \in \mathscr{S}$ and	Negation
$\qquad p \vee \sim p = T \qquad$ and $\qquad p \wedge \sim p = F$	

We have already shown the validity of one of the distributive properties in our preceding work (see the paragraph just before Example 4.2.5). Assuming the other properties have been verified in the same fashion, then we can say that

$$(\mathscr{S}, \vee, \wedge)$$

is a Boolean algebra, called the ***Boolean algebra of sentences.*** Thus, we can list the six additional properties that apply to any Boolean algebra as stated in Section 3.3.

Additional Sentence Facts

For any $p, q, r \in \mathscr{S}$:	Name of Property
5P. $p \vee (q \vee r) = (p \vee q) \vee r$	Associative
and	
$p \wedge (q \wedge r) = (p \wedge q) \wedge r$	
6P. $p \vee p = p$ and $p \wedge p = p$	Idempotent
7P. $p \vee T = T$ and $p \wedge F = F$	Domination
8P. $\sim \sim p = p$	Double negation
9P. $\sim T = F$ and $\sim F = T$	Negation of T and F

10P. $\sim(p \vee q) = \sim p \wedge \sim q$ De Morgan[†]

and

$\sim(p \wedge q) = \sim p \vee \sim q$

Now let's turn our attention to electronic circuits. By the word *circuit*, we mean either a single input or the result of combining inputs by finitely many AND, OR, NOT, NAND, and NOR gates to produce a single output. Since such circuits have one output that is either on or off, we shall hereafter use uppercase letters P, Q, R, and so on, to represent *either* inputs or circuits. Thus, given a nonempty set of inputs, let $\mathscr{C}$ be the collection of *all* circuits formed from these inputs. The method of proving circuit equality demonstrated in the preceding section can be applied to verify each of the following statements. When no ambiguity results, we shall subsequently write the AND gate as PQ rather than $P \cdot Q$. Also, for expressions like

$P + (QR)$

we shall drop the parentheses and write it as

$P + QR$

The suffix C stands for circuit.

Circuit Facts

	Name of Property
1C. For any $P, Q \in \mathscr{C}$,	Commutative
$\quad P + Q = Q + P \quad$ and $\quad PQ = QP$	
2C. For any $P, Q, R \in \mathscr{C}$,	Distributive
$\quad P + QR = (P + Q)(P + R)$	
and	
$\quad P(Q + R) = PQ + PR$	
3C. For off input $0 \in \mathscr{C}$ and on input $1 \in \mathscr{C}$	Identity
$\quad P + 0 = P \quad$ and $\quad P \cdot 1 = P$	
for all $P \in \mathscr{C}$.	
4C. For each $P \in \mathscr{C}$, then $\bar{P} \in \mathscr{C}$ and	Inversion
$\quad P + \bar{P} = 1 \quad$ and $\quad P\bar{P} = 0$	

The circuit diagrams corresponding to each of these properties are given in Figure 8.

[†] After English mathematician Augustus De Morgan (1806–1871), who was a friend and colleague of George Boole.

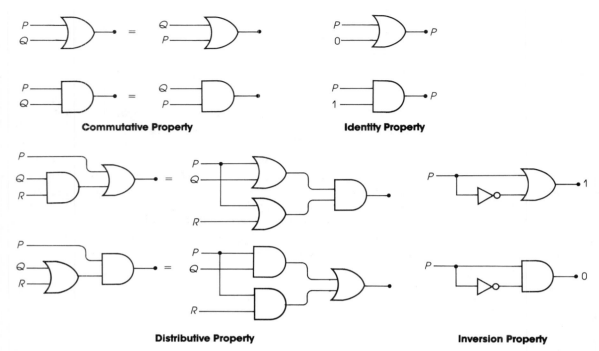

Commutative Property   **Identity Property**

Distributive Property   **Inversion Property**

Figure 8

Once all of the circuit facts have been verified, we can say that

$$(\mathscr{C}, +, \cdot)$$

is a Boolean algebra, called the **Boolean algebra of circuits**. Thus, the next six properties are valid; actually, one of the De Morgan properties was proved in Example 4.2.6. Figure 9 gives the corresponding circuit diagrams.

Additional Circuit Facts

For any $P, Q, R \in \mathscr{C}$:	Name of Property
5C. $P + (Q + R) = (P + Q) + R$	Associative
and	
$P(QR) = (PQ)R$	
6C. $P + P = P$ and $PP = P$	Idempotent
7C. $P + 1 = 1$ and $P \cdot 0 = 0$	Domination
8C. $\bar{\bar{P}} = P$	Double inversion
9C. $\bar{1} = 0$ and $\bar{0} = 1$	Inversion of 1 and 0
10C. $\overline{P + Q} = \bar{P}\bar{Q}$	De Morgan
$\overline{PQ} = \bar{P} + \bar{Q}$	

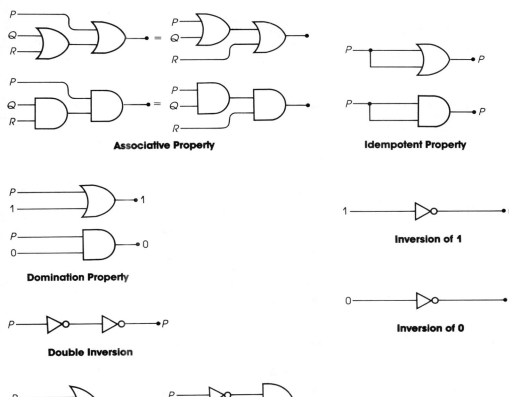

Associative Property

Idempotent Property

Domination Property

Inversion of 1

Inversion of 0

Double Inversion

De Morgan Properties

Figure 9

In Figure 9, notice that the De Morgan properties give alternate representations for NOR and NAND gates. Also, because of the associative property, it is now unnecessary to place parentheses in the expressions

$$P + Q + R \quad \text{or} \quad PQR$$

and we can replace the top and bottom circuit in the associative property diagram with the respective three-input gates.

The properties we have listed in this section are called the **Boolean properties** of sentences and circuits. How (1C) through (10C) can be used to simplify complex circuits will be illustrated in Section 4.4. We shall conclude this section with an example to show how these properties might be applied to check some already known results.

Example 4.3.1

Boolean Properties of Circuits

The three circuit equalities given in Example 4.2.9 to demonstrate the universality of NAND gates can be verified as follows:

$$\overline{PP} = \bar{P} + \bar{P} \qquad \text{De Morgan}$$
$$= \bar{P} \qquad \text{Idempotent}$$

$$\overline{(\overline{PQ})(\overline{PQ})} = \overline{\overline{PQ} + \overline{PQ}} \qquad \text{De Morgan}$$
$$= PQ + PQ \qquad \text{Double inversion}$$
$$= PQ \qquad \text{Idempotent}$$

$$\overline{(\overline{PP})(\overline{QQ})} = \overline{\overline{PP} + \overline{QQ}} \qquad \text{De Morgan}$$
$$= PP + QQ \qquad \text{Double inversion}$$
$$= P + Q \qquad \text{Idempotent} \qquad \blacksquare$$

One final comment. The commutative, distributive, associative, and De Morgan properties stated above can be generalized in obvious ways to include more terms (see exercises 15 to 19). References to one of these properties in succeeding sections may involve such a generalization.

EXERCISES 4.3

In exercises 1 through 8, use the definitions of sentence and circuit equality to prove the indicated fact(s).

1. 1P, 2P, 3P, 4P

2. 1C, 2C, 3C, 4C

3. 5C

4. 5P

5. 6P, 7P, 8P

6. 6C, 7C, 8C

7. 10C

8. 10P

9. Write the symbolic form of the circuit given in Figure 10.

10. Write the symbolic form of the circuit given in Figure 11.

11. Use Boolean properties to verify each of the following:

  a. $\overline{\bar{P}Q} = P + \bar{Q}$
  b. $\bar{P} + Q\bar{R} = \overline{(\bar{P} + Q)\,\overline{P}R}$
  c. $P + \bar{P}Q = P + Q$

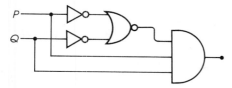

Figure 10

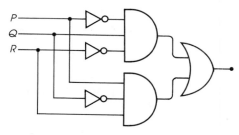

Figure 11

12. Use Boolean properties to verify each of the following:

a. $\overline{P + \bar{Q}} = \bar{P}Q$
b. $\bar{P}(\bar{Q} + R) = \overline{(P + Q)} + \bar{P}R$
c. $P(\bar{P} + Q) = PQ$

13. Use Boolean properties to verify the universality of the OR gate by proving the following:

a. $\overline{P + P} = \bar{P}$
b. $\overline{(P + Q) + (P + Q)} = P + Q$
c. $\overline{(P + P) + (Q + Q)} = PQ$

14. Do the distributive properties (2C) hold if we interpret $+$ and $\cdot$ as usual addition and multiplication of numbers P, Q, and R?

15. Use the commutative property to show the following;

a. For any $p, q, r \in \mathcal{S}$,

$$p \vee q \vee r = p \vee r \vee q = q \vee p \vee r = r \vee p \vee q$$
$$= q \vee r \vee p = r \vee q \vee p$$

and

$$p \wedge q \wedge r = p \wedge r \wedge q = q \wedge p \wedge r = r \wedge p \wedge q$$
$$= q \wedge r \wedge p = r \wedge q \wedge p$$

b. For any $P, Q, R \in \mathcal{C}$,

$$P + Q + R = P + R + Q = Q + P + R$$
$$= R + P + Q = Q + R + P$$
$$= R + Q + P$$

and

$$PQR = PRQ = QPR = RPQ = QRP = RQP$$

16. Use the distributive property to verify the following:

a. For any $p, q, r, s \in \mathcal{S}$,

$$p \vee (q \wedge r \wedge s) = (p \vee q) \wedge (p \vee r) \wedge (p \vee s)$$

and

$$p \wedge (q \vee r \vee s) = (p \wedge q) \vee (p \wedge r) \vee (p \wedge s)$$

b. For any $P, Q, R, S \in \mathcal{C}$,

$$P + QRS = (P + Q)(P + R)(P + S)$$
$$P(Q + R + S) = PQ + PR + PS$$

17. Use the distributive and commutative properties to show the following:

a. For any $p, q, r \in \mathcal{S}$,

$$(p \wedge q) \vee r = (p \vee r) \wedge (q \vee r)$$

and

$$(p \vee q) \wedge r = (p \wedge r) \vee (q \wedge r)$$

b. For any $P, Q, R \in \mathcal{C}$,

$$PQ + R = (P + R)(Q + R)$$

and

$$(P + Q)R = PR + QR$$

18. Use the associative property to verify the following:

a. For any $p, q, r, s \in \mathcal{S}$

$$(p \vee q) \vee r \vee s = p \vee (q \vee r) \vee s = p \vee q \vee (r \vee s)$$

and

$$(p \wedge q) \wedge r \wedge s = p \wedge (q \wedge r) \wedge s = p \wedge q \wedge (r \wedge s)$$

b. For any $P, Q, R, S \in \mathcal{C}$

$$(P + Q) + R + S = P + (Q + R) + S$$
$$= P + Q + (R + S)$$

and

$$(PQ)RS = P(QR)S = PQ(RS)$$

[Exercise 18(b) justifies the use of four-input OR gates and AND gates, shown in Figure 12.]

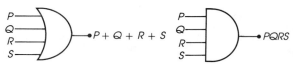

Figure 12

19. Use the De Morgan property to show the following:

a. For any $p, q, r \in \mathcal{S}$,

$$\sim(p \vee q \vee r) = \sim p \wedge \sim q \wedge \sim r$$

and

$$\sim(p \wedge q \wedge r) = \sim p \vee \sim q \vee \sim r$$

b. For any $P, Q, R \in \mathcal{C}$,

$$\overline{P + Q + R} = \bar{P}\bar{Q}\bar{R}$$

and

$$\overline{PQR} = \bar{P} + \bar{Q} + \bar{R}$$

4.4

Simplification of Circuits

The Boolean properties that we have just studied offer one possible method for simplifying circuits. The real-world value of such simplification is evident: The simpler the circuit, the fewer gates and connections required, and the less expensive to build!

In this section we present several examples of circuit simplification. However, before reading these, you should be aware of several points. First, the particular steps used in a certain simplification need not be unique; there may be alternate ways to simplify that are just as good or better. Second, some moves may appear to be mysterious ("Why did he apply *that* property?"), but with practice and increased familiarity with the Boolean properties, you will have fewer such reactions. Third, often valid applications of a Boolean property to a given expression may yield an equivalent expression that is no simpler and perhaps more complicated. In this situation, back up and try another property; trial and error is part of the learning process here. Nevertheless, by carefully reading various illustrations and practicing on related exercises, you should develop some facility for circuit simplification using the Boolean properties.

Whenever a property name is listed to the right of a line in the examples of this section, it explains the reason for the equal sign to the left on the same line.

Example 4.4.1

Simplification

Consider the circuit

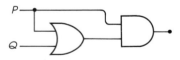

The symbolic form is

$$P(P + Q)$$

which can be simplified as follows:

$$
\begin{aligned}
P(P + Q) &= PP + PQ & &\text{Distributive} \\
&= P + PQ & &\text{Idempotent} \\
&= P \cdot 1 + PQ & &\text{Identity} \\
&= P(1 + Q) & &\text{Distributive} \\
&= P \cdot 1 & &\text{Domination} \\
&= P & &\text{Identity}
\end{aligned}
$$

Thus, the entire given circuit can be replaced by the single input P. ∎

Example 4.4.2   **Simplification**

The symbolic form of the circuit

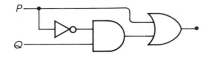

is

$$P + \bar{P}Q$$

which can be simplified as follows:

$$
\begin{aligned}
P + \bar{P}Q &= (P + \bar{P})(P + Q) &&\text{Distributive} \\
&= 1 \cdot (P + Q) &&\text{Inversion} \\
&= P + Q &&\text{Identity}
\end{aligned}
$$

So the OR gate $P + Q$ can be used in place of the given circuit. ∎

Example 4.4.3   **Simplification**

The circuit

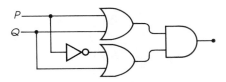

has symbolic form $(P + Q)(\bar{P} + Q)$; let's simplify it.

$$
\begin{aligned}
(P + Q)(\bar{P} + Q) &= (P + Q)\bar{P} + (P + Q)Q &&\text{Distributive} \\
&= P\bar{P} + Q\bar{P} + PQ + QQ &&\text{Distributive} \\
&= 0 + Q\bar{P} + PQ + QQ &&\text{Inversion} \\
&= Q\bar{P} + PQ + QQ &&\text{Identity} \\
&= Q\bar{P} + PQ + Q &&\text{Idempotent} \\
&= \bar{P}Q + PQ + Q &&\text{Commutative} \\
&= (\bar{P} + P)Q + Q &&\text{Distributive} \\
&= 1 \cdot Q + Q &&\text{Inversion} \\
&= Q + Q &&\text{Identity} \\
&= Q &&\text{Idempotent}
\end{aligned}
$$

Hence, the given circuit can be replaced by the single input Q. ∎

Example 4.4.4   **Simplification**

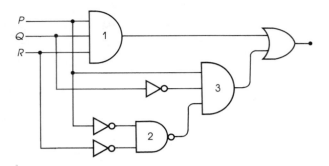

Here gates 1, 2, and 3 are represented by

$$PQR \qquad \overline{\overline{P}\overline{R}} \qquad P\overline{Q}(\overline{\overline{P}\overline{R}})$$

respectively, so the given circuit is

$$PQR + P\overline{Q}(\overline{\overline{P}\overline{R}})$$

which we now simplify.

$$
\begin{aligned}
PQR + P\overline{Q}(\overline{\overline{P}\overline{R}}) &= PQR + P\overline{Q}(\overline{\overline{P}} + \overline{\overline{R}}) & \text{De Morgan} \\
&= PQR + P\overline{Q}(P + R) & \text{Double inversion} \\
&= PQR + P\overline{Q}P + P\overline{Q}R & \text{Distributive} \\
&= PRQ + PP\overline{Q} + PR\overline{Q} & \text{Commutative} \\
&= PRQ + P\overline{Q}\ \ + PR\overline{Q} & \text{Idempotent} \\
&= PRQ + PR\overline{Q} + P\overline{Q} & \text{Commutative} \\
&= \ \ PR(Q + \overline{Q}) + P\overline{Q} & \text{Distributive} \\
&= \ \ \ \ \ PR \cdot 1 \ \ + P\overline{Q} & \text{Inversion} \\
&= \ \ \ \ \ \ \ \ PR \ \ \ + P\overline{Q} & \text{Identity} \\
&= \ \ \ \ \ \ P(R + \overline{Q}) & \text{Distributive}
\end{aligned}
$$

Thus, the given complex circuit can be replaced by

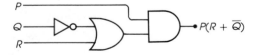

The Boolean properties can also be applied to determine alternate symbolic forms of given expressions or to derive further algebraic properties, as the next three examples indicate.

Example 4.4.5   **Alternate Form of NOR Gate**

In Example 4.4.2 we derived an alternate form for an OR gate, namely

$$P + Q = P + \overline{P}Q$$

So using this result, we have

$$\overline{P + Q} = \overline{P + \bar{P}Q} \qquad \text{Example 4.4.2}$$
$$= \bar{P}(\overline{\bar{P}Q}) \qquad \text{De Morgan}$$
$$= \bar{P}(\bar{\bar{P}} + \bar{Q}) \qquad \text{De Morgan}$$
$$= \bar{P}(P + \bar{Q}) \qquad \text{Double inversion}$$

Hence, $\bar{P}(P + \bar{Q})$ is an alternate representation for a NOR gate.   ■

Example 4.4.6

Alternate Forms of Exclusive OR Gate

In Section 4.2 we discussed the exclusive OR gate $\oplus$ and characterized it in terms of the AND gate and inclusive OR gate.

$$P \oplus Q = \qquad \bar{P}Q + P\bar{Q} \qquad \text{Example 4.2.7}$$
$$= 0 + \bar{P}Q + P\bar{Q} + 0 \qquad \text{Identity}$$
$$= P\bar{P} + \bar{P}Q + P\bar{Q} + Q\bar{Q} \qquad \text{Inversion}$$
$$= P\bar{P} + Q\bar{P} + P\bar{Q} + Q\bar{Q} \qquad \text{Commutative}$$
$$= (P + Q)\bar{P} + (P + Q)\bar{Q} \qquad \text{Distributive}$$
$$= (P + Q)(\bar{P} + \bar{Q}) \qquad \text{Distributive}$$
$$= (P + Q)\overline{PQ} \qquad \text{De Morgan}$$

Therefore, either of the last two expressions

$$(P + Q)(\bar{P} + \bar{Q}) \quad \text{or} \quad (P + Q)\overline{PQ}$$

can be used as an alternate representation of the exclusive OR gate.   ■

Example 4.4.7

Associativity of

We can prove the associative property for $\oplus$

$$P \oplus (Q \oplus R) = (P \oplus Q) \oplus R$$

by working on each side separately and reducing them to the same expression.

$$P \oplus (Q \oplus R) = P \oplus (\bar{Q}R + Q\bar{R}) \qquad \text{Example 4.2.7}$$
$$= \bar{P}(\bar{Q}R + Q\bar{R}) + P(\overline{\bar{Q}R + Q\bar{R}}) \qquad \text{Example 4.2.7}$$
$$= \qquad '' \qquad + P(\overline{\bar{Q}R})(\overline{Q\bar{R}}) \qquad \text{De Morgan}$$
$$= \qquad '' \qquad + P(\bar{\bar{Q}} + \bar{R})(\bar{Q} + \bar{\bar{R}}) \qquad \text{De Morgan}$$
$$= \qquad '' \qquad + P(Q + \bar{R})(\bar{Q} + R) \qquad \text{Double inversion}$$
$$= \qquad '' \qquad + P[(Q + \bar{R})\bar{Q} + (Q + \bar{R})R] \qquad \text{Distributive}$$
$$= \qquad '' \qquad + P(Q\bar{Q} + \bar{R}\bar{Q} + QR + \bar{R}R) \qquad \text{Distributive}$$
$$= \qquad '' \qquad + P(0 + \bar{R}\bar{Q} + QR + 0) \qquad \text{Inversion}$$
$$= \qquad '' \qquad + P(\bar{R}\bar{Q} + QR) \qquad \text{Identity}$$
$$= \qquad '' \qquad + P(\bar{Q}\bar{R} + QR) \qquad \text{Commutative}$$
$$= \bar{P}\bar{Q}R + \bar{P}Q\bar{R} + P\bar{Q}\bar{R} + PQR \qquad \text{Distributive}$$

and

$$
\begin{aligned}
(P \oplus Q) \oplus R &= & (\bar{P}Q + P\bar{Q}) \oplus R & \qquad \text{Example 4.2.7} \\
&= & (\overline{\bar{P}Q + P\bar{Q}})R + (\bar{P}Q + P\bar{Q})\bar{R} & \qquad \text{Example 4.2.7} \\
&= & (\overline{\bar{P}Q})(\overline{P\bar{Q}})R + \quad\text{"} & \qquad \text{De Morgan} \\
&= & (\bar{\bar{P}} + \bar{Q})(\bar{P} + \bar{\bar{Q}})R + \quad\text{"} & \qquad \text{De Morgan} \\
&= & (P + \bar{Q})(\bar{P} + Q)R + \quad\text{"} & \qquad \text{Double inversion} \\
&= & [(P + \bar{Q})\bar{P} + (P + \bar{Q})Q]R + \quad\text{"} & \qquad \text{Distributive} \\
&= & (P\bar{P} + \bar{Q}\bar{P} + PQ + \bar{Q}Q)R + \quad\text{"} & \qquad \text{Distributive} \\
&= & (0 + \bar{Q}\bar{P} + PQ + 0)R + \quad\text{"} & \qquad \text{Inversion} \\
&= & (\bar{Q}\bar{P} + PQ)R + \quad\text{"} & \qquad \text{Identity} \\
&= & (\bar{P}\bar{Q} + PQ)R + \quad\text{"} & \qquad \text{Commutative} \\
&= & \bar{P}\bar{Q}R + PQR \quad + \bar{P}Q\bar{R} + P\bar{Q}\bar{R} & \qquad \text{Distributive} \\
&= & \bar{P}\bar{Q}R + \bar{P}Q\bar{R} + P\bar{Q}\bar{R} + PQR & \qquad \text{Commutative}
\end{aligned}
$$

This completes the proof of the associativity of $\oplus$. Hereafter, then, we can write

$$P \oplus Q \oplus R$$

without any parentheses and use the multiinput exclusive OR gate symbol

In a similar but shorter and less difficult fashion, one can show that

$$P \oplus Q = Q \oplus P$$

so $\oplus$ is also commutative (see exercise 23 on next page).

One natural, important question arises here: How do we know that a simplified circuit cannot be simplified further? Systematic procedures can be used to determine minimal circuit forms (the Quine–McCluskey method and Karnaugh maps are examples), which are usually studied in courses on digital systems and logic design.

EXERCISES **4.4**

In exercises 1 through 20, draw the given circuit and then use Boolean properties to find a simpler circuit.

1. $P + PQ$

2. $Q(P + Q)$

3. $P(\bar{P} + Q)$

4. $P\bar{Q} + Q$

5. $\bar{P}Q + PQ$

6. $(P + Q)(P + \bar{Q})$

7. $PQ + P\bar{Q} + \bar{P}Q + \bar{P}\bar{Q}$

8. $(\bar{P} + \bar{Q})(\bar{P} + Q)(P + \bar{Q})(P + Q)$

9. $PPQ(P + \bar{Q})$

10. $Q + Q + P + \bar{P}Q$

11. $(P\bar{Q} + \bar{P}Q)P$

12. $PQ(\bar{P} + Q)$

13. $PQR + P\bar{Q}R + \bar{P}\bar{Q}R$

14. $P\bar{Q}R + P\bar{Q}\bar{R}$

15. $PQR + PQ\bar{R} + P\bar{Q}R$

16. $(P + Q)R + (\bar{P} + \bar{R})Q$

17. $(P + Q)(\bar{P} + R)(Q + R)$

18. $\overline{P\bar{Q} + R}$

19. $\overline{(\bar{P} + R)(Q + \bar{S})}$

20. $PQ + \bar{P}R + QR$

21. Use the expression

$$PQ = (P + \bar{Q})Q$$

and Boolean properties to find an alternate form for the NAND gate $\overline{PQ}$.

22. Use the expression

$$P \oplus Q = \bar{P}Q + P\bar{Q}$$

and Boolean properties to find an alternate form for the exclusive NOR gate

$$\overline{P \oplus Q}$$

23. Verify that the commutative property holds for $\oplus$;

that is,

$$P \oplus Q = Q \oplus P$$

24. Do both of the distributive properties hold if $+$ is replaced by $\oplus$? That is, does

$$P \oplus QR = (P \oplus Q)(P \oplus R)?$$
$$P(Q \oplus R) = PQ \oplus PR?$$

4.5

Design of Circuits

In previous sections we have taken given circuits, either in symbolic or diagrammatic form, and proceeded to analyze or simplify them. Now we shall illustrate a method that we can use to construct circuits that accomplish particular tasks.

To begin with, recall the general nature of our discussion of a parity bit system in Section 3.8.

Example 4.5.1

Parity Check Circuit

For a parity system to function, one needs at both the transmitting and receiving ends a circuit that will take a given binary string and determine whether it contains an even or odd number of 1 bits. In Section 3.8 our examples dealt with the attachment of a parity bit to four-bit strings. So let's construct a circuit that will take in a four-bit string and determine whether it has even or odd parity—that is, an even or odd number of 1 bits—so that the appropriate 0 or 1 bit can be affixed.

Ordinarily, one would handle such a four-bit string by scanning bit by bit from left to right, counting the 1 bits. The running parity (output) at any particular instant depends on two things (inputs); namely, the value of the current bit scanned, and the parity of the preceding bits. For instance, using the string

1101

and working from left to right we have:

$$
\begin{array}{rl}
\text{First bit} \;=\; & 1 \\
& +\text{even} \quad \text{(no preceding parity)} \\
\hline
& \text{odd} = \text{parity through first bit} \\[4pt]
\text{Second bit} \;=\; & 1 \\
& +\text{odd} \quad \text{(parity of preceding bits)} \\
\hline
& \text{even} = \text{parity through second bit} \\[4pt]
\text{Third bit} \;=\; & 0 \\
& +\text{even} \quad \text{(parity of preceding bits)} \\
\hline
& \text{even} = \text{parity through third bit} \\[4pt]
\text{Fourth bit} \;=\; & 1 \\
& +\text{even} \quad \text{(parity of preceding bits)} \\
\hline
& \text{odd} = \text{parity through fourth bit}
\end{array}
$$

Now let's generalize this procedure by assigning

B = value (0 or 1) of current bit b scanned

P = parity (0 for even, 1 for odd) of bits preceding b

Par = parity (0 for even, 1 for odd) of bits up to and including b

Therefore, *when a particular bit b in the string is scanned*, we want a circuit that will produce the following results:

B	P	Par
1	1	0
1	0	1
0	1	1
0	0	0

Note that we use the symbols 0 and 1 in two different ways here. On the one hand, they represent the possible bit values B; on the other hand, 0 and 1 represent even and odd respectively for P and Par.

From our work in preceding sections, we should recognize the above as the output table for the exclusive OR gate

$$B \oplus P$$

It is the circuit needed *to handle each bit in the string.*

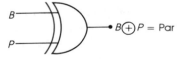

To process a four-bit string $B_1 B_2 B_3 B_4$, we need to apply four $\oplus$ gates successively:

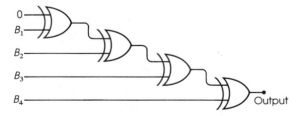

Here the top input is 0 since the initial parity is even. However, it is easy to verify that

$$0 \oplus B = B$$

so the topmost $\oplus$ gate can be replaced by the B_1 input

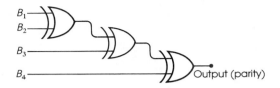

The output is 0 if $B_1 B_2 B_3 B_4$ contains an even number of 1 bits, and 1 if it contains an odd number. Due to the associativity of $\oplus$, (see Example 4.4.7), we can symbolize this 4-bit *parity check circuit* by

$$B_1 \oplus B_2 \oplus B_3 \oplus B_4$$

and draw it as

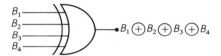

This circuit could alternately be represented as shown in Figure 13.

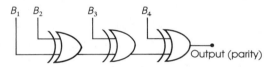

Figure 13

This is an example of an *iterated network*, since it consists of a number of identical cells that are interconnected. The primary inputs B_1, B_2, B_3, and B_4 are applied at the same time, and electrical signals are propagated down the line of cells. ∎

In our first example we were fortunate to be able to recognize the output table as that of $B \oplus P$. The next example illustrates one method that can be used when we're not so lucky.

Example **4.5.2** Comparitor Circuit

We wish to build a circuit that will take in two four-bit strings and determine whether the first is less than the second. One way of proceeding is to compare bit by bit from right to left, where at each step of comparing particular digits the result (output) depends upon two things (inputs), namely, the values of the two bits currently being compared and the result of the comparison of the two strings preceding the current bits. Let's look at an example using strings

1010   and   1101

We'll use a and b to represent bits of 1010 and 1101, respectively.

							Comparison through This Bit

First Bits: $\quad$ 1 $\quad$ 0 $\quad$ 1 $\quad$ | 0 | $\quad$ $a = 0 < 1 = b$ $\qquad$ $0 < 1$
$\qquad\qquad\qquad$ 1 $\quad$ 1 $\quad$ 0 $\quad$ | 1 |

Second Bits: $\quad$ 1 $\quad$ 0 $\quad$ | 1 | $\quad$ 0 $\quad$ $a = 1 \not< 0 = b$ $\qquad$ $10 \not< 01$
$\qquad\qquad\qquad$ 1 $\quad$ 1 $\quad$ | 0 | $\quad$ 1

Third Bits: $\quad$ 1 $\quad$ | 0 | $\quad$ 1 $\quad$ 0 $\quad$ $a = 0 < 1 = b$ $\qquad$ $010 < 101$
$\qquad\qquad\qquad$ 1 $\quad$ | 1 | $\quad$ 0 $\quad$ 1

Fourth Bits: | 1 | $\quad$ 0 $\quad$ 1 $\quad$ 0 $\quad$ $a = 1 = 1 = b$ *and* $\qquad$ $1010 < 1101$
$\qquad\qquad\qquad$ | 1 | $\quad$ 1 $\quad$ 0 $\quad$ 1 $\quad$ for preceding strings
$\qquad\qquad\qquad\qquad\qquad\qquad\qquad\qquad$ $010 < 101$

This last line of work yields the conclusion

$\qquad$ $1010 < 1101$

Now let's generalize this process; let

$\qquad$ A = value of bit in first four-bit string

$\qquad$ B = value of corresponding bit in second four-bit string

$$P = \begin{cases} 0 & \text{if string preceding } A \not< \text{string preceding } B \\ 1 & \text{if string preceding } A < \text{string preceding } B \end{cases}$$

$$\text{CMP} = \begin{cases} 0 & \text{if string through } A \not< \text{string through } B \\ 1 & \text{if string through } A < \text{string through } B \end{cases}$$

Here we use the word *through* in the sense of "up to and including." Then *at each step of comparing one bit with another*, we desire a circuit that will produce the result shown in Table 23.

In this table, the first row says that when bits $A = B = 1$ and the string preceding A is less than the string preceding B (since $P = 1$), then the string through A is less than the string through B (that is, CMP = 1). Again, as with the table in the preceding example, 0 and 1 are used in two different ways.

Our task now is to find a circuit whose outputs match the outputs CMP above. This can be done as follows:

Table 23

A	B	P	CMP
1	1	1	1
1	1	0	0
1	0	1	0
1	0	0	0
0	1	1	1
0	1	0	1
0	0	1	1
0	0	0	0

Sum of Products Method

Step 1: Locate each row in the table with output 1.

Step 2: For each of these rows, precisely one of the "products" ABP, $\bar{A}BP$, $A\bar{B}P$, $AB\bar{P}$, $\bar{A}\bar{B}P$, $\bar{A}B\bar{P}$, $A\bar{B}\bar{P}$, and $\bar{A}\bar{B}\bar{P}$ will have value equal to 1. Write it down.

Step 3: Connect the results of step 2 with OR gate symbols +.

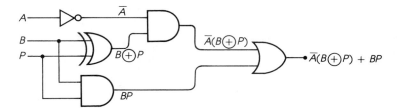

Figure 14

The resulting expression is the symbolic form of a circuit whose output table equals the given one.

To apply this method to the example at hand, we use Table 23 with output CMP and proceed as follows:

Step 1: Rows 1, 5, 6, and 7 have output equal to 1.
Step 2: For row 1, $ABP = 1$; row 5, $\bar{A}BP = 1$; row 6, $\bar{A}B\bar{P} = 1$; and row 7, $\bar{A}\bar{B}P = 1$.
Step 3: $ABP + \bar{A}BP + \bar{A}B\bar{P} + \bar{A}\bar{B}P$ is the desired symbolic form.

Observe that the derived form has one term "added" for each row with an output equal to 1. Rather than stop here, let's apply the simplification method of Section 4.4.

$$
\begin{aligned}
ABP + \bar{A}BP + \bar{A}B\bar{P} + \bar{A}\bar{B}P &= (A + \bar{A})BP + \bar{A}(B\bar{P} + \bar{B}P) \quad \text{Distributive} \\
&= 1 \cdot BP + \bar{A}(B\bar{P} + \bar{B}P) \quad \text{Inversion} \\
&= BP + \bar{A}(B\bar{P} + \bar{B}P) \quad \text{Identity} \\
&= BP + \bar{A}(\bar{B}P + B\bar{P}) \quad \text{Commutative} \\
&= BP + \bar{A}(B \oplus P) \quad \text{Example 4.2.7}
\end{aligned}
$$

Therefore, the circuit shown in Figure 14 is one that will compare two digits in the desired manner.

If we abbreviate this circuit by the box

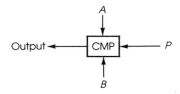

then the ***comparitor circuit*** to compare two four-bit strings $A_4A_3A_2A_1$ and $B_4B_3B_2B_1$ is shown in Figure 15; it is another example of an ***iterated network***. We shall leave it to the reader to verify (see exercise 4 of this section)

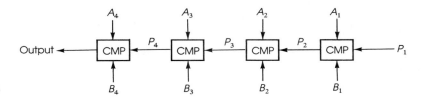

Figure 15

that the initial input $P_1 = 0$ yields a strict $<$ comparison of the two strings, while setting $P_1 = 1$ initially yields a $\leq$ comparison.   ◼

The procedure just described for generating the circuit's symbolic form from the output table is an illustration of using the *sum-of-products method*. It is easily adapted to tables with a different number of inputs:

If the table has n inputs, then in step 2 for each row with output 1, find that product of the n inputs (or inverted inputs) that equals 1.

Other methods of generating such circuits are studied in courses on digital systems or logic design.

The last example of this section relates to a circuit with two desired outputs where each must be handled separately and the results combined.

Example 4.5.3   **Full Adder Circuit**

The objective here is to construct a circuit that will add two four-bit strings. From our work in Chapter 2 we are quite familiar with the mechanics of binary addition.

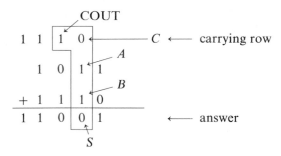

We've highlighted one step in the illustration above in order to generalize. Let

$$A = \text{bit to be added}$$
$$B = \text{bit to be added}$$
$$C = \text{carry bit to be added to } A \text{ and } B \text{ (carry-}in)$$
$$S = \text{right digit of sum of } A, B, \text{ and } C$$
$$\text{COUT} = \text{left digit of sum of } A, B, \text{ and } C \text{ (carry-}out)$$

Notice that at each step there are always three input bits A, B, and C and two output bits S and COUT; also, the COUT of one step is the C of the next. The desired table, shown in Table 24, is easy to construct, for we merely add A, B, and C and tally the results. For instance, the second row summarizes the calculation

$$1 + 1 + 0 = 10$$
$$\uparrow \quad \uparrow \quad \uparrow \quad \uparrow \searrow$$
$$A \quad B \quad C \quad \text{COUT} \quad S$$

Table 24

A	B	C	COUT	S
1	1	1	1	1
1	1	0	1	0
1	0	1	1	0
1	0	0	0	1
0	1	1	1	0
0	1	0	0	1
0	0	1	0	1
0	0	0	0	0
Input			Output	

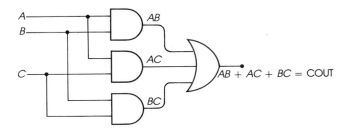

Figure 16

Since there are two output columns in this table, we apply the sum-of-products method to each separately. The COUT rows 1, 2, 3, and 5 have 1 outputs, giving the sum

$$ABC + AB\bar{C} + A\bar{B}C + \bar{A}BC$$

$= ABC + ABC + ABC + AB\bar{C} + A\bar{B}C + \bar{A}BC$	Idempotent
$= ABC + AB\bar{C} + ABC + A\bar{B}C + ABC + \bar{A}BC$	Commutative
$= ABC + AB\bar{C} + ACB + AC\bar{B} + BCA + BC\bar{A}$	Commutative
$= AB(C + \bar{C}) + AC(B + \bar{B}) + BC(A + \bar{A})$	Distributive
$= \quad AB \cdot 1 \quad + \quad AC \cdot 1 \quad + \quad BC \cdot 1$	Inversion
$= \quad\quad AB \quad + \quad\quad AC \quad + \quad\quad BC$	Identity

This circuit for COUT is drawn in Figure 16.

For output column S, the rows 1, 4, 6, and 7 have outputs equaling 1, yielding the sum

$$ABC + A\bar{B}\bar{C} + \bar{A}B\bar{C} + \bar{A}\bar{B}C$$

whose simplification is tricky:

$\quad ABC + A\bar{B}\bar{C} \quad\quad + \bar{A}B\bar{C} + \bar{A}\bar{B}C$	
$= A(BC + \bar{B}\bar{C}) \quad\quad + \bar{A}(B\bar{C} + \bar{B}C)$	Distributive
$= \quad\quad " \quad\quad\quad\quad + \bar{A}(\bar{B}C + B\bar{C})$	Commutative
$= \quad\quad " \quad\quad\quad\quad + \bar{A}(B \oplus C)$	Example 4.2.7
$= A(\overline{\overline{BC + \bar{B}\bar{C}}}) \quad\quad + \quad\quad "$	Double inversion
$= A(\overline{\overline{BC})(\overline{\bar{B}\bar{C}}}) \quad\quad + \quad\quad "$	De Morgan
$= A(\bar{B} + \bar{C})(\bar{\bar{B}} + \bar{\bar{C}}) + \quad\quad "$	De Morgan
$= A(\bar{B} + \bar{C})(B + C) + \quad\quad "$	Double inversion
$= A(B + C)(\bar{B} + \bar{C}) + \quad\quad "$	Commutative
$= \quad A(\overline{B \oplus C}) \quad + \bar{A}(B \oplus C)$	Example 4.4.6
$= \quad \bar{A}(B \oplus C) \quad + A(\overline{B \oplus C})$	Commutative
$= A \oplus (B \oplus C)$	Example 4.2.7
$= A \oplus B \oplus C$	Example 4.4.7

Thus a three-input exclusive OR gate will suffice as a circuit for the S output.

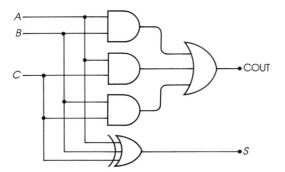

Figure 17

Combining this with the COUT circuit drawn in Figure 16, we find a representation of the *full adder circuit*, shown in Figure 17. If we abbreviate the full adder by the box

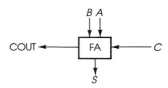

then the circuit to add two four-bit strings $A_4A_3A_2A_1$ and $B_4B_3B_2B_1$ is drawn in Figure 18, which is another example of an *iterated network*. Notice that the initial carry-in is 0 (see exercise 14), and the answer is $S_5S_4S_3S_2S_1$.

∎

In this section we have merely scratched the surface of the topic of circuit design. The examples presented were those of *combinatorial logic networks*, meaning that they do not contain any feedback or memory, and the output at any instant depends only on the input values at that instant. Most digital systems incorporate this type of combinatorial network with memory devices, so that the output at any instant depends on both the input values and the information stored in memory.

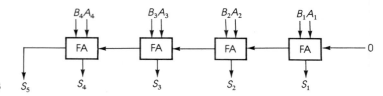

Figure 18

EXERCISES **4.5**

1. Work through the parity check iterated network of Figure 13 with each of the following four-bit strings, and in each case interpret the output:

a. 0000   **b.** 1000   **c.** 0110   **d.** 1011

2. Work through the comparitor iterated network of Figure 15 by setting the initial input P_1 at 0 and taking four-bit strings $A_4A_3A_2A_1$ and $B_4B_3B_2B_1$ to be, respectively, the following:

a. 0110, 0101   **b.** 1001, 1010
c. 0010, 0001   **d.** 1011, 1011

3. Repeat exercise 2, but this time set P_1 at 1.

4. Compare the answers to exercises 2(d) and 3(d). From this comparison draw a conclusion regarding the value of initial input P_1.

5. Work through the iterated network for addition shown in Figure 18 taking four-bit strings $A_4A_3A_2A_1$ and $B_4B_3B_2B_1$, respectively, to be the following:

a. 1011, 0100   **b.** 1010, 0011
c. 1000, 1000   **d.** 0111, 1011

In each case interpret the meaning of the output.

6. Construct an output table for two input bits A and B that will indicate whether $A < B$ (represented by 1) or $A \not< B$ (represented by 0). Then find a circuit to implement this table—that is, whose outputs match this table.

7. Repeat exercise 6, but replace $<$ by $\le$, and $\not<$ by $\not\le$.

8. Construct an output table for two input bits A and B that will indicate whether the algebraic difference $A - B$ is negative (represented by 1) or not (represented by 0). Then find a circuit to implement this table.

9. Construct an output table for three-bit inputs A, B, and C that will indicate whether the algebraic expression $A * B - C$ is negative (represented by 1) or not (represented by 0). Then find a circuit to implement this table.

10. For each case, find circuits that implement the given table and simplify if possible.

a.

P	Q	Output
1	1	1
1	0	0
0	1	1
0	0	0

b.

P	Q	R	Output
1	1	1	1
1	1	0	0
1	0	1	0
1	0	0	0
0	1	1	1
0	1	0	0
0	0	1	0
0	0	0	1

11. For each case, find circuits that implement the given table and simplify if possible.

a.

P	Q	Output
1	1	0
1	0	1
0	1	0
0	0	1

b.

P	Q	R	Output
1	1	1	0
1	1	0	1
1	0	1	0
1	0	0	1
0	1	1	1
0	1	0	0
0	0	1	1
0	0	0	0

12. For the full adder circuit (see Example 4.5.3):

a. Verify that $(A \oplus B)C + AB$ is an alternate circuit for the COUT bit.
b. Draw a circuit diagram for $(A \oplus B)C + AB$
c. Find an alternate circuit diagram by replacing the COUT circuit of Figure 17 with that of part (b).

13. Construct a table that will take two input bits A and B and output *both* bits C and S of their sum. Then find separate circuits to implement each of these bits C (for carry) and S (for sum), and combine them into one circuit diagram. This circuit is called the ***half-adder circuit***.

14. Use the half-adder circuit of exercise 13 to simplify one part of the iterated circuit for addition drawn in Figure 18.

15. Letting HA represent the half-adder circuit described in exercise 13, verify that the circuit of Figure 19 is an alternate representation of the full adder circuit.

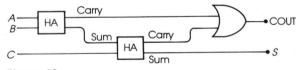

Figure 19

16. Design a circuit that will control a light by two switches in different locations.

17. Design a circuit that will control a light by three switches in different locations.

18. Design a circuit that will be on if

a. at least two of the three inputs are on;
b. precisely two of the three inputs are on.

19. Design a circuit that will be on if

   a. precisely two of the four inputs are on;

   b. at least three of the four inputs are on.

20. The president of a small company implements a proposal only if she and a majority of her three advisors approve the proposal. Design a circuit so that each advisor and the president can vote by pressing a button, and a light goes on when a proposal is to be implemented.

4.6

Conditional and Biconditional Connectives

In the last two sections of this chapter, we shall move away from the study of electronic gates and circuitry and complete our discussion of sentential logic.

We often use sentences with the word combination "if . . . then . . . " in everyday language (as well as in math books!); for instance,

If it rains, then I will take my umbrella.

The symbol $\rightarrow$ is ordinarily used in this context. If p and q are propositions, then the form

$$p \rightarrow q$$

represents the sentence

If p then q.

Let's consider for a moment the sense in which we use

$$p \rightarrow q$$

It is in the sense that

If p happens, then q necessarily happens.

But of course, q can happen independently of p. So

$$p \rightarrow q$$

is the same as saying

Either q happens or p does not happen or both.

That is,

$$q \quad \text{or} \quad \sim p \quad \text{or both}$$

which is written symbolically

$$q \vee \sim p$$

The result is that

$$p \rightarrow q \quad = \quad q \vee \sim p$$

Table 25

p	q	$\sim p$	$q \vee \sim p$
T	T	F	T
T	F	F	F
F	T	T	T
F	F	T	T

Table 26

p	q	$p \to q$
T	T	T
T	F	F
F	T	T
F	F	T

Therefore, we can construct the truth table for $q \vee \sim p$ in the usual fashion (see Table 25) and use the output column to write the **truth table for the conditional connective** $\to$ (see Table 26). In this table note that $p \to q$ is false *only* in the case that p is true and q is false.

Thus we have another connective at our disposal when translating sentences into symbolic form and constructing their truth tables.

Example 4.6.1

Sentence Involving →

Consider the sentence

> If Anne doesn't get the ticket, then either I go alone or we both stay home.

Letting

$p =$ Anne gets the ticket

$q =$ I go alone

$r =$ we stay home

we get for this sentence the symbolic form

$$\sim p \to (q \vee r)$$

Observe that parentheses are needed here to avoid ambiguity; the truth table for this expression is given in Table 27. ∎

Example 4.6.2

Symbolic Form Involving →

For the expression

$$p \vee [\sim q \to (r \wedge s)]$$

Table 27

p	q	r	$\sim p$	$q \vee r$	$\sim p \to (q \vee r)$
T	T	T	F	T	T
T	T	F	F	T	T
T	F	T	F	T	T
T	F	F	F	F	T
F	T	T	T	T	T
F	T	F	T	T	T
F	F	T	T	T	T
F	F	F	T	F	F

Table 28

p	q	r	s	$\sim q$	$r \wedge s$	$\sim q \rightarrow (r \wedge s)$	$p \vee [\sim q \rightarrow (r \wedge s)]$
T	T	T	T	F	T	T	T
T	T	T	F	F	F	T	T
T	T	F	T	F	F	T	T
T	T	F	F	F	F	T	T
T	F	T	T	T	T	T	T
T	F	T	F	T	F	F	T
T	F	F	T	T	F	F	T
T	F	F	F	T	F	F	T
F	T	T	T	F	T	T	T
F	T	T	F	F	F	T	T
F	T	F	T	F	F	T	T
F	T	F	F	F	F	T	T
F	F	T	T	T	T	T	T
F	F	T	F	T	F	F	F
F	F	F	T	T	F	F	F
F	F	F	F	T	F	F	F

the truth table is that given in Table 28. Since there are four propositions p, q, r, and s involved, this table has $2^4 = 16$ rows. ∎

The sentences

> p is a sufficient condition that q.
>
> p only if q.
>
> p implies q.

are sometimes used as alternate ways of expressing

> *If p then q.*

Example 4.6.3   **Alternate Expression of If/Then**

The sentence

> I will go and take Clare only if Kerry is not driving.

has the symbolic form

$$(p \wedge q) \rightarrow \sim r$$

Table 29

p	q	r	$p \wedge q$	$\sim r$	$(p \wedge q) \rightarrow \sim r$
T	T	T	T	F	F
T	T	F	T	T	T
T	F	T	F	F	T
T	F	F	F	T	T
F	T	T	F	F	T
F	T	F	F	T	T
F	F	T	F	F	T
F	F	F	F	T	T

where

$$p = \text{I will go}$$
$$q = \text{I will take Clare}$$
$$r = \text{Kerry is driving}$$

The truth table is shown in Table 29.   ■

Definitions

> The statement
>
> $$p \rightarrow q$$
>
> is called a ***conditional sentence***, and
>
> $$q \rightarrow p$$
>
> is the ***converse of $p \rightarrow q$***,
>
> $$\sim p \rightarrow \sim q$$
>
> is the ***inverse of $p \rightarrow q$***, and
>
> $$\sim q \rightarrow \sim p$$
>
> is the ***contrapositive of $p \rightarrow q$***.

The truth tables for these are combined into one table given in Table 30. Notice that the columns for the conditional and its converse do *not* coincide. However, the columns for the conditional and its contrapositive *do* coincide, as do the columns for the converse and inverse. Therefore, the following statements are valid.

Fact

> $$(p \rightarrow q) = (\sim q \rightarrow \sim p)$$
> $$(q \rightarrow p) = (\sim p \rightarrow \sim q)$$

Example 4.6.4

Contrapositive, Converse, and Inverse

The conditional sentence

If Maura arrives on time,
   then we will go to the ballet. $p \rightarrow q$

Table 30

p	q	$\sim p$	$\sim q$	$p \rightarrow q$	$\sim q \rightarrow \sim p$	$q \rightarrow p$	$\sim p \rightarrow \sim q$
T	T	F	F	T	T	T	T
T	F	F	T	F	F	T	T
F	T	T	F	T	T	F	F
F	F	T	T	T	T	T	T

Table 31

p	q	$p \rightarrow q$	$q \rightarrow p$	$(p \rightarrow q) \wedge (q \rightarrow p)$
T	T	T	T	T
T	F	F	T	F
F	T	T	F	F
F	F	T	T	T

Table 32

p	q	$p \leftrightarrow q$
T	T	T
T	F	F
F	T	F
F	F	T

has the same truth value as its contrapositive

> If we don't go to the ballet,
>   then Maura did not arrive on time. $\sim q \rightarrow \sim p$

And the converse of the given conditional

> If we go to the ballet,
>   then Maura did arrive on time. $q \rightarrow p$

has the same truth value as the inverse

> If Maura does not arrive on time,
>   then we will not go to the ballet. $\sim p \rightarrow \sim q$

The *biconditional connective* $\leftrightarrow$ is defined by the relationship

$$(p \leftrightarrow q) = [(p \rightarrow q) \wedge (q \rightarrow p)]$$

where p and q are propositions. In Table 31 we have constructed the truth table for

$$(p \rightarrow q) \wedge (q \rightarrow p)$$

We can use this to write the *truth table for the biconditional connective* $\leftrightarrow$ in Table 32. In this latter table observe that $p \leftrightarrow q$ is true when p and q have the same truth values.

In colloquial English, the biconditional symbol $\leftrightarrow$ is usually translated as

> "means the same as"
> "is the same as saying" and
> "is identical to"

or more formally as

> "is equivalent to"
> "is necessary and sufficient that" and
> "if and only if"

Example 4.6.5   **Biconditional Connective**

Consider the sentence

> To say that Chatham is a dog is the same as saying she has four legs
>   and does not chew bubble gum.

Table 33

p	q	r	$\sim r$	$q \wedge \sim r$	$p \leftrightarrow (q \wedge \sim r)$
T	T	T	F	F	F
T	T	F	T	T	T
T	F	T	F	F	F
T	F	F	T	F	F
F	T	T	F	F	T
F	T	F	T	T	F
F	F	T	F	F	T
F	F	F	T	F	T

Letting

p = Chatham is a dog

q = Chatham has four legs

r = Chatham chews bubble gum

we get for this sentence the symbolic form

$p \leftrightarrow (q \wedge \sim r)$

The truth table is given in Table 33. In four cases, namely rows 2, 5, 7, and 8, this sentence is true, which means

p and $q \wedge \sim r$

have the same truth values. ■

The phrase "if ..., then ..." is a part of the structure of many high-level computer programming languages, and it can be used to translate into a particular language the diamond symbol of a flowchart. For instance, the diamond

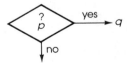

can be expressed as "if p then q." Here, when the question asked by proposition p is answered *yes* or *true*, then the proposition q is executed; otherwise, the *no* or *false* path is followed. Some languages have an extended version

If p then q else r

which can be applied to a diamond of the form

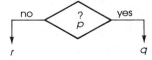

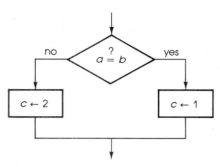

Figure 20

Figure 21

When the answer to question p is *true*, then q is executed; when *false*, proposition r is executed.

Example 4.6.6

If/Then/Else

a. The flowchart portion shown in Figure 20 can be expressed by the phrase

If $a = b$ then $c \leftarrow 1$ else $c \leftarrow 2$.

b. When one diamond (question) is followed by another as in Figure 21, then for clarity it is advisable to express these **nested conditional statements** by vertically aligning the matching *if* and *else* words, as

If $x < 100$ then
               if $x < 10$ then r
               else q
else p. ∎

EXERCISES 4.6

1. Construct a truth table for each of the following sentence forms:

  a. $(\sim p \lor q) \to p$ **b.** $(p \to q) \to (p \land q)$
  c. $(p \land q) \to \sim r$ **d.** $(p \lor q) \to (r \to p)$
  e. $(p \land \sim r) \leftrightarrow (r \lor \sim s)$

2. Construct a truth table for each of the following sentence forms:

  a. $\sim p \to (q \land p)$ **b.** $(p \to q) \lor \sim(p \leftrightarrow \sim q)$
  c. $p \to (\sim q \lor r)$ **d.** $(p \to \sim q) \to (r \land p)$
  e. $(p \land \sim r) \to (r \lor \sim s)$

3. In each of the following, translate the given sentence into symbolic form and construct a truth table. The letters in parentheses indicate the number of propositions you should use.

  a. If I leave, then I will not return. (p, q)
  b. You will succeed if you work hard or hit the lottery. (p, q, r)

c. Increased productivity implies increased wages and improved sales. (p, q, r)
d. If you remain in that job, you will be accepted if you don't complain. (p, q, r)
e. All athletes are healthy. (p, q)
f. If today is Saturday and the weather is mild, then I'll go to the beach or wax the car. (p, q, r, s)
g. For us to win, it is necessary and sufficient that we don't fumble the ball. (p, q)

4. In each of the following, translate the given sentence into symbolic form and construct a truth table. The letters in parentheses indicate the number of propositions you should use.

  a. If he doesn't go, then he'll be angry. (p, q)
  b. You will win first prize or second prize if you compete in the exhibition. (p, q, r)

c. For us to win, it is sufficient that we have no turnovers and play flawless defense. (p, q, r)
d. If I drive we'll be early, but if she drives we'll be late. (p, q, r)
e. If you are willing and able, you will succeed. (p, q, r)
f. If you eat, drink, and be merry, then tomorrow you shall die. (p, q, r, s)
g. To say that he'll be elected president is equivalent to saying that I'm nine feet tall and have green hair. (p, q, r)

5. Express each of the conditional sentences in parts (a), (b), and (c) of exercise 3 as an either/or sentence.

6. Express each of the conditional sentences in parts (a), (b), and (c) of exercise 4 as an either/or sentence.

7. For each of the following conditional sentences, write the contrapositive, converse, and inverse:

a. If John wins the primary, then he'll win the election.
b. You will lose weight only if you diet.
c. Promotion implies retention.

8. For each of the following conditional sentences, write the contrapositive, converse, and inverse:

a. If $x^2 = 0$ then $x = 0$.
b. It is necessary to practice to become an accomplished cellist.
c. Since Smith won, it follows that Jones will resign.

9. Translate each of the flowchart portions of Figure 22 into if/then or if/then/else statements.

a.

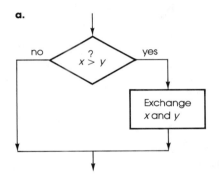

b.

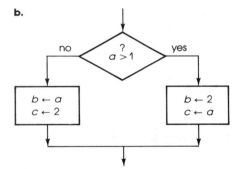

c.

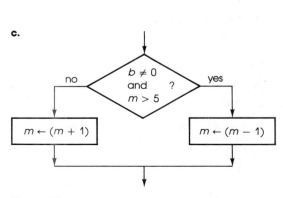

d.

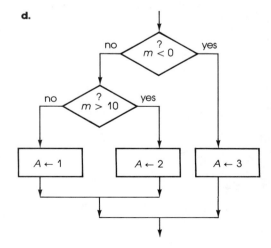

Figure 22

10. Express each of the following statements in flow-chart form:

   a. If $i = 5$, then $m = n = 2$.

   b. If $a > 1$ and $b \le 10$, then $c \leftarrow c + 1$; else $c \leftarrow c + 2$.

   c. If $j < 10$, then $P \leftarrow 10 * P$; otherwise $P \leftarrow 5 * P$.

   d. If $B = 0$ then $A \leftarrow 0$;

      else if $B > 0$ then $A \leftarrow 1$;

          else $A \leftarrow -1$.

11. Verify that

$$(p \leftrightarrow q) = [(\sim p \vee q) \wedge (p \vee \sim q)]$$

12. Verify that

$$(p \leftrightarrow q) = [(p \vee q) \rightarrow (p \wedge q)]$$

13. Is the connective $\rightarrow$

   a. associative in the sense that

$$[p \rightarrow (q \rightarrow r)] = [(p \rightarrow q) \rightarrow r]$$

   for propositions p, q, and r?

   b. commutative in the sense that

$$(p \rightarrow q) = (q \rightarrow p)$$

   for propositions p and q?

14. Is the connective $\leftrightarrow$

   a. associative in the sense that

$$[p \leftrightarrow (q \leftrightarrow r)] = [(p \leftrightarrow q) \leftrightarrow r]$$

   for propositions p, q, and r?

   b. commutative in the sense that

$$(p \leftrightarrow q) = (q \leftrightarrow p)$$

   for propositions p and q?

15. In the truth table for the connective $\rightarrow$, in order to show that statement $p \rightarrow q$ is true, it suffices to show that q is true whenever p is true. (Why?)

   a. Verify that the statement

     If x is a multiple of 5, then x^2 is a multiple of 5.

   is true.

   b. Using the contrapositive form, verify that

     If x^2 is an odd integer, then x is an odd integer.

   is true.

4.7

Deductive Arguments

The type of reasoning by which a ***conclusion*** is drawn from a finite set of statements called ***premises*** is called ***deductive argumentation***. Sometimes such arguments can be represented in symbolic form.

Example 4.7.1   **Deductive Argument**

    Premises: $\begin{cases} \text{Every glitch is a gorch.} \\ \text{Every gorch is a gump.} \end{cases}$

    Conclusion: Every glitch is a gump.

If we let

    $p = $ object is a glitch

    $q = $ object is a gorch

    $r = $ object is a gump

then the given argument can be expressed symbolically as

    Premises: $\begin{cases} p \rightarrow q \\ q \rightarrow r \end{cases}$

    Conclusion: $p \rightarrow r$

For a reason we shall explain shortly, we have constructed in Table 34 a truth table containing each of the two premises and the conclusion as column headings.

Table 34

			Premises		Conclusion
p	q	r	$p \to q$	$q \to r$	$p \to r$
T	T	T	***T***	***T***	T
T	T	F	T	F	F
T	F	T	F	T	T
T	F	F	F	T	F
F	T	T	***T***	***T***	T
F	T	F	T	F	T
F	F	T	***T***	***T***	T
F	F	F	***T***	***T***	T

■

Definition

> A deductive argument is called **valid** if the conclusion is true whenever all the premises are true.

In the preceding example, both premises $p \to q$ and $q \to r$ are true *only* in the four instances marked in boldface; in each of these cases, the conclusion $p \to r$ is true. Therefore, the deductive argument in Example 4.7.1 is valid. (This type of valid argument is called a *hypothetical syllogism*.) It is important to note that *validity pertains only to the symbolic form of the argument* and not to the truth value or context of the component propositions, which in this example are senseless. Then why bother to determine the validity of an argument? Because *if* an argument is valid and *if* we can show that the premises are true, then the conclusion *must* be true. In the real world, much debate is on the truth value of premises!

Example 4.7.2 **Valid Argument**

Premises: $\begin{cases} \text{If it rains, then I carry my umbrella.} \\ \text{It rains.} \end{cases}$

Conclusion: I carry my umbrella.

Letting

$p =$ it rains

$q =$ I carry my umbrella

we get for the symbolic form of this argument

Premises: $\begin{cases} p \to q \\ p \end{cases}$

Conclusion: q

Table 35

Conclusion	Premises	
q	p	$p \to q$
T	T	T
F	T	F
T	F	T
F	F	T

Table 35 shows the truth table for $p \to q$ with the columns permuted so that premises $p \to q$ and p are adjacent. There is only one case in which both premises are true (as marked with boldface); in this instance, the conclusion q is also true. Hence, this argument is valid. ∎

Example 4.7.3 **Invalid Argument**

Premises:
$\begin{cases} \text{If you are sophisticated, then you buy Brand X.} \\ \text{You buy Brand X.} \end{cases}$

Conclusion: You are sophisticated.

Letting

p = you are sophisticated

q = you buy Brand X

we get for this argument the form

Premises: $\begin{cases} p \to q \\ q \end{cases}$

Conclusion: p

But the truth table in Table 36 shows that there is a case in which both premises $p \to q$ and q are true, but conclusion p is *false* (see the third row). Thus, this argument is invalid. ∎

Table 36

Conclusion	Premises	
p	q	$p \to q$
T	*T*	*T*
T	F	F
F	*T*	*T*
F	F	T

Example 4.7.4 **Invalid Argument**

Consider the following argument:

Premises:
$\begin{cases} \text{If I go, then Barbara stays.} \\ \text{If Anne stays, then Kerry goes.} \\ \text{Either Anne or I go.} \end{cases}$

Conclusion: Either Barbara or Kerry goes.

Table 37

						Premises			Conclusion
p	q	r	s	$\sim q$	$\sim r$	$p \to \sim q$	$\sim r \to s$	$r \vee p$	$q \vee s$
T	T	T	T	F	F	F	T	T	T
T	T	T	F	F	F	F	T	T	T
T	T	F	T	F	T	F	T	T	T
T	T	F	F	F	T	F	F	T	T
T	F	T	T	T	F	*T*	*T*	*T*	T
T	F	T	F	T	F	*T*	*T*	*T*	F
T	F	F	T	T	T	*T*	*T*	*T*	T
T	F	F	F	T	T	T	F	T	F.
F	T	T	T	F	F	*T*	*T*	*T*	T
F	T	T	F	F	F	*T*	*T*	*T*	T
F	T	F	T	F	T	T	T	F	T
F	T	F	F	F	T	T	F	F	T
F	F	T	T	T	F	*T*	*T*	*T*	T
F	F	T	F	T	F	*T*	*T*	*T*	F
F	F	F	T	T	T	T	T	F	T
F	F	F	F	T	T	T	F	F	F

Allowing letters p, q, r, and s to represent the respective propositions "I go," "Barbara goes," "Anne goes," and "Kerry goes," we have the symbolic form

$$\text{Premises:} \begin{cases} p \to \sim q \\ \sim r \to s \\ r \vee p \end{cases}$$

Conclusion: $q \vee s$

The truth table in Table 37 indicates that the argument is invalid, since there are instances (namely, rows 6 and 14) in which the premises are true but the conclusion false. ■

When one can express an argument in symbolic form as we have above, then there is another, more mechanical way to determine validity.

Fact

> If an argument has premises P_1, $P_2, \ldots, P_n$ and conclusion C, then to say that the argument is valid is the same as saying that the conditional sentence
>
> $$(P_1 \wedge P_2 \wedge \cdots \wedge P_n) \to C$$
>
> is a tautology.

Example 4.7.5   **Validity/Tautology**

Each of the first three examples of this section has two premises, so the conditional sentence

$$(P_1 \wedge P_2) \to C$$

for these arguments becomes

$$[(p \to q) \wedge (q \to r)] \to (p \to r)$$
$$[(p \to q) \wedge p] \to q$$
$$[(p \to q) \wedge q] \to p$$

respectively. Their truth tables are given in Tables 38, 39, and 40. The first

Table 38

p	q	r	$p \to q$	$q \to r$	$p \to r$	$(p \to q) \wedge (q \to r)$	$[(p \to q) \wedge (q \to r)] \to (p \to r)$
T	T	T	T	T	T	T	T
T	T	F	T	F	F	F	T
T	F	T	F	T	T	F	T
T	F	F	F	T	F	F	T
F	T	T	T	T	T	T	T
F	T	F	T	F	T	F	T
F	F	T	T	T	T	T	T
F	F	F	T	T	T	T	T

Table 39

p	q	$p \to q$	$(p \to q) \wedge p$	$[(p \to q) \wedge p] \to q$
T	T	T	T	T
T	F	F	F	T
F	T	T	F	T
F	F	T	F	T

Table 40

p	q	$p \to q$	$(p \to q) \wedge q$	$[(p \to q) \wedge q] \to p$
T	T	T	T	T
T	F	F	F	T
F	T	T	T	F
F	F	T	F	T

two are tautologies and the third is not; thus, by the fact above, the first two arguments are valid and the third is not valid. This corroborates our previous conclusions. ∎

Sometimes the premises of an argument can easily be translated into statements concerning sets of objects, and so they can naturally be represented by Venn diagrams. For instance, the following statements have their Venn diagram equivalents shown at the right:

Every senior citizen
is eligible for
deduction *B*.

Those eligible for deduction *B*

No one under age 20
may drink alcoholic
beverages.

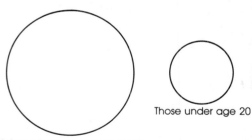

Drinkers of alcoholic beverages   Those under age 20

John plays the violin.

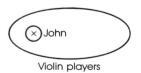

Violin players

Some basketball players
are over 7 feet tall.

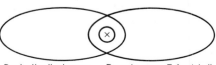

Basketball players   People over 7 feet tall

Fact

> If a Venn diagram (or diagrams) is used to represent the premises of a deductive argument, then to say that the argument is valid is the same as saying that the conclusion follows from the diagram (or *each* of the diagrams).

In using this method, you must be careful to draw *all possible diagrams* that could apply to the premises.

Example 4.7.6   **Valid Argument/Venn Diagram**

Premises: $\begin{cases} \text{Every glitch is a gorch.} \\ \text{Bozo is a glitch.} \end{cases}$

Conclusion: Bozo is a gorch.

The Venn diagrams of the individual premises are

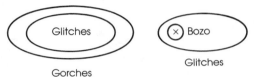

and the diagram of both premises looks like

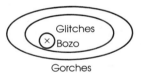

from which the conclusion necessarily follows. Hence, the argument is valid. ∎

Example 4.7.7   **Invalid Argument/Venn Diagram**

Premises: $\begin{cases} \text{Every glitch is a gorch.} \\ \text{Bozo is a gorch.} \end{cases}$

Conclusion: Bozo is a glitch.

Here there are two possible diagrams that apply to the premises:

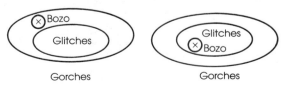

Since the conclusion does *not* follow from each of these, the argument is invalid. ∎

EXERCISES 4.7

In exercises 1 through 8, use a truth table to determine whether the given argument is valid or invalid.

1. Premises: $\begin{cases} p \to r \\ q \to r \end{cases}$

Conclusion: $(p \lor q) \to r$

2. Premises: $\begin{cases} p \to r \\ q \to r \end{cases}$

Conclusion: $(p \land q) \to r$

3. Premises: $\begin{cases} p \to q \\ \sim p \end{cases}$

Conclusion: $\sim q$

4. Premises: $\begin{cases} p \to q \\ r \to s \\ p \lor r \end{cases}$

Conclusion: $q \land s$

5. Premises: $\begin{cases} \sim p \to (q \to \sim r) \\ \sim p \lor \sim s \\ \sim u \to q \end{cases}$

Conclusion: $r \to u$

6. Premises: $\begin{cases} p \to q \\ r \to \sim q \\ r \end{cases}$

Conclusion: p

7. Premises: $\begin{cases} p \leftrightarrow q \\ q \end{cases}$

Conclusion: p

8. Premises: $\begin{cases} p \leftrightarrow \sim q \\ p \end{cases}$

Conclusion: $p \to q$

In exercises 9 through 14, first translate the given argument into symbolic form and then use a truth table to determine its validity.

9. Premises: $\begin{cases} \text{If the temperature rises then it explodes.} \\ \text{Either the temperature rises or it explodes.} \end{cases}$

Conclusion: The temperature rises.

10. Premises: $\begin{cases} \text{Either bankruptcy implies bad} \\ \text{management or bankruptcy} \\ \text{implies bad luck.} \\ \text{Bankruptcy occurs.} \end{cases}$

Conclusion: Either bad management or bad luck.

11. Premises: $\begin{cases} \text{If Jack wins then Paul loses.} \\ \text{If Kim wins then Jane loses.} \\ \text{Either Jack or Kim wins.} \end{cases}$

Conclusion: Either Paul or Jane loses.

12. Premises: $\begin{cases} \text{If Peg passes then Pat passes.} \\ \text{If Mike fails then Pat fails.} \\ \text{If Mike passes then Jim passes.} \end{cases}$

Conclusion: If Jim passes then Peg passes.

13. Premises: $\begin{cases} \text{To say that Katie will not win the} \\ \text{title is the same as saying the moon} \\ \text{is made of cheese.} \\ \text{The moon is made of cheese.} \end{cases}$

Conclusion: Katie wins the title.

14. Premises: $\begin{cases} \text{If Catherine studies, she will pass} \\ \text{the course.} \\ \text{If Catherine does not play softball, she} \\ \text{will study.} \\ \text{Catherine fails the course.} \end{cases}$

Conclusion: Catherine played softball.

In exercises 15 through 20, translate the given statement into a Venn diagram.

15. No late arrivals will be admitted.

16. All florks are blue.

17. Some workers are over age 65.

18. Some whales are mammals.

19. Left-handers are creative.

20. Eric is a great basketball player.

In exercises 21 through 26, use Venn diagrams to determine the validity of the given arguments.

21. Premises: $\begin{cases} \text{All teachers are wealthy.} \\ \text{Lorraine is a teacher.} \end{cases}$

Conclusion: Lorraine is wealthy.

22. Premises: $\begin{cases} \text{All farmers are wise.} \\ \text{All wise people are happy.} \end{cases}$

Conclusion: All farmers are happy.

23. Premises: $\begin{cases} \text{No math teachers are nasty.} \\ \text{No computer science teachers are} \\ \text{nasty.} \end{cases}$

Conclusion: All math teachers are computer science teachers.

24. Premises: $\begin{cases} \text{No truck drivers eat asparagus.} \\ \text{Millie is a truck driver.} \end{cases}$

Conclusion: Millie does not eat asparagus.

25. Premises: $\begin{cases} \text{Some politicians are liars.} \\ \text{All liars should be ignored.} \end{cases}$

Conclusion: Some politicians should be ignored.

26. Premises: $\begin{cases} \text{Some politicians are liars.} \\ \text{All liars should be ignored.} \end{cases}$

Conclusion: Some politicians should not be ignored.

CHAPTER 5

Computer-Related Linear Mathematics

Many problem-solving situations involve data to be analyzed and processed that naturally fall into linear or rectangular patterns. For instance, if we wish to develop an algorithm to alphabetize a collection of 100 names, it is helpful to use the designation

NAME(1), NAME(2), . . . , NAME(100)

for the given list. This is an example of ordering data in a **_linear array_**, or **_one-dimensional array_**; in the language of mathematics, this is called a **_vector_**.

In other applications, like the one cited in Figure 1 taken from another textbook,[†] the data can be organized naturally into a **_rectangular array_**, or **_two-dimensional array_**; this is called a **_matrix_** in mathematical terminology.

There is a vast mathematical literature on the study of vectors and matrices, each being a part of the broad field called **_linear mathematics_**. In this chapter we present the rudiments of these two topics and emphasize those aspects that are of immediate application within the area of computer science. In particular, Section 5.2 presents techniques for sorting and searching one-dimensional arrays. If you wish to postpone, omit, or skim the material of this section, you may do so without impairing your ability to follow the remainder of the chapter.

5.1

One-Dimensional Arrays: Vectors

In some applications, a large set of data is processed one piece at a time with no need to recall any particular input datum already processed. When this is the case, the same computer memory locations can be utilized for the data;

[†] J.-P. Tremblay and R. B. Bunt, *An Introduction to Computer Science: An Algorithmic Approach* [New York: McGraw-Hill, 1979], p. 197. Reprinted with permission.

Table 4-15 Family Allowance Payments

Yearly Income	Number of children						
	0	1	2	3	4	5	6 and up
Less than $3,000	0	17	19	20	22	24	25
$3,000–$3,999	0	16	18	19	21	23	24
$4,000–$4,999	0	15	17	18	20	22	23
$5,000–$5,999	0	14	16	17	19	21	22
$6,000–$6,999	0	13	15	16	18	20	21
$7,000–$7,999	0	12	14	15	17	19	20
$8,000–$8,999	0	11	13	14	16	18	19
$9,000–$9,999	0	10	12	13	15	17	18
$10,000 and over	0	9	11	12	14	16	17

Algorithm BENEFITS. This algorithm computes the monthly family allowance payments according to the schedule shown in Table 4-15. Assume that this information is stored in the integer array SCHED. The integer variables R and C are used to determine the appropriate row and column subscripts, respectively, from the values read into the integer variables INCOME and CHILDREN.

1. [Repeat for all families]
   Repeat thru step 5 while there still remains a family
2. [Input data for next family]
   Read (INCOME, CHILDREN)
   If there is no more input then Exit
3. [Determine appropriate row subscript]
   If INCOME < 3000
   then R ← 1
   else If INCOME ≥ 10000
       then R ← 9
       else R ← TRUNC((INCOME − 1000)/1000)
4. [Determine appropriate column subscript]
   If CHILDREN ≥ 6
   then C ← 6
   else C ← CHILDREN
5. [Select correct payment]
   Write ('PAYMENT IS', SCHED[R, C]).

Figure 1

the incoming data to be processed merely replaces the outgoing data already processed. The algorithm in the first example illustrates this situation.

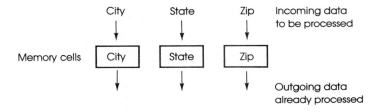

Example **5.1.1**

A company wishes to determine which of its 675 employees are over age 50. Using the variables NAME and AGE for the name and age of a particular employee, the flowchart in Figure 2 will output a list of all such employees, along with their respective ages. Here the variable k is used as a counter. Note that only two internal memory cells are required to store the changing values of NAME and AGE. ∎

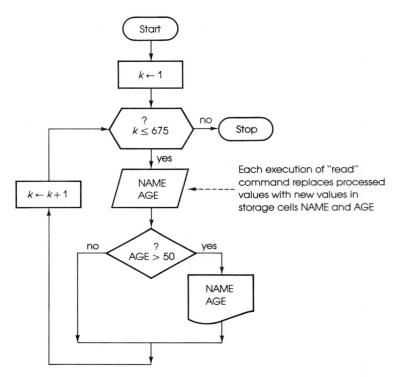

Figure 2

The next example shows that in some applications, each piece of input datum must be stored in a separate, identifiable memory cell for later use in the algorithm.

Example **5.1.2**

The same company mentioned above also wants to determine the average age of its 675 employees and produce a list of all those over that average age. The algorithm in Figure 3 accomplishes this objective, where

$$k = \text{counter of employees}$$
$$\text{NAME}(k) = \text{name of } k\text{th employee}$$
$$\text{AGE}(k) = \text{age of } k\text{th employee}$$
$$\text{SUM} = \text{accumulating sum of ages}$$
$$\text{AVG} = \text{average age of all employees}$$
∎

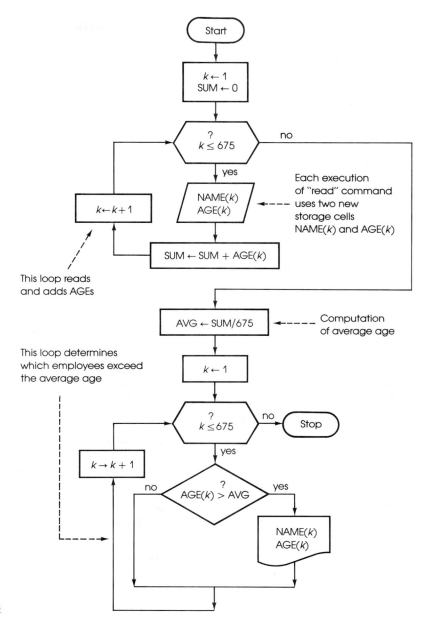

Figure 3

Observe the difference between this example and the preceding one. Here the name and age of each employee are given separate variable names and hence separate computer storage locations. Therefore,

$$2 * 675 = 1350$$

separate memory cells are required to store this set of input data.

In the preceding illustration, the lists NAME(1), NAME(2), ..., NAME(675) and AGE(1), AGE(2), ..., AGE(675) are examples of *one-dimensional arrays* or *vectors*.

Definition

Let n be a positive integer. An ordered collection of data

$$a(1), a(2), \ldots, a(n)$$

is called a ***vector***, or ***one-dimensional array, of length n***. Each individual datum $a(k)$ is called a ***component*** of the vector.

Notation: **a** or $(a(k))$ for the vector

We must make two comments on notation here. First, in mathematics you would ordinarily use the subscript notation a_k instead of $a(k)$. However, in computer programming one cannot use subscript or superscript notation, so we shall not use it here. Second, in instances like Example 5.1.2, it is desirable to use full variable names like NAME and AGE instead of single letters when expressing the arrays (NAME(k)) and (AGE(k)).

Example 5.1.3 **Vectors**

$$\mathbf{a} = (\text{S. Carcieri, S. Greene, M. Hauser, K. Pesta})$$

is a vector of length 4 with components

$$a(1) = \text{S. Carcieri} \qquad a(2) = \text{S. Greene}$$
$$a(3) = \text{M. Hauser} \qquad a(4) = \text{K. Pesta}$$

And

$$\mathbf{b} = (14, 25.2, 64.01, -19, 8.5, -1.5, 15.9)$$

is a vector of length 7 with components

$$b(1) = 14 \qquad b(2) = 25.2 \qquad b(3) = 64.01 \qquad b(4) = -19$$
$$b(5) = 8.5 \qquad b(6) = -1.5 \qquad b(7) = 15.9$$

∎

Definition

Let $\mathbf{a} = (a(k))$ and $\mathbf{b} = (b(k))$ be vectors each of length n. Then we say that **a** *equals* **b** if

$$a(k) = b(k)$$

for each $k = 1, 2, \ldots, n$.

Notation: $\mathbf{a} = \mathbf{b}$

Example 5.1.4   **Vector Equality**

(Erving, \$100.52) = (Erving, \$100.52)

(Erving, \$100.52) = (Erving, \$51.52 + \$49)

(Erving, \$100.52) $\neq$ (Bird, \$100.52) because the first components are not equal;

(Erving, \$100.52) $\neq$ (Erving, \$100) because the second components are not equal;

(1, 2, 3, 4, 5) = (1, 2, 3, 10 − 6, 5)

(6, 0, 1) $\neq$ (6, 1, 0) because the second and third components are not equal.   ▰

Vectors like the last two presented, all of whose components are numbers, are called **numeric vectors**.

Notation

For a numeric vector

$$\mathbf{a} = (a(k))$$

of length n, the sum of the components

$$a(1) + a(2) + \cdots + a(n)$$

can be expressed as

$$\sum_{k=1}^{n} a(k)$$

Here the Greek letter $\sum$ (read sigma) is called a **summation symbol** and variable k is called a **summation index**.

Observe that other letters can be used for the summation index when the summation symbol is used:

$$\sum_{k=1}^{n} a(k) = \sum_{i=1}^{n} a(i) = \sum_{j=1}^{n} a(j)$$

In any of these expressions, we consider the given index to have initial value 1, and it increases successively by 1 until the upper limit of n is reached. If you wish to begin the sum at a different component, for instance $a(3)$, then alter the initial value of the index accordingly, as

$$\sum_{k=3}^{n} a(k) = a(3) + a(4) + \cdots + a(n)$$

Example 5.1.5   **Summation Symbol**

a. For vector

$$\mathbf{a} = (0, 2, 4, 6, 8, 10)$$

we have

$$\sum_{k=1}^{6} a(k) = a(1) + a(2) + a(3) + a(4) + a(5) + a(6)$$

$$= \ 0 \ + \ 2 \ + \ 4 \ + \ 6 \ + \ 8 \ + \ 10$$

$$= 30$$

and

$$\sum_{k=4}^{6} a(k) = a(4) + a(5) + a(6)$$

$$= \ 6 \ + \ 8 \ + \ 10$$

$$= 24$$

b. For

$$\mathbf{b} = (14, 25.2, 64.01, -19, 8.5, -1.5, 15.9)$$

we have

$$\sum_{k=1}^{7} b(k) = b(1) + b(2) + \ b(3) \ + b(4) + b(5) + b(6) + b(7)$$

$$= \ 14 \ + 25.2 + 64.01 - \ 19 \ + 8.5 - \ 1.5 + 15.9$$

$$= 107.11$$

c. If the vector

$$\mathbf{SALES} = (1562.16, 982.05, 220.00, 1235.10, 1852.90)$$

represents the weekly sales of five marketing agents at a particular office, then

$$\sum_{k=1}^{5} \text{SALES}(k) = 1562.16 + 982.05 + 220.00 + 1235.10 + 1852.90$$

$$= 5852.21$$

equals the total sales of those five agents for that week.

d. In the algorithm of Figure 3 in this chapter, the final value of the variable **SUM** is

$$\sum_{k=1}^{675} \text{AGE}(k) = \text{AGE}(1) + \text{AGE}(2) + \cdots + \text{AGE}(675)$$

Of course, the numeric value of the answer here depends on the ages $\text{AGE}(k)$ that are input.

e. $\displaystyle\sum_{k=1}^{4} (-1)^k(k+2) = (-1)^1(1+2) + (-1)^2(2+2) + (-1)^3(3+2) + (-1)^4(4+2)$

$$= \quad -3 \quad + \quad 4 \quad - \quad 5 \quad + \quad 6$$

$$= \quad 2 \qquad\qquad\qquad\qquad\qquad\qquad \blacksquare$$

Part (e) illustrates the use of summation notation outside the context of vectors.

Now that we have the concepts of vector equality and numeric vectors in hand, we can define in natural ways certain fundamental operations with vectors. In this context the word *scalar* is often used to distinguish an ordinary number, such as the numbers we've dealt with so far in this text, from a vector or one of its components.

Definitions

Let $\mathbf{a} = (a(k))$ and $\mathbf{b} = (b(k))$ be numeric vectors of the same length, and let t be a scalar. Then the vector

1. $(a(k) + b(k))$ is called the *sum of* $\mathbf{a}$ *and* $\mathbf{b}$ and is denoted $\mathbf{a} + \mathbf{b}$;
2. $(a(k) - b(k))$ is called the *difference of* $\mathbf{a}$ *and* $\mathbf{b}$ and is denoted $\mathbf{a} - \mathbf{b}$; and
3. $(t * a(k))$ is called the *scalar multiple of* $\mathbf{a}$ *by* t and is denoted $t\mathbf{a}$.

These operations are called vector *addition*, *subtraction*, and *scalar multiplication*, respectively.

Observe that you can only add or subtract vectors of the same length, and the result is a vector of that length. Likewise, a scalar multiple of a given vector has the same length as the given vector.

Example 5.1.6   **Addition, Subtraction, and Scalar Multiplication**

a. If $\mathbf{a} = (2, 0, -3)$, $\mathbf{b} = (.5, -1.2, 7)$, $\mathbf{c} = (-9, 4)$, $t = 4$, and $s = -6$, then

$$\mathbf{a} + \mathbf{b} = (2, 0, -3) + (.5, -1.2, 7)$$
$$= (2 + .5, 0 - 1.2, -3 + 7) = \boxed{(2.5, -1.2, 4)}$$

$$\mathbf{a} - \mathbf{b} = (2, 0, -3) - (.5, -1.2, 7)$$
$$= (2 - .5, 0 + 1.2, -3 - 7) = \boxed{(1.5, 1.2, -10)}$$

$$\mathbf{b} - \mathbf{a} = (.5, -1.2, 7) - (2, 0, -3)$$
$$= (.5 - 2, -1.2 - 0, 7 + 3) = \boxed{(-1.5, -1.2, 10)}$$

$$t\mathbf{a} = 4(2, 0, -3)$$
$$= (4 * 2, 4 * 0, 4 * (-3)) = \boxed{(8, 0, -12)}$$

$$s\mathbf{b} = -6(.5, -1.2, 7)$$
$$= (-6 * .5, (-6) * (-1.2), -6 * 7) = \boxed{(-3, 7.2, -42)}$$

$$t\mathbf{c} = 4(-9, 4)$$
$$= (4 * (-9), 4 * 4) = \boxed{(-36, 16)}$$

$$s\mathbf{c} = -6(-9, 4)$$
$$= ((-6) * (-9), (-6) * 4) = \boxed{(54, -24)}$$

$$2\mathbf{b} - 3\mathbf{a} = 2(.5, -1.2, 7) - 3(2, 0, -3)$$
$$= (2 * .5, 2 * (-1.2), 2 * 7) - (3 * 2, 3 * 0, 3 * (-3))$$
$$= (1, -2.4, 14) - (6, 0, -9)$$
$$= (1 - 6, -2.4 - 0, 14 + 9) = \boxed{(-5, -2.4, 23)}$$

$$.3(\mathbf{b} - \mathbf{a}) = .3[(.5, -1.2, 7) - (2, 0, -3)]$$
$$= .3(.5 - 2, -1.2 - 0, 7 + 3)$$
$$= .3(-1.5, -1.2, 10)$$
$$= (.3 * (-1.5), .3 * (-1.2), .3 * 10) = \boxed{(-.45, -.36, 3)}$$

And finally, *none* of the expressions

$$\mathbf{a} + \mathbf{c} \qquad \mathbf{a} - \mathbf{c} \qquad \mathbf{b} + \mathbf{c} \qquad \mathbf{b} - \mathbf{c}$$

are defined because in each case the two vectors have different lengths.

b. Suppose the vector

$$\mathbf{FIR} = (75, 62, 49, 17, 81)$$

represents the number of units sold by each of five salespersons during the first week of a month. Similarly,

$$\mathbf{SEC} = (82, 79, 80, 63, 68)$$
$$\mathbf{THR} = (64, 81, 76, 79, 72)$$
$$\mathbf{FOR} = (77, 75, 63, 58, 74)$$

represent the sales during the second, third, and fourth weeks, respectively, where the salespersons are always listed in the same order. Then the monthly sales for each of the five salespersons are represented by the vector

$$\mathbf{FIR} + \mathbf{SEC} + \mathbf{THR} + \mathbf{FOR} = (75 + 82 + 64 + 77, 62 + 79 + 81 + 75,$$
$$49 + 80 + 76 + 63, 17 + 63 + 79 + 58,$$
$$81 + 68 + 72 + 74) \cdot$$
$$= (298, 297, 268, 217, 295)$$

Similarly, the change in sales during the second week as compared with the first week is

$$\mathbf{SEC} - \mathbf{FIR} = (82, 79, 80, 63, 68) - (75, 62, 49, 17, 81)$$
$$= (7, 17, 31, 46, -13)$$

Also, the vector

$$4\mathbf{THR} = 4(64, 81, 76, 79, 72)$$
$$= (256, 324, 304, 316, 288)$$

represents what the monthly sales *would have been* had the sales reps worked all month at the level of the third week. ∎

If you are wondering why we did not include in the preceding definition the "obvious" way of multiplying vectors

$$(a(k) * b(k))$$

it's because that type of multiplication is not very useful. Rather, there are other, more important kinds of vector multiplication, one of which we present now and shall use later in this chapter.

Definition

Let $\mathbf{a} = (a(k))$ and $\mathbf{b} = (b(k))$ be numeric vectors of length n. Then the number

$$a(1) * b(1) + \cdots + a(n) * b(n) = \sum_{k=1}^{n} a(k) * b(k)$$

is called the ***inner product of a and b***.

Notation: a · b

Note that, unlike vector addition, subtraction, and scalar multiplication, this type of operation with vectors produces a scalar and *not* a vector.

Example 5.1.7

Inner Product

a. If $\mathbf{a} = (2, 0, -3)$ and $\mathbf{b} = (.5, -1.2, 7)$, then

$$\mathbf{a} \cdot \mathbf{b} = 2 * .5 + 0 * (-1.2) + (-3) * 7$$
$$= 1 + 0 - 21 = \boxed{-20}$$

and

$$\mathbf{b} \cdot \mathbf{a} = .5 * 2 + (-1.2) * 0 + 7 * (-3)$$
$$= 1 + 0 - 21 = \boxed{-20}$$

b. $(7.9, 8) \cdot (100, -14.5) = (7.9) * 100 + 8 * (-14.5)$
$$= 790 - 116 = \boxed{674}$$

c. $(1, 2, 3, 4) \cdot (-2, 4, 0, 3) = 1 * (-2) + 2 * 4 + 3 * 0 + 4 * 3$
$$= -2 + 8 + 0 + 12 = \boxed{18} \qquad \blacksquare$$

Sometimes it is advantageous to write the right-hand factor in an inner product as a ***column vector*** $\begin{pmatrix} b(1) \\ b(2) \\ \vdots \\ b(n) \end{pmatrix}$ rather than as a ***row vector***

$$(b(1), b(2), \ldots, b(n))$$

In this case, the inner product takes the form

$$(a(1), a(2), \ldots a(n)) \cdot \begin{pmatrix} b(1) \\ b(2) \\ \vdots \\ b(n) \end{pmatrix} = \sum_{k=1}^{n} a(k) * b(k)$$

Example 5.1.8   **Inner Product**

a. $(7.9, 8) \cdot \begin{pmatrix} 100 \\ -14.5 \end{pmatrix} = 7.9 * 100 + 8 * (-14.5)$

$$= 790 - 116 = \boxed{674}$$

b. $(.3, 1, -10) \cdot \begin{pmatrix} 4 \\ -11 \\ 2 \end{pmatrix} = .3 * 4 + 1 * (-11) + (-10) * 2$

$$= 1.2 - 11 - 20 = \boxed{-29.8}$$

c. $(4, -7, -15, 6, 0) \cdot \begin{pmatrix} 6 \\ -2 \\ 0 \\ -5 \\ 4 \end{pmatrix}$

$$= 4 * 6 + (-7) * (-2) + (-15) * 0 + 6 * (-5) + 0 * 4$$

$$= 24 + 14 + 0 - 30 + 0 = \boxed{8} \qquad ■$$

EXERCISES 5.1

1. In each of the following, specify the length of the given vector and the third component, if any:

  a. $n = (2, 4, 6, 8, 10)$
  b. $AGE = (50, 62)$
  c. $NAME = $ (W. Miley, B. McCann, R. Hokenson)
  d. $WAGE = $ ($100, $110.50, $150, $108.25)

2. In each of the following, specify the length of the given vector and the fourth component, if any:

  a. $m = (0, 1, 2, 3, 4, 5)$
  b. $WGT = (123, 146, 210)$
  c. $TEAM = $ (F. Sneesby, R. Hayman)
  d. $L = (6.7, -4, 5.01, -17)$

3. In each case, determine if the given vectors are equal.

  a. $(1, 2, 3)$ $(1, 2, 3)$

  b. $(0, 0, 0)$ $(0, 0, 0, 0)$
  c. (McCloskey, $100)   (McCloskey, $99 + $1)
    (McCloskey, $99)

4. In each case, determine if the given vectors are equal.

  a. $(2, 4, 8)$ $(2, 4, 4 + 4)$
  b. (A, B, C, D) (B, C, D, A)
  c. $(2, .1)$ $(2, \frac{1}{10})$ $(2, \frac{2}{5})$

5. For each of the following, compute $\sum_{k=1}^{7} a(k)$ where

  a. $a = (-3, 6, 0, 1, .4, -5, 2)$
  b. $a(k) = 3k$
  c. $a(k) = 2$
  d. $a(k) = (-1)^k(k - 3)$

6. For each of the following, compute $\sum_{k=1}^{5} a(k)$ where

  a. $a = (\frac{7}{2}, -3, 1, -1, 4)$ **b.** $a(k) = k + 1$
  c. $a(k) = k^2$ **d.** $a(k) = (-k)^2$

7. For the vector $\mathbf{AMT} = (17, 22, 14, 51, 37)$, compute the following:

a. $\displaystyle\sum_{k=1}^{5} \mathrm{AMT}(k)$ **b.** $\displaystyle\sum_{k=2}^{4} \mathrm{AMT}(k)$ **c.** $\displaystyle\sum_{k=3}^{5} \mathrm{AMT}(k)$

8. For the vector $\mathbf{NUM} = (-3, 0, 1, 8, -11, 22, 5)$, compute the following:

a. $\displaystyle\sum_{k=1}^{7} \mathrm{NUM}(k)$ **b.** $\displaystyle\sum_{k=3}^{6} \mathrm{NUM}(k)$ **c.** $\displaystyle\sum_{k=4}^{7} \mathrm{NUM}(k)$

9. Given the vectors

$\mathbf{a} = (1, 2, 3)$ $\mathbf{b} = (-6, 0, 5)$ $\mathbf{c} = (4, -7, 0, -3)$

compute, if possible, each of the following:

$\mathbf{a} + \mathbf{b}$	$\mathbf{a} - \mathbf{b}$	$\mathbf{a} + \mathbf{c}$	$\mathbf{b} - \mathbf{c}$
$3\mathbf{a}$	$.25\mathbf{c}$	$-3(\mathbf{a} + \mathbf{b})$	$4\mathbf{b} - 3\mathbf{a}$
$\mathbf{a} \cdot \mathbf{b}$	$\mathbf{a} \cdot \mathbf{c}$	$(5\mathbf{a}) \cdot \mathbf{b}$	$5(\mathbf{a} \cdot \mathbf{b})$

10. Given the vectors

$\mathbf{e} = (4, 8)$ $\mathbf{f} = (0, 0, -1, 1)$ $\mathbf{g} = (-4, -3, -2, -1)$

compute, if possible, each of the following:

$\mathbf{e} + \mathbf{f}$	$\mathbf{f} + \mathbf{g}$	$\mathbf{f} - \mathbf{g}$	$\mathbf{g} - \mathbf{f}$
$3\mathbf{e}$	$.4\mathbf{g}$	$6(\mathbf{f} - \mathbf{g})$	$\mathbf{g} - 4\mathbf{f}$
$\mathbf{e} \cdot \mathbf{f}$	$\mathbf{g} \cdot \mathbf{f}$	$(-2\mathbf{f}) \cdot \mathbf{g}$	$(3\mathbf{f}) \cdot (4\mathbf{g})$

11. Compute each of the following inner products:

a. $(1, 0, -1, 0) \cdot \begin{pmatrix} 10 \\ -2 \\ -1 \\ 0 \end{pmatrix}$ **b.** $(.6, -1) \cdot \begin{pmatrix} .5 \\ .02 \end{pmatrix}$

12. Compute each of the following inner products:

a. $(6, -2) \cdot \begin{pmatrix} 4 \\ 1 \end{pmatrix}$ **b.** $(2, -1, 3) \cdot \begin{pmatrix} 0 \\ 5 \\ .5 \end{pmatrix}$

13. For the vectors

$\mathbf{a} = (3, 1, -2)$ $\mathbf{b} = (-1, 0, 4)$ $\mathbf{c} = (1, -2, 5)$

and scalar $t = 7$, verify each of the following algebraic properties:

a. $\mathbf{a} + \mathbf{b} = \mathbf{b} + \mathbf{a}$
b. $\mathbf{a} + (\mathbf{b} + \mathbf{c}) = (\mathbf{a} + \mathbf{b}) + \mathbf{c}$
c. $t(\mathbf{a} + \mathbf{b}) = t\mathbf{a} + t\mathbf{b}$

14. For the vectors

$\mathbf{a} = (3, 1, -2)$ $\mathbf{b} = (-1, 0, 4)$ $\mathbf{c} = (1, -2, 5)$

and scalar $t = 7$, verify each of the following algebraic properties:

a. $\mathbf{a} \cdot \mathbf{b} = \mathbf{b} \cdot \mathbf{a}$
b. $t(\mathbf{a} \cdot \mathbf{b}) = (t\mathbf{a}) \cdot \mathbf{b} = \mathbf{a} \cdot (t\mathbf{b})$
c. $\mathbf{a} \cdot (\mathbf{b} + \mathbf{c}) = (\mathbf{a} \cdot \mathbf{b}) + (\mathbf{a} \cdot \mathbf{c})$

The next three exercises relate to an exam for 357 students that has already been graded. In each case, construct a flowchart that will take in the name NAME and raw exam score RAW for each student and perform the specified tasks.

15. Print out the names and grades of all those students who passed the exam (that is, achieved raw score $RAW \geq 60$).

16. Compute the class average score and print out the names and grades of all those who achieved at least this average score.

17. Compute the class average and scaling factor so that the adjusted class average is 75 and then print out the name and scaled grade for each student.

Exercises 18 through 20 concern a particular company with 20 salespeople. A transaction is recorded by entering the name NAME of the salesperson, that person's sales number NUM (1 through 20), and the amount AMT of the sale. In each case, construct an algorithm that will take in 500 such transactions and perform the specified tasks.

18. Print out all transactions over \$100 in amount, along with the name of the appropriate salesperson.

19. Compute the average transaction and print out all transactions exceeding that average, along with the name of the appropriate salesperson.

20. Compute and print out the total sales amounts (in dollars) for each of the 20 salespeople (identified on the printout by sales number NUM).

21. A certain multiple-choice exam contains ten questions, each having five possible answers, 1, 2, 3, 4, or 5. Let KEY be the array of length 10 containing the correct answers, and let ANS be an array of length 10 containing the answers given by a particular student. Construct a flowchart that will take in NAME and ANS for each of 850 students, and print out the NAME and SCORE (out of a possible 10) for each student.

5.2

Sorting and Searching

Of great importance in data processing is the ability to order, or sort, large collections of numeric or alphabetic information quickly and then search this sorted information for a particular entry. For instance, in the first example of the preceding section, we sought a list of employees over age 50. Now if the list of employees had initially been *sorted by age* from youngest to oldest, we could have quickly skipped to those of age 51 and printed out each name thereafter. Such order allows a more efficient procedure, particularly if several searches of the same data are required.

In this section we first present a few methods for *sorting* (or ordering) data and then two methods of *searching* the sorted data to determine whether a particular entry appears. To keep notation simple, we shall deal only with numeric data, but the procedures presented are suitable for sorting and searching alphabetic data. As mentioned in the introduction to this chapter, the remainder of this chapter can be studied without covering the material on sorting and searching in this section.

To begin with, let's consider the mechanics of one version of a *selection sort*. Throughout this section, the word *pass* will mean going through a given list and making desired changes.

Example 5.2.1   Selection Sort

Here is one method for placing the following numbers in ascending order:[†]

$$4 \quad 9 \quad 3 \quad 8 \quad 6$$

Original List	Pass 1		Pass 2		Pass 3		Pass 4		Sorted List
4	4	**3**	3	3	3	3	3	3	3
9	9	9	9	**4**	4	4	4	4	4
3	3	4	4	9	9	**6**	6	6	6
8	8	8	8	8	8	8	8	**8**	8
6	6	6	6	6	6	9	9	9	9

In pass 1, we go through the original list to find the smallest number and then exchange it with the first array number. In pass 2, we take this partially ordered list and, *beginning with the second element*, search downward for the least element. When found, this element is exchanged with the second array element. In general, for pass k we go down the list starting at $a(k)$ to find the least element and then exchange it with $a(k)$. Note that after four such passes, the greatest element is necessarily at the bottom, so that a fifth pass is not needed.   ∎

With this example in hand, we can now take the first of several steps in developing a general algorithm for the selection sort.

[†] In this section, all the tables and traces should be read column by column from left to right.

Example 5.2.2   **Preliminary Algorithm for Selection Sort**

A flowchart (soon to be refined) that describes the selection sort is drawn in Figure 4. It takes in an array $(a(k))$ of length N and produces a sorted array. For instance, if the array

$$a(1) = 4 \qquad a(2) = 9 \qquad a(3) = 3 \qquad a(4) = 8 \qquad a(5) = 6$$

is input, then the output is the sorted array

$$a(1) = 3 \qquad a(2) = 4 \qquad a(3) = 6 \qquad a(4) = 8 \qquad a(5) = 9 \qquad ■$$

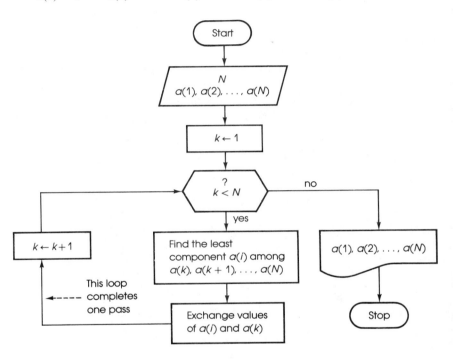

Figure 4

Two parts of the above flowchart need to be refined, so let's take the simpler one first. In many sort algorithms, it is necessary to exchange the values of two variables—that is, to exchange the contents of two memory cells. Conceptually, this is the same as switching the contents of two soda bottles, which requires a third bottle for temporary storage. Here (see Figure 5) we need a third memory cell TEMP to hold the contents of the smallest

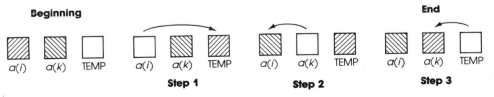

Figure 5

component $a(i)$ so that the contents of $a(k)$ can be emptied into $a(i)$. Thus, the switching of $a(i)$ and $a(k)$ can be accomplished by the instructions

$$\text{TEMP} \leftarrow a(i)$$
$$a(i) \quad \leftarrow a(k)$$
$$a(k) \quad \leftarrow \text{TEMP}$$

The difficult part to refine in the flowchart of Figure 4 is finding the least component $a(i)$ among the elements

$$a(k), a(k + 1), \ldots, a(N)$$

which we shall call a ***tail of the array*** $a(1), a(2), \ldots, a(N)$.

For a given value of k, the algorithm drawn in Figure 6 will examine the appropriate tail and determine the smallest value $a(i)$ among the tail elements. Note that no start or stop instructions are included in this figure. This is because we are working on a refinement of a portion of the flowchart given in Figure 4.

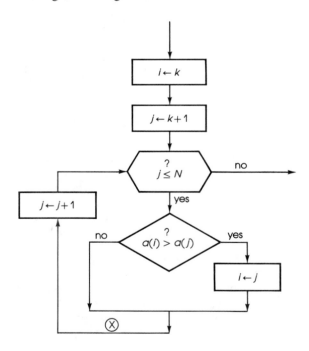

Figure 6

Example **5.2.3** Finding the Smallest Component: Selection Sort

Let's perform a trace at point $\circledX$ of the flowchart segment in Figure 6, beginning with the array

$$a(1) = 4 \qquad a(2) = 9 \qquad a(3) = 3 \qquad a(4) = 8 \qquad a(5) = 6$$

so here $N = 5$.

For $k = 1$ we have $i \leftarrow 1, j \leftarrow 2$, and (remember to read column by column from left to right!) at point (X)

j	2	3	4	5
$a(j)$	9	3	8	6
i	1	3	3	3
$a(i)$	4	3	3	③

← Least component

Thus, when we exit the loop in Figure 6 (that is, receive a no answer to the question "$6 \leq 5$?"), the value of $a(i)$ is 3, the smallest value among the given array.

One word of caution here! Before we trace this same algorithm for $k = 2$, the values of $a(1)$ and $a(3)$ must be exchanged, since this is the next step in the larger algorithm of Figure 4. After this is done, we have

$$a(1) = 3 \qquad a(2) = 9 \qquad a(3) = 4 \qquad a(4) = 8 \qquad a(5) = 6$$

and we can apply Figure 6 when $k = 2$. In this case, $i \leftarrow 2, j \leftarrow 3$, and at point (X)

j	3	4	5
$a(j)$	4	8	6
i	3	3	3
$a(i)$	4	4	④

← Least component

So 4 is the smallest among the tail values

$$a(2) = 9 \qquad a(3) = 4 \qquad a(4) = 8 \qquad a(5) = 6$$

Once again a switch must be made, this time between $a(2)$ and $a(3)$, yielding

$$a(1) = 3 \qquad a(2) = 4 \qquad a(3) = 9 \qquad a(4) = 8 \qquad a(5) = 6$$

Now for $k = 3$ we have $i \leftarrow 3, j \leftarrow 4$, and at point (X)

j	4	5
$a(j)$	8	6
i	4	5
$a(i)$	8	⑥

← Least component

So 6 is the smallest among the tail values

$$a(3) = 9 \qquad a(4) = 8 \qquad a(5) = 6$$

After exchanging the values of $a(3)$ and $a(5)$, the array becomes

$$a(1) = 3 \qquad a(2) = 4 \qquad a(3) = 6 \qquad a(4) = 8 \qquad a(5) = 9$$

Finally, when $k = 4$ we get $i \leftarrow 4, j \leftarrow 5$, and at point (X)

j	5
$a(j)$	9
i	4
$a(i)$	(8)	$\leftarrow$ Lesser component

That is, 8 is the smaller of the tail values

$$a(4) = 8 \qquad a(5) = 9$$

An exchange in this case means that the values of $a(4)$ and $a(4)$ are switched, which of course does not alter the array values. The net result, then, is the sorted array

$$a(1) = 3 \qquad a(2) = 4 \qquad a(3) = 6 \qquad a(4) = 8 \qquad a(5) = 9 \qquad ∎$$

Now we are at the stage where we can refine the flowchart in Figure 4.

Example 5.2.4   **Selection Sort Algorithm**

The flowchart in Figure 7 will take in an array $a(1), a(2), \ldots, a(N)$ and produce a sorted array. Observe that this flowchart has the same general shape as that of Figure 4, but this one includes more detail about the important processes of finding the smallest component and exchanging, which we worked on separately. This method is an example of what is called *modular algorithm development*, meaning that a rough algorithm was initially presented and then gradually refined to include all important details.

Figure 7 also includes a trace of the flowchart at points (X) and (Y) for the given array

$$a(1) = 4 \qquad a(2) = 9 \qquad a(3) = 3 \qquad a(4) = 8 \qquad a(5) = 6$$

The reader might find it interesting to compare these trace columns with the work done in Example 5.2.1 on the same array.   ∎

There is a disadvantage to the selection sort just described. Namely, for an array of length N, there are *always* $(N - 1)$ passes through the array, even if the array is completely sorted after the first few passes or was completely sorted from the start. Fortunately, there are other types of sorts, like the one we are about to present, which halt once the array is ordered.

Example 5.2.5   **Bubble Sort**

Starting with the (now very familiar) array

$$\mathbf{a} = (4, 9, 3, 8, 6)$$

of length 5, each pass in the bubble sort consists of the four comparisons

$$a(k) > a(k + 1)$$

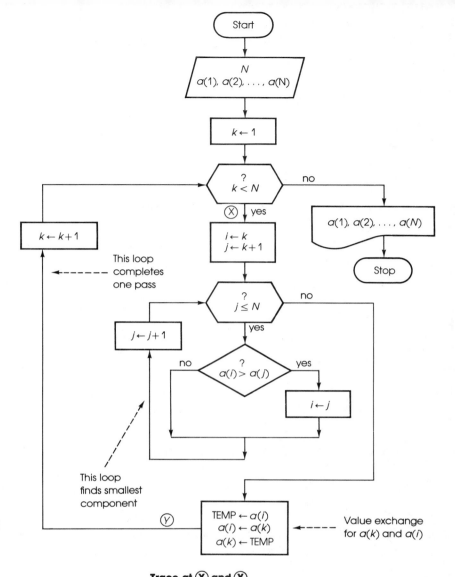

Figure 7 Selection sort

Trace at ⊗ and ⓨ

	Pass 1		Pass 2		Pass 3		Pass 4	
	⊗	ⓨ	⊗	ⓨ	⊗	ⓨ	⊗	ⓨ
k	1	1	2	2	3	3	4	4
a(1)	4	3	3	3	3	3	3	3
a(2)	9	9	9	4	4	4	4	4
a(3)	3	4	4	9	9	6	6	6
a(4)	8	8	8	8	8	8	8	8
a(5)	6	6	6	6	6	9	9	9

Initial array ⌐ ⌐ ⌐ ↑

↑ ⌐ ⌐ ⌐ Sorted array

where $k = 1, 2, 3, 4$. When the answer to this comparison is no, we leave the values of $a(k)$ and $a(k + 1)$ alone; when the answer is yes, their values are exchanged.

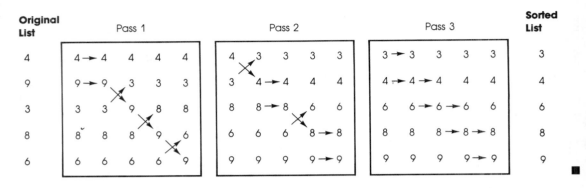

The name of this particular sort is derived from the observation that at each pass the smaller numbers "bubble up" toward the top of the array and the larger ones fall down. The big advantage of this type of sort is that when a pass is made in which *no* array values are exchanged, the sorting process terminates. In this example, the ordering was accomplished in three passes, one fewer than the selection sort required. If the given array had already been sorted, the bubble sort would have terminated after just one pass.

Example 5.2.6 **Bubble Sort Algorithm**

Figure 8 contains a flowchart for the bubble sort procedure. The inside loop is somewhat similar to the inside loop of the selection sort in that array values are compared and exchanged if necessary. The big difference, however, lies in the use of the variable FLAG to signal whether an exchange *was* made (FLAG = 0) within the inner loop or not (FLAG remains equal to 1). In the former case (FLAG = 0), we want to enter the outside loop again to perform another pass. In the latter case (FLAG = 1), no exchange of array values has been made on the last pass, and the sorting process can be ended.

At the bottom of Figure 8 is a trace of the flowchart at points $\textcircled{X}$ and $\textcircled{Y}$ for the array

$$a(1) = 4 \qquad a(2) = 9 \qquad a(3) = 3 \qquad a(4) = 8 \qquad a(5) = 6$$

The reader should note that the columns in this trace coincide with the first and last columns of each pass shown in Example 5.2.5.

Also, this algorithm can be modified to reduce the number of comparisons and thereby improve its efficiency (see exercise 25). ∎

Now that we have two methods of sorting data in arrays (and there are others!), let's turn to the problem of searching to find a particular entry. *Here we shall assume that the given arrays are already sorted.*

The most obvious but least efficient way to search ordered data is to start at the beginning and examine each item.

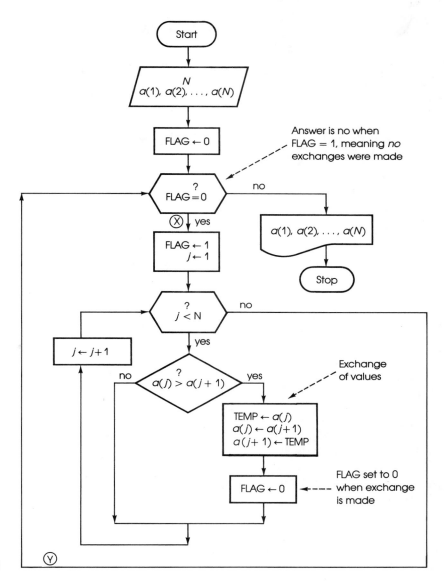

Figure 8 Bubble sort

Trace at Ⓧ and Ⓨ

	Pass 1		Pass 2		Pass 3	
	Ⓧ	Ⓨ	Ⓧ	Ⓨ	Ⓧ	Ⓨ
FLAG	0	0	0	0	0	①	← ───── Terminates sort
$a(1)$	4	4	4	3	3	3
$a(2)$	9	3	3	4	4	4
$a(3)$	3	8	8	6	6	6
$a(4)$	8	6	6	8	8	8
$a(5)$	6	9	9	9	9	9

Initial array ── ↑   Sorted array ── ↑

Example 5.2.7 **Linear Search**

The flowchart in Figure 9a will receive as input an ordered array

$$a(1), a(2), \ldots, a(N)$$

and VALUE, which represents the value for which the array is to be

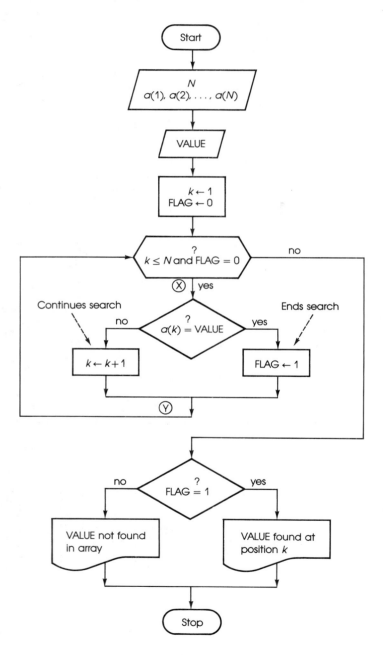

Figure 9a Linear search

Traces at (X) and (Y)

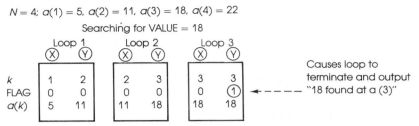

$N = 4$; $a(1) = 5$, $a(2) = 11$, $a(3) = 18$, $a(4) = 22$

Searching for VALUE = 18

Causes loop to terminate and output "18 found at a (3)"

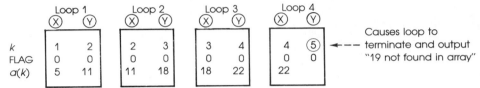

Searching for VALUE = 19

Causes loop to terminate and output "19 not found in array"

Figure 9b

searched. It will output one of two results:

"VALUE not found in array."

or

"VALUE found at position k."

In Figure 9b are traces for two different searches of the array 5, 11, 18, 22.

When you want to perform a linear search of only a portion of an array, the flowchart of Figure 9 can easily be modified. Suppose we wish to search the array $(a(k))$ from starting position BEGIN to final position END where

$$1 \leq \text{BEGIN} < \text{END} \leq N$$

so

$$a(1), a(2), \ldots, \underbrace{a(\text{BEGIN}), \ldots, a(\text{END})}, \ldots, a(N)$$

Portion to be searched

In this situation, the algorithm in Figure 9 needs only two modifications, changing the initialization $k \leftarrow 1$ to $k \leftarrow \text{BEGIN}$ and changing the loop-controlling question from $k \leq N$ to $k \leq \text{END}$. We shall use such a partial search of an array in the next example.

It is evident that a linear search is an extremely inefficient way to search *large* collections of data. Imagine using a telephone book by starting at page one and proceeding name by name to the name you seek! Of course, the usual way to use a telephone book is to skip through pages, noting the index

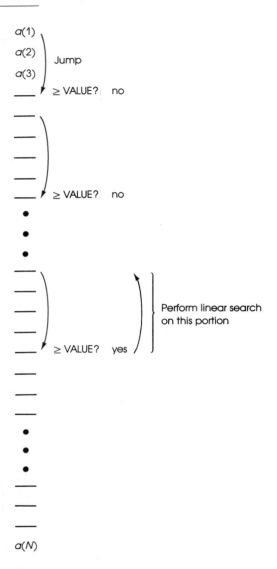

Figure 10 Jump linear search

at the tops of the pages until you arrive in the vicinity of the desired name; then you perform a linear search. This process is an illustration of what we'll call a *jump linear search* (see Figure 10).

Example 5.2.8

Jump 50 Linear Search

Figure 11 contains a flowchart that will read an ordered array of length 2000

$$a(1), a(2), \ldots, a(2000)$$

and search for VALUE in jumps of 50. If at any jump the array value equals or exceeds VALUE, the linear search algorithm is executed, or "called," for that portion of the array just jumped over. Once that search is completed,

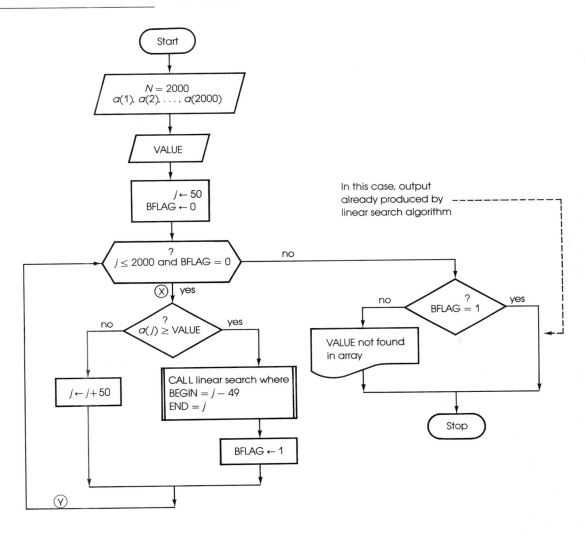

In this case, output
already produced by
linear search algorithm

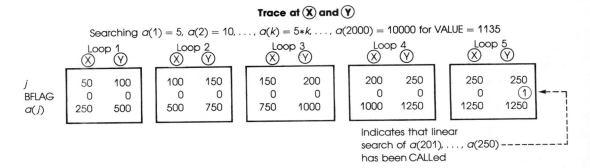

Trace at Ⓧ and Ⓨ

Searching $a(1) = 5$, $a(2) = 10, \ldots, a(k) = 5*k, \ldots, a(2000) = 10000$ for VALUE = 1135

Indicates that linear
search of $a(201), \ldots, a(250)$
has been CALLed

Figure 11 Jump 50 linear search

the algorithm of Figure 11 continues at the position immediately following where the call was made, namely at BFLAG $\leftarrow$ 1.

Also, at the bottom of Figure 11 is a trace of the search of the array

$$a(1) = 5, \ a(2) = 10, \ldots, \ a(k) = 5 * k, \ldots, \ a(2000) = 10000$$

for VALUE = 1135. During the fifth time through the loop, the question

$$a(250) = 1250 \geq 1135?$$

is answered yes. So the linear search algorithm is called for the array portion

$$a(201), \ a(202), \ldots, \ a(250)$$

the result of which is the output "1135 found at position $a(227)$." ■

Note that the algorithm of the preceding example should be used only on arrays of length 2000. If we were to execute it on an array of length 2039, then the array tail

$$a(2001), \ a(2002), \ldots, \ a(2039)$$

would not be searched. However, an addition to the flowchart to include a search of the tail would handle such an array (see exercises 21–24).

The two searches presented in this section are elementary ones. More sophisticated and efficient procedures, encountered in advanced studies, include the **quadratic search** and the **binary search**.

EXERCISES **5.2**

The first ten exercises below refer to the arrays

$$\mathbf{a} = (6, \ 1, \ 9, \ 2, \ 7, \ 4) \quad and \quad \mathbf{b} = (7, \ 9, \ 6, \ 4, \ 1, \ 2)$$

1. Apply the selection sort to array **a**.

2. Apply the selection sort to array **b**.

In exercises 3 and 4, the work of Example 5.2.3 can be used as a model.

3. For array **a** perform a trace at point $\widehat{X}$ of the flowchart in Figure 6, identifying the element $a(i)$ after each pass.

4. For array **b** perform a trace at point $\widehat{X}$ of the flowchart in Figure 6, identifying the element $a(i)$ after each pass.

5. Beginning with array **a**, do a trace at points $\widehat{X}$ and $\widehat{Y}$ of the selection sort algorithm given in Figure 7.

6. Beginning with array **b**, do a trace at points $\widehat{X}$ and $\widehat{Y}$ of the selection sort algorithm given in Figure 7.

7. Apply the bubble sort to array **a**.

8. Apply the bubble sort to array **b**.

9. Using array **a**, perform a trace at points $\widehat{X}$ and $\widehat{Y}$ of the bubble sort algorithm given in Figure 8.

10. Using array **b**, perform a trace at points $\widehat{X}$ and $\widehat{Y}$ of the bubble sort algorithm given in Figure 8.

In the next six exercises, complete a trace at points $\widehat{X}$ and $\widehat{Y}$ of the linear search algorithm shown in Figure 9, given $N = 5$, the array

$$a(1) = 2 \quad a(2) = 9 \quad a(3) = 14 \quad a(4) = 25 \quad a(5) = 33$$

and the specified VALUE.

11. VALUE = 25   **12.** VALUE = 14

13. VALUE = 10   **14.** VALUE = 28

15. VALUE = 1   **16.** VALUE = 45

17. Improve the efficiency of the linear search flowchart in Figure 9 so that the loop terminates when $a(k) >$ VALUE. (Remember that the given array **a** is sorted.)

18. How would the trace of the search for VALUE = 1136 differ from the trace given in Figure 11?

19. Perform a trace of Figure 11 at points $\widehat{X}$ and $\widehat{Y}$ while searching for VALUE = 1000.

20. Perform a trace of Figure 11 at points $\widehat{X}$ and $\widehat{Y}$ while searching for VALUE = 1005.

21. Modify the jump 50 linear search algorithm so that it will correctly search any ordered array of length N, where $2000 \leq N \leq 2049$.

22. Modify the jump 50 linear search algorithm so that it will correctly search any ordered array of length N, where N is a multiple of 50.

23. Modify the jump 50 linear search algorithm so that it will correctly search any ordered array of length N, where $N > 0$. (*Hint:* Combine the modifications of exercises 21 and 22.)

24. Given a positive integer J, modify the algorithm of Figure 11 so that it becomes a jump J linear search that processes any ordered array of length N, where $N > 0$. (*Hint:* Modify the result of exercise 23.)

25. In the bubble sort, one pass causes the largest element *not* in its correct position in the array to fall to its correct position. Use this observation to reduce the number of the comparisons $a(j) > a(j + 1)$ made in the bubble sort algorithm in Figure 8.

26. In light of the comment made in the first sentence of the preceding exercise, revise the flowchart in Figure 8 so that one pass causes the *smallest* element *not* in its correct position in the array to rise to its correct position.

27. Figure 12 is a variation of the selection sort algorithm presented in the text. At each pass through the array, the least element is determined, replaced by the bottom array element, and stored in a separate array. Construct a flowchart for this type of sort that will take in array $(a(k))$ and produce sorted array $(b(k))$.

28. Figure 13 is a variation of the selection sort algorithm presented in the text. At each pass through the array, the least element is determined, replaced by a predetermined number known to be out of the range of the array values (here we'll choose 999), and stored in a separate array. Construct a flowchart for this **insertion sort**, which will take in any array $(a(k))$ consisting of integers with at most three-digits and produce sorted array $(b(k))$.

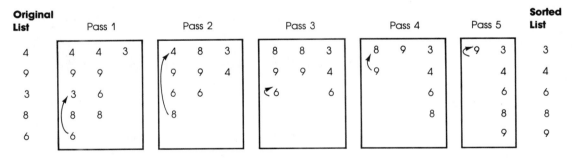

Figure 12

Figure 13

5.3

Two-Dimensional Arrays: Matrices

Just as linear lists, or one-dimensional arrays, provide a natural way of organizing data in certain instances, so do ***two-dimensional arrays***, or ***matrices***, in other situations.

Definition

Let m and n be positive integers. An ordered rectangular array of data

$$\begin{bmatrix} a_{11} & a_{12} & \cdots & a_{1n} \\ a_{21} & a_{22} & \cdots & a_{2n} \\ \vdots & \vdots & & \vdots \\ a_{m1} & a_{m2} & \cdots & a_{mn} \end{bmatrix}$$

where entry a_{ij} is placed in row i and column j, is called a ***matrix*** (or ***two-dimensional array***) of ***dimension $m \times n$***.

Notation: $A = [a_{ij}]$

Once again some comments on notation are in order here. First, we'll use the double-subscript notation a_{ij} rather than the more cumbersome $a(i, j)$ that must be used in computer programming. In either form, the letter i represents the row number and j the column number. Second, it is sometimes useful to consider the $m \times n$ matrix written above as being built from the ***m row vectors***,

$$(a_{11}, a_{12}, \cdots, a_{1n})$$
$$(a_{21}, a_{22}, \cdots, a_{2n})$$
$$\vdots$$
$$(a_{m1}, a_{m2}, \cdots, a_{mn})$$

or as being constructed from the ***n column vectors***,

$$\begin{pmatrix} a_{11} \\ a_{21} \\ \vdots \\ a_{m1} \end{pmatrix} \begin{pmatrix} a_{12} \\ a_{22} \\ \vdots \\ a_{m2} \end{pmatrix} \cdots \begin{pmatrix} a_{1n} \\ a_{2n} \\ \vdots \\ a_{mn} \end{pmatrix}$$

These are vectors of the type discussed in Section 6.1.

Example 5.3.1   **Matrices**

a. $A = \begin{bmatrix} \text{Nixon} & 72 \\ \text{Carter} & 76 \\ \text{Reagan} & 80 \end{bmatrix}$

is a 3 × 2 matrix built from the three row vectors

(Nixon,   72)

(Carter,   76)

(Reagan, 80)

Using the matrix entry notation $A = [a_{ij}]$, we get

$$a_{11} = \text{Nixon} \qquad a_{12} = 72$$
$$a_{21} = \text{Carter} \qquad a_{22} = 76$$
$$a_{31} = \text{Reagan} \qquad a_{32} = 80$$

b. The 4 × 5 matrix

$$S = \begin{bmatrix} 7 & 5 & 2 & 3 & 4 \\ 4 & 5 & 3 & 0 & 7 \\ 5 & 6 & 6 & 4 & 2 \\ 1 & 4 & 4 & 5 & 3 \end{bmatrix}$$

consists of integer entries that represent the number of new automobiles sold by five salespersons (each associated with a separate column) over a particular four-week period (each week associated with a separate row). For instance, $S_{32} = 6$ means that during week 3, salesperson 2 sold six autos; $S_{45} = 3$ means that during week 4, salesperson 5 sold three autos. The four row vectors

(7, 5, 2, 3, 4)

(4, 5, 3, 0, 7)

(5, 6, 6, 4, 2)

(1, 4, 4, 5, 3)

represent the weekly sales, and the five column vectors

$$\begin{pmatrix} 7 \\ 4 \\ 5 \\ 1 \end{pmatrix} \begin{pmatrix} 5 \\ 5 \\ 6 \\ 4 \end{pmatrix} \begin{pmatrix} 2 \\ 3 \\ 6 \\ 4 \end{pmatrix} \begin{pmatrix} 3 \\ 0 \\ 4 \\ 5 \end{pmatrix} \begin{pmatrix} 4 \\ 7 \\ 2 \\ 3 \end{pmatrix}$$

represent the sales for each of the five salespersons.   ■

Example 5.3.2   **Payroll Matrix**

Let $E = (e_{ij})$ be a 576 × 7 matrix, where each row vector contains the name, social security number, gross salary, federal tax deduction, state tax deduction, retirement deduction, and net salary for one of a company's 576 employees. The algorithm in Figure 14 takes in this matrix E and prints out the company totals for federal tax TOTFED and net salary TOTNET. In essence, this algorithm just adds the numbers in the fourth and seventh columns of matrix E.

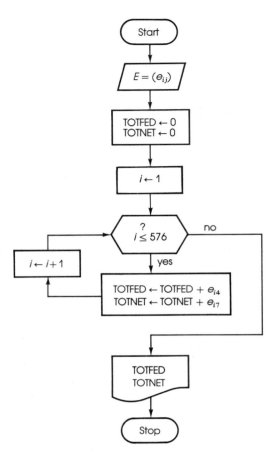

Figure 14

Actually, the entire, large matrix E need not be stored in computer memory to produce the desired printout here (see exercise 44), but this may be desirable if the employee salary data is to be processed further. ∎

Definition

Let $A = [a_{ij}]$ and $B = [b_{ij}]$ each be $m \times n$ matrices. Then we say that **A equals B** if

$$a_{ij} = b_{ij}$$

for all $i = 1, 2, \ldots, m$ and $j = 1, 2, \ldots, n$.

Notation: $A = B$

Example 5.3.3   **Matrix Equality**

a. $\begin{bmatrix} 1 & 1+1 \\ 2+1 & 4 \end{bmatrix} = \begin{bmatrix} 1 & 2 \\ 3 & 4 \end{bmatrix} \neq \begin{bmatrix} 1 & 3 \\ 2 & 4 \end{bmatrix}$

b. $\begin{bmatrix} \text{Nixon} & 70 + 2 \\ \text{Carter} & 76 \\ \text{Reagan} & 84 - 4 \end{bmatrix} = \begin{bmatrix} \text{Nixon} & 72 \\ \text{Carter} & 76 \\ \text{Reagan} & 80 \end{bmatrix} \neq \begin{bmatrix} \text{Reagan} & 80 \\ \text{Carter} & 76 \\ \text{Nixon} & 72 \end{bmatrix}$ ∎

Matrices of the type in (a) of the preceding example, all of whose entries are numbers, are called *numeric matrices*. (When it is obvious that matrices are numeric, the modifier *numeric* is often dropped.) Again, the word *scalar* is used in this context to distinguish an ordinary number from a matrix or matrix entry.

Definitions

Let $A = [a_{ij}]$ and $B = [b_{ij}]$ be numeric matrices of the same dimension $m \times n$, and let t be a scalar. Then the matrix

1. $[a_{ij} + b_{ij}]$ is called the *sum of A and B* and is denoted $A + B$;
2. $[a_{ij} - b_{ij}]$ is called the *difference of A and B* and is denoted $A - B$; and
3. $[t * a_{ij}]$ is called the *scalar multiple of A by t* and is denoted tA.

These operations are called matrix *addition*, *subtraction*, and *scalar multiplication*, respectively.

Note that we can only add or subtract matrices of the same dimension, and the result is a matrix of that dimension. Also, a scalar multiple of a given matrix has the same dimension as that given matrix.

Example 5.3.4   **Addition, Subtraction, and Scalar Multiplication**

a. If

$$A = \begin{bmatrix} 7 & -12 \\ -3 & 4 \\ 0 & 6.5 \end{bmatrix} \qquad B = \begin{bmatrix} -1 & 6 \\ 14 & 0 \\ .3 & -5 \end{bmatrix} \qquad C = \begin{bmatrix} 4 & -1 \\ 2 & 5 \end{bmatrix}$$

and $t = 4$ and $s = -3$, then

$$A + B = \begin{bmatrix} 7 & -12 \\ -3 & 4 \\ 0 & 6.5 \end{bmatrix} + \begin{bmatrix} -1 & 6 \\ 14 & 0 \\ .3 & -5 \end{bmatrix} = \begin{bmatrix} 7-1 & -12+6 \\ -3+14 & 4+0 \\ 0+.3 & 6.5-5 \end{bmatrix} = \begin{bmatrix} 6 & -6 \\ 11 & 4 \\ .3 & 1.5 \end{bmatrix}$$

$$A - B = \begin{bmatrix} 7 & -12 \\ -3 & 4 \\ 0 & 6.5 \end{bmatrix} - \begin{bmatrix} -1 & 6 \\ 14 & 0 \\ .3 & -5 \end{bmatrix} = \begin{bmatrix} 7+1 & -12-6 \\ -3-14 & 4-0 \\ 0-.3 & 6.5+5 \end{bmatrix} = \begin{bmatrix} 8 & -18 \\ -17 & 4 \\ -.3 & 11.5 \end{bmatrix}$$

$$B - A = \begin{bmatrix} -1 & 6 \\ 14 & 0 \\ .3 & -5 \end{bmatrix} - \begin{bmatrix} 7 & -12 \\ -3 & 4 \\ 0 & 6.5 \end{bmatrix} = \begin{bmatrix} -1-7 & 6+12 \\ 14+3 & 0-4 \\ .3-0 & -5-6.5 \end{bmatrix} = \begin{bmatrix} -8 & 18 \\ 17 & -4 \\ .3 & -11.5 \end{bmatrix}$$

$$tA = 4\begin{bmatrix} 7 & -12 \\ -3 & 4 \\ 0 & 6.5 \end{bmatrix} = \begin{bmatrix} 4*7 & 4*(-12) \\ 4*(-3) & 4*4 \\ 4*0 & 4*6.5 \end{bmatrix} = \begin{bmatrix} 28 & -48 \\ -12 & 16 \\ 0 & 26 \end{bmatrix}$$

$$sB = -3\begin{bmatrix} -1 & 6 \\ 14 & 0 \\ .3 & -5 \end{bmatrix} = \begin{bmatrix} (-3)*(-1) & (-3)*6 \\ (-3)*14 & (-3)*0 \\ (-3)*.3 & (-3)*(-5) \end{bmatrix} = \begin{bmatrix} 3 & -18 \\ -42 & 0 \\ -.9 & 15 \end{bmatrix}$$

$$tC = 4\begin{bmatrix} 4 & -1 \\ 2 & 5 \end{bmatrix} = \begin{bmatrix} 4*4 & 4*(-1) \\ 4*2 & 4*5 \end{bmatrix} = \begin{bmatrix} 16 & -4 \\ 8 & 20 \end{bmatrix}$$

$$2B - 5A = 2\begin{bmatrix} -1 & 6 \\ 14 & 0 \\ .3 & -5 \end{bmatrix} - 5\begin{bmatrix} 7 & -12 \\ -3 & 4 \\ 0 & 6.5 \end{bmatrix} = \begin{bmatrix} -2 & 12 \\ 28 & 0 \\ .6 & -10 \end{bmatrix} - \begin{bmatrix} 35 & -60 \\ -15 & 20 \\ 0 & 32.5 \end{bmatrix} = \begin{bmatrix} -37 & 72 \\ 43 & -20 \\ .6 & -42.5 \end{bmatrix}$$

$$6(B - A) = 6\left(\begin{bmatrix} -1 & 6 \\ 14 & 0 \\ .3 & -5 \end{bmatrix} - \begin{bmatrix} 7 & -12 \\ -3 & 4 \\ 0 & 6.5 \end{bmatrix}\right) = 6\begin{bmatrix} -8 & 18 \\ 17 & -4 \\ .3 & -11.5 \end{bmatrix} = \begin{bmatrix} -48 & 108 \\ 102 & -24 \\ 1.8 & -69 \end{bmatrix}$$

Also, *none* of the expressions

$$A + C \qquad A - C \qquad B + C \qquad B - C$$

are defined because in each case the two matrices have different dimensions.

b. Recall the 4×5 matrix

$$S = \begin{bmatrix} 7 & 5 & 2 & 3 & 4 \\ 4 & 5 & 3 & 0 & 7 \\ 5 & 6 & 6 & 4 & 2 \\ 1 & 4 & 4 & 5 & 3 \end{bmatrix}$$

given in Example 5.3.1(b), and let

$$T = \begin{bmatrix} 6 & 6 & 4 & 5 & 3 \\ 5 & 7 & 5 & 5 & 6 \\ 0 & 4 & 6 & 7 & 4 \\ 4 & 3 & 5 & 6 & 7 \end{bmatrix}$$

be a matrix having the same interpretation but for the succeeding four-week sales period. Then

$$S + T = \begin{bmatrix} 13 & 11 & 6 & 8 & 7 \\ 9 & 12 & 8 & 5 & 13 \\ 5 & 10 & 12 & 11 & 6 \\ 5 & 7 & 9 & 11 & 10 \end{bmatrix}$$

gives two-month totals for each salesperson;

$$T - S = \begin{bmatrix} -1 & 1 & 2 & 2 & -1 \\ 1 & 2 & 2 & 5 & -1 \\ -5 & -2 & 0 & 3 & 2 \\ 3 & -1 & 1 & 1 & 4 \end{bmatrix}$$

represents the change in sales the second month as compared with the first; and

$$12T = \begin{bmatrix} 72 & 72 & 48 & 60 & 36 \\ 60 & 84 & 60 & 60 & 72 \\ 0 & 48 & 72 & 84 & 48 \\ 48 & 36 & 60 & 72 & 84 \end{bmatrix}$$

gives the yearly totals if the sales level represented by matrix T were maintained for 12 consecutive months. ∎

As was the case with vectors, the "multiplication" of two matrices A and B of the same dimension defined by multiplying corresponding matrix entries

$$[a_{ij} * b_{ij}]$$

is not very useful. However, the following kind of matrix multiplication *is* important:

Definition

Let $A = [a_{ij}]$ and $B = [b_{ij}]$ be numeric matrices of dimensions $m \times n$ and $n \times r$, respectively. Then the matrix

$$C = [c_{ij}]$$

of dimension $m \times r$ defined by

$$c_{ij} = a_{i1}b_{1j} + a_{i2}b_{2j} + a_{i3}b_{3j} + \cdots + a_{in}b_{nj}$$

$$= \sum_{k=1}^{n} a_{ik}b_{kj}$$

is called the **matrix product of A and B**.

Notation: AB

This operation is called **matrix multiplication**.

If you're wondering where such a crazy definition came from, the answer is not easy. Briefly, a matrix can be interpreted as representing a certain type of function called a linear transformation, and the multiplication of two matrices as defined above is the composition as functions of the associated linear transformations. This idea is examined in detail in a course on linear algebra.

Observe the relationship between the dimensions of A and B so that the definition makes sense:

Dimension A   **Dimension B**

$m \times n$ $n \times r$

must be equal

Dimension of product AB

Also, while the formula for c_{ij} looks formidable, it's really just the inner product of the ith row vector of A with the jth column vector of B:

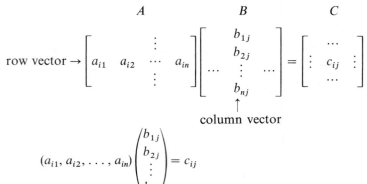

$$\left(a_{i1}, a_{i2}, \ldots, a_{in}\right)\begin{pmatrix} b_{1j} \\ b_{2j} \\ \vdots \\ b_{nj} \end{pmatrix} = c_{ij}$$

$$a_{i1}b_{1j} + a_{i2}b_{2j} + \cdots + a_{in}b_{nj} = c_{ij}$$

Example 5.3.5   **Matrix Multiplication**

Let

$$A = \begin{bmatrix} 1 & 2 \\ 0 & -3 \end{bmatrix} \qquad B = \begin{bmatrix} -5 & -1 \\ 2 & 6 \end{bmatrix} \qquad C = \begin{bmatrix} -4 & 3 & 2 \\ 1 & 0 & -5 \end{bmatrix}$$

$$D = \begin{bmatrix} 2 & -1 & 4 \\ -1 & 5 & 0 \\ 0 & 2 & 3 \end{bmatrix} \qquad E = \begin{bmatrix} -6 & 2 \end{bmatrix} \qquad I_3 = \begin{bmatrix} 1 & 0 & 0 \\ 0 & 1 & 0 \\ 0 & 0 & 1 \end{bmatrix}$$

Then

$$AB = \begin{bmatrix} 1 & 2 \\ 0 & -3 \end{bmatrix}\begin{bmatrix} -5 & -1 \\ 2 & 6 \end{bmatrix} = \begin{bmatrix} 1(-5) + 2(2) & 1(-1) + 2(6) \\ 0(-5) + (-3)(2) & 0(-1) + (-3)(6) \end{bmatrix}$$

$$= \begin{bmatrix} -1 & 11 \\ -6 & -18 \end{bmatrix}$$

$$BA = \begin{bmatrix} -5 & -1 \\ 2 & 6 \end{bmatrix}\begin{bmatrix} 1 & 2 \\ 0 & -3 \end{bmatrix} = \begin{bmatrix} (-5)(1) + (-1)(0) & (-5)(2) + (-1)(-3) \\ 2(1) + 6(0) & 2(2) + 6(-3) \end{bmatrix}$$

$$= \begin{bmatrix} -5 & -7 \\ 2 & -14 \end{bmatrix}$$

$$CD = \begin{bmatrix} -4 & 3 & 2 \\ 1 & 0 & -5 \end{bmatrix} \begin{bmatrix} 2 & -1 & 4 \\ -1 & 5 & 0 \\ 0 & 2 & 3 \end{bmatrix}$$

$$= \begin{bmatrix} (-4)(2) + 3(-1) + 2(0) & (-4)(-1) + 3(5) + 2(2) & (-4)(4) + 3(0) + 2(3) \\ 1(2) + 0(-1) + (-5)(0) & 1(-1) + 0(5) + (-5)(2) & 1(4) + 0(0) + (-5)(3) \end{bmatrix}$$

$$= \begin{bmatrix} -11 & 23 & -10 \\ 2 & -11 & -11 \end{bmatrix}$$

$$EC = \begin{bmatrix} -6 & 2 \end{bmatrix} \begin{bmatrix} -4 & 3 & 2 \\ 1 & 0 & -5 \end{bmatrix}$$

$$= \begin{bmatrix} (-6)(-4) + 2(1) & (-6)(3) + 2(0) & (-6)(2) + 2(-5) \end{bmatrix}$$

$$= \begin{bmatrix} 26 & -18 & -22 \end{bmatrix}$$

$$B^2 = BB = \begin{bmatrix} -5 & -1 \\ 2 & 6 \end{bmatrix} \begin{bmatrix} -5 & -1 \\ 2 & 6 \end{bmatrix} = \begin{bmatrix} (-5)(-5) + (-1)(2) & (-5)(-1) + (-1)(6) \\ 2(-5) + 6(2) & 2(-1) + 6(6) \end{bmatrix}$$

$$= \begin{bmatrix} 23 & -1 \\ 2 & 34 \end{bmatrix}$$

$$DI_3 = \begin{bmatrix} 2 & -1 & 4 \\ -1 & 5 & 0 \\ 0 & 2 & 3 \end{bmatrix} \begin{bmatrix} 1 & 0 & 0 \\ 0 & 1 & 0 \\ 0 & 0 & 1 \end{bmatrix}$$

$$= \begin{bmatrix} 2(1) + (-1)(0) + 4(0) & 2(0) + (-1)(1) + 4(0) & 2(0) + (-1)(0) + 4(1) \\ (-1)(1) + 5(0) + 0(0) & (-1)(0) + 5(1) + 0(0) & (-1)(0) + 5(0) + 0(1) \\ 0(1) + 2(0) + 3(0) & 0(0) + 2(1) + 3(0) & 0(0) + 2(0) + 3(1) \end{bmatrix}$$

$$= \begin{bmatrix} 2 & -1 & 4 \\ -1 & 5 & 0 \\ 0 & 2 & 3 \end{bmatrix}$$

$$I_3D = \begin{bmatrix} 1 & 0 & 0 \\ 0 & 1 & 0 \\ 0 & 0 & 1 \end{bmatrix} \begin{bmatrix} 2 & -1 & 4 \\ -1 & 5 & 0 \\ 0 & 2 & 3 \end{bmatrix}$$

$$= \begin{bmatrix} 1(2) + 0(-1) + 0(0) & 1(-1) + 0(5) + 0(2) & 1(4) + 0(0) + 0(3) \\ 0(2) + 1(-1) + 0(0) & 0(-1) + 1(5) + 0(2) & 0(4) + 1(0) + 0(3) \\ 0(2) + 0(-1) + 1(0) & 0(-1) + 0(5) + 1(2) & 0(4) + 0(0) + 1(3) \end{bmatrix}$$

$$= \begin{bmatrix} 2 & -1 & 4 \\ -1 & 5 & 0 \\ 0 & 2 & 3 \end{bmatrix}$$

$$(A + B)C = \left(\begin{bmatrix} 1 & 2 \\ 0 & -3 \end{bmatrix} + \begin{bmatrix} -5 & -1 \\ 2 & 6 \end{bmatrix} \right) \begin{bmatrix} -4 & 3 & 2 \\ 1 & 0 & -5 \end{bmatrix}$$

$$= \begin{bmatrix} -4 & 1 \\ 2 & 3 \end{bmatrix} \begin{bmatrix} -4 & 3 & 2 \\ 1 & 0 & -5 \end{bmatrix}$$

$$= \begin{bmatrix} (-4)(-4) + 1(1) & (-4)(3) + 1(0) & (-4)(2) + 1(-5) \\ 2(-4) + 3(1) & 2(3) + 3(0) & 2(2) + 3(-5) \end{bmatrix}$$

$$= \begin{bmatrix} 17 & -12 & -13 \\ -5 & 6 & -11 \end{bmatrix}$$

Finally, notice that the product DC is not defined because

Dimension D   **Dimension C**

3×3 2×3

Not equal

For similar reasons, the products CE, CA, CB, AD, BD, C^2, and E^2 are not defined. ∎

In the preceding example, the matrix I_3, which has all 1's on the **main diagonal** and all 0's elsewhere, is called the **3 × 3 identity matrix** because it has the property

$$MI_3 = I_3M = M$$

for any 3×3 matrix M. In general, the $n \times n$ matrix I_n that has all 1's on the main diagonal and all 0's elsewhere is called the **n × n identity matrix**; it has the property that

$$MI_n = I_nM = M$$

for any $n \times n$ matrix M.

Also, it is important to observe in the above example that

$$AB \neq BA$$

This shows that *matrix multiplication does not possess the property of commutativity*; that is, for some matrices A and B, the products AB and BA are equal, and for others these products are not equal. (Recall that ordinary multiplication of numbers is commutative!)

Nevertheless, matrix multiplication does share many of the algebraic properties of our familiar number systems (see exercises 34, 35, and 37 to 40). One such property that we shall use in a later section of this chapter is the property that matrix multiplication is associative; that is,

$$A(BC) = (AB)C$$

whenever these products are defined.

Example 5.3.6

Associativity of Matrix Multiplication

Here we shall verify that the associative property for matrix multiplication holds for the matrices

$$A = \begin{bmatrix} -7 & 2 \end{bmatrix} \quad B = \begin{bmatrix} 0 & 4 \\ -8 & 5 \end{bmatrix} \quad C = \begin{bmatrix} 3 & -2 & 1 \\ -4 & -6 & 0 \end{bmatrix}$$

First,

$$BC = \begin{bmatrix} 0 & 4 \\ -8 & 5 \end{bmatrix} \begin{bmatrix} 3 & -2 & 1 \\ -4 & -6 & 0 \end{bmatrix}$$

$$= \begin{bmatrix} 0(3) + 4(-4) & 0(-2) + 4(-6) & 0(1) + 4(0) \\ (-8)(3) + 5(-4) & (-8)(-2) + 5(-6) & (-8)(1) + 5(0) \end{bmatrix}$$

$$= \begin{bmatrix} -16 & -24 & 0 \\ -44 & -14 & -8 \end{bmatrix}$$

so

$$A(BC) = \begin{bmatrix} -7 & 2 \end{bmatrix} \begin{bmatrix} -16 & -24 & 0 \\ -44 & -14 & -8 \end{bmatrix}$$

$$= \begin{bmatrix} (-7)(-16) + 2(-44) & (-7)(-24) + 2(-14) & (-7)(0) + 2(-8) \end{bmatrix}$$

$$= \begin{bmatrix} 24 & 140 & -16 \end{bmatrix}$$

Also,

$$AB = \begin{bmatrix} -7 & 2 \end{bmatrix} \begin{bmatrix} 0 & 4 \\ -8 & 5 \end{bmatrix}$$

$$= \begin{bmatrix} (-7)(0) + 2(-8) & (-7)(4) + 2(5) \end{bmatrix} = \begin{bmatrix} -16 & -18 \end{bmatrix}$$

and

$$(AB)C = \begin{bmatrix} -16 & -18 \end{bmatrix} \begin{bmatrix} 3 & -2 & 1 \\ -4 & -6 & 0 \end{bmatrix}$$

$$= \begin{bmatrix} (-16)(3) + (-18)(-4) & (-16)(-2) + (-18)(-6) & (-16)(1) + (-18)0 \end{bmatrix}$$

$$= \begin{bmatrix} 24 & 140 & -16 \end{bmatrix}$$

Therefore, we see that

$$A(BC) = (AB)C \qquad \blacksquare$$

While we could linger here to discuss other algebraic properties of matrix addition, subtraction, scalar multiplication, and multiplication, we shall resist that temptation and leave the statements of several such properties to the exercises (see exercises 33 to 40).

EXERCISES 5.3

Exercises 1 through 42 below use one or more of the matrices

$$A = \begin{bmatrix} 1 & 2 & 3 \\ 0 & -5 & 6 \end{bmatrix} \quad B = \begin{bmatrix} .5 & 0 & -2 \\ -4 & 1 & -1 \end{bmatrix}$$

$$C = \begin{bmatrix} 1 & 0 & -1 \\ 5 & 8 & 0 \\ -3 & -4 & 1 \end{bmatrix} \quad D = \begin{bmatrix} -6 & 1 & 4 \\ 2 & 3 & 7 \\ 0 & 0 & -5 \end{bmatrix}$$

$$E = \begin{bmatrix} 2 & 0 & -1 & 4 & 3 & 1 \\ 0 & 0 & 1 & 5 & 6 & -2 \\ 7 & 3 & 0 & -1 & 0 & 1 \end{bmatrix}$$

and scalars $t = 3$ and $s = -2$. In the first 32 exercises, compute the desired matrix, if possible.

1. $A + B$ **2.** $A - B$ **3.** $B - A$

4. $C + D$ **5.** $C - D$ **6.** $A + C$

7. $D + B$ **8.** $C + I_3$ **9.** tA

10. sC **11.** $-4D$ **12.** $6D - C$

13. $A + 2D$ **14.** AB **15.** CD

16. DC **17.** AC **18.** CA

19. DB **20.** BD **21.** AE

22. EB **23.** ED **24.** CE

25. CI_3 **26.** $2I_3D$ **27.** $-5CI_3$

28. A^2 **29.** I_4^2 **30.** C^2

31. $(A + B)C$ **32.** $A(C - I_3)$

The algebraic properties stated in exercises 33 to 40 are valid in general, provided the operations are defined. Verify each for the matrices A, B, C, D, and E and scalar t given above.

33. $A + B = B + A$

34. $A(BC) = (AB)C$

35. $t(BC) = (tB)C = B(tC)$

36. $t(A + B) = tA + tB$

37. $A(C + D) = AC + AD$

38. $A(C - D) = AC - AD$

39. $(A + B)C = AC + BC$

40. $(A - B)C = AC - BC$

41. For matrices C and D above, which of the following are true?

a. $CD = DC$ **b.** $I_3C = CI_3$

42. The *transpose of matrix* $A = [a_{ij}]$ is defined as that matrix $T = [t_{ij}]$ with the property that

$$t_{ij} = a_{ji} \quad \text{for all } i \text{ and } j$$

In effect, then, the rows of T are the columns of A. Compute the transposes of matrices A, B, C, D, E, and I_3 given above.

43. Verify that $I_4M = MI_4$ for any 4×4 matrix M.

44. Revise the flowchart in Figure 14 so that the same printout is produced without initially storing the entire matrix E in memory.

45. In a particular course, each of 850 students has taken four exams, and their names and scores have been assembled into an 850×5 matrix S of the form

		Score			
Name		**Exam 1**	**Exam 2**	**Exam 3**	**Exam 4**
850 Rows	$\begin{bmatrix} \vdots \\ \text{D. GREENE} \\ \vdots \end{bmatrix}$	$\begin{matrix} \vdots \\ 88 \\ \vdots \end{matrix}$	$\begin{matrix} \vdots \\ 92 \\ \vdots \end{matrix}$	$\begin{matrix} \vdots \\ 79 \\ \vdots \end{matrix}$	$\begin{matrix} \vdots \\ 95 \\ \vdots \end{bmatrix}$

Construct a flowchart that will initially read in matrix S and produce the following:

a. Each student's name and average grade on the four exams taken

b. The average class grade on each of the four exams

46. For each of 275 marketing representatives, the three largest sales are recorded in descending order next to the person's name. The 275 rows of data are used to form the 275×4 matrix S illustrated below.

		Sales		
Name		**Largest**	**2nd Largest**	**3rd Largest**
275 Rows	$\begin{bmatrix} \vdots \\ \text{W. BRADY} \\ \vdots \end{bmatrix}$	$\begin{matrix} \vdots \\ 2490 \\ \vdots \end{matrix}$	$\begin{matrix} \vdots \\ 2061 \\ \vdots \end{matrix}$	$\begin{matrix} \vdots \\ 1938 \\ \vdots \end{bmatrix}$

Construct an algorithm that will initially read in matrix S and produce the following:

a. For each representative, the name and average of the two largest sales

b. The average largest sale for all reps

5.4

Determinants

A matrix in which the number of rows equals the number of columns is called a *square matrix*; for instance, the 2×2 and 3×3 matrices

$$\begin{bmatrix} 2 & -.3 \\ 0 & 4 \end{bmatrix} \qquad \begin{bmatrix} -5 & 0 & 2 \\ 1 & 7 & -6 \\ 4 & .1 & 3 \end{bmatrix}$$

are square matrices, whereas the 2×3 matrix

$$\begin{bmatrix} -6 & 1 & .5 \\ 4 & .2 & -7 \end{bmatrix}$$

is not square. One important tool associated with square matrices is the *determinant*. There are several ways of presenting the general definition of the determinant of an $n \times n$ matrix; in computer-related math, we shall take an approach that requires several preliminary steps.

Definition

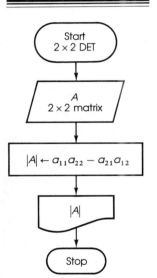

Figure 15

> The *determinant of a 2 × 2 matrix*
>
> $$A = \begin{bmatrix} a_{11} & a_{12} \\ a_{21} & a_{22} \end{bmatrix}$$
>
> is the number
>
> $$a_{11}a_{22} - a_{21}a_{12}$$
>
> **Notation:** $|A|$ or det A

The phrase "determinant of a 2×2 matrix" is often abbreviated as **2 × 2 determinant**. The flowchart in Figure 15 gives the procedure for evaluating a 2×2 determinant. Note that we have placed the label

$$2 \times 2 \text{ DET}$$

within the start symbol, since we shall call this algorithm in succeeding flowcharts.

Example 5.4.1

2 × 2 Determinants

a. If

$$A = \begin{bmatrix} 2 & -.3 \\ 0 & 4 \end{bmatrix} \qquad \text{and} \qquad B = \begin{bmatrix} -5 & -1 \\ 2 & 6 \end{bmatrix}$$

then

$$|A| = \begin{vmatrix} 2 & -.3 \\ 0 & 4 \end{vmatrix} = 2(4) - (0)(-.3) = \boxed{8}$$

and

$$|B| = \begin{vmatrix} -5 & -1 \\ 2 & 6 \end{vmatrix} = (-5)(6) - (2)(-1) = \boxed{-28}$$

b. $\det \begin{bmatrix} 1 & 0 \\ 0 & 1 \end{bmatrix} = \begin{vmatrix} 1 & 0 \\ 0 & 1 \end{vmatrix} = 1(1) - 0(0) = \boxed{1}$

$\det \begin{bmatrix} 2 & 4 \\ -1 & -2 \end{bmatrix} = \begin{vmatrix} 2 & 4 \\ -1 & -2 \end{vmatrix} = 2(-2) - (-1)(4) = \boxed{0}$ ∎

Before proceeding further, we should introduce the terms *minor* and *cofactor* of a matrix entry.

Definitions

Let $A = (a_{ij})$ be an $n \times n$ matrix and a_{kp} an arbitrary entry, where $n \geq 3$.

a. The *minor of a_{kp}* is the determinant of the matrix remaining after the deletion of the kth row and pth column of A.

Notation: $\mathrm{MIN}(a_{kp})$

b. The *cofactor of a_{kp}* is the number

$$(-1)^{k+p} \, \mathrm{MIN}(a_{kp})$$

Notation: $\mathrm{COF}(a_{kp})$

Observe that at the moment we can apply these definitions *only* to 3×3 matrices, since we have defined only 2×2 determinants.

Example 5.4.2   **Minors and Cofactors**

For the 3×3 matrix

$$A = \begin{bmatrix} -2 & 1 & 7 \\ 5 & -3 & 0 \\ -4 & 6 & 8 \end{bmatrix}$$

we have

$$\mathrm{MIN}(a_{11}) = \det \begin{bmatrix} \cancel{-2} & \cancel{1} & \cancel{7} \\ \cancel{5} & -3 & 0 \\ \cancel{-4} & 6 & 8 \end{bmatrix} = \begin{vmatrix} -3 & 0 \\ 6 & 8 \end{vmatrix}$$

$$= (-3)(8) - (6)(0) = \boxed{-24}$$

$$\text{COF}(a_{11}) = (-1)^{1+1} \text{ MIN}(a_{11}) = (-1)^2(-24) = \boxed{-24}$$

$$\text{MIN}(a_{21}) = \det \begin{bmatrix} -2 & 1 & 7 \\ 5 & -3 & 0 \\ -4 & 6 & 8 \end{bmatrix} = \begin{vmatrix} 1 & 7 \\ 6 & 8 \end{vmatrix} = 1(8) - (6)(7) = \boxed{-34}$$

$$\text{COF}(a_{21}) = (-1)^{2+1} \text{ MIN}(a_{21}) = (-1)^3(-34) = \boxed{34}$$

$$\text{MIN}(a_{31}) = \det \begin{bmatrix} -2 & 1 & 7 \\ 5 & -3 & 0 \\ -4 & 6 & 8 \end{bmatrix} = \begin{vmatrix} 1 & 7 \\ -3 & 0 \end{vmatrix}$$

$$= (1)(0) - (-3)(7) = \boxed{21}$$

$$\text{COF}(a_{31}) = (-1)^{3+1} \text{ MIN}(a_{31}) = (-1)^4(21) = \boxed{21}$$

$$\text{MIN}(a_{22}) = \det \begin{bmatrix} -2 & 1 & 7 \\ 5 & -3 & 0 \\ -4 & 6 & 8 \end{bmatrix} = \begin{vmatrix} -2 & 7 \\ -4 & 8 \end{vmatrix}$$

$$= (-2)(8) - (-4)(7) = \boxed{12}$$

$$\text{COF}(a_{22}) = (-1)^{2+2} \text{ MIN}(a_{22}) = (-1)^4(12) = \boxed{12}$$

$$\text{MIN}(a_{13}) = \det \begin{bmatrix} -2 & 1 & 7 \\ 5 & -3 & 0 \\ -4 & 6 & 8 \end{bmatrix} = \begin{vmatrix} 5 & -3 \\ -4 & 6 \end{vmatrix}$$

$$= 5(6) - (-4)(-3) = \boxed{18}$$

$$\text{COF}(a_{13}) = (-1)^{1+3} \text{ MIN}(a_{13}) = (-1)^4(18) = \boxed{18}$$

$$\text{MIN}(a_{23}) = \det \begin{bmatrix} -2 & 1 & 7 \\ 5 & -3 & 0 \\ -4 & 6 & 8 \end{bmatrix} = \begin{vmatrix} -2 & 1 \\ -4 & 6 \end{vmatrix}$$

$$= (-2)(6) - (-4)(1) = \boxed{-8}$$

$$\text{COF}(a_{23}) = (-1)^{2+3} \text{ MIN}(a_{23}) = (-1)^5(-8) = \boxed{8}$$

The minors and cofactors of the remaining three matrix entries a_{12}, a_{32}, and a_{33} are computed similarly. ∎

Now we are ready to extend our definition of a 2×2 determinant.

Definition

The *determinant of a 3 × 3 matrix*

$$A = \begin{bmatrix} a_{11} & a_{12} & a_{13} \\ a_{21} & a_{22} & a_{23} \\ a_{31} & a_{32} & a_{33} \end{bmatrix}$$

is the number

$$a_{11} \, \text{COF}(a_{11}) + a_{21} \, \text{COF}(a_{21}) + a_{31} \, \text{COF}(a_{31})$$

Notation: $|A|$ or det A

This statement uses cofactors of the first column entries a_{11}, a_{21}, and a_{31}. However, there is nothing special about the first column; equivalent definitions could be given using the cofactors of any single row or column of the given matrix (see exercises 17 and 18).

Example 5.4.3   **3 × 3 Determinants**

a. Using the matrix

$$A = \begin{bmatrix} -2 & 1 & 7 \\ 5 & -3 & 0 \\ -4 & 6 & 8 \end{bmatrix}$$

and results of the preceding example, we have

$$\begin{aligned} |A| &= a_{11} \, \text{COF}(a_{11}) + a_{21} \, \text{COF}(a_{21}) + a_{31} \, \text{COF}(a_{31}) \\ &= (-2)(-24) + (5)(34) + (-4)(21) \\ &= 48 + 170 - 84 = \boxed{134} \end{aligned}$$

b. In the next three examples we use the same definition but minimize notation while computing the determinants.

$$\begin{vmatrix} 1 & 0 & -3 \\ 3 & -1 & 2 \\ -2 & 1 & 0 \end{vmatrix} = 1(-1)^{1+1} \begin{vmatrix} -1 & 2 \\ 1 & 0 \end{vmatrix} + 3(-1)^{2+1} \begin{vmatrix} 0 & -3 \\ 1 & 0 \end{vmatrix} + (-2)(-1)^{3+1} \begin{vmatrix} 0 & -3 \\ -1 & 2 \end{vmatrix}$$

$$= (1)(-2) + (-3)(3) + (-2)(-3) = \boxed{-5}$$

$$\begin{vmatrix} -3 & 6 & 9 \\ 5 & 0 & 1 \\ 1 & -2 & -3 \end{vmatrix} = (-3)(-1)^{1+1} \begin{vmatrix} 0 & 1 \\ -2 & -3 \end{vmatrix} + 5(-1)^{2+1} \begin{vmatrix} 6 & 9 \\ -2 & -3 \end{vmatrix} + 1(-1)^{3+1} \begin{vmatrix} 6 & 9 \\ 0 & 1 \end{vmatrix}$$

$$= (-3)(2) + (-5)(0) + (1)(6) = \boxed{0}$$

$$\begin{vmatrix} 1 & 0 & 0 \\ 0 & 1 & 0 \\ 0 & 0 & 1 \end{vmatrix} = 1(-1)^{1+1} \begin{vmatrix} 1 & 0 \\ 0 & 1 \end{vmatrix} + 0(-1)^{2+1} \begin{vmatrix} 0 & 0 \\ 0 & 1 \end{vmatrix} + 0(-1)^{3+1} \begin{vmatrix} 0 & 0 \\ 1 & 0 \end{vmatrix}$$

$$= \quad (1)(1) \quad + \quad (0)(0) \quad + \quad (0)(0) \quad = \boxed{1} \qquad \blacksquare$$

Figure 16 contains a flowchart of the algorithm involved in computing a 3×3 determinant. Notice that it uses three calls (or executions) of the algorithm to compute a 2×2 determinant. Remember that after a call is completed, the execution of the main algorithm resumes at that point immediately following the call. You should compare the numbers in the trace included in Figure 16 with the work in Example 5.4.3(a).

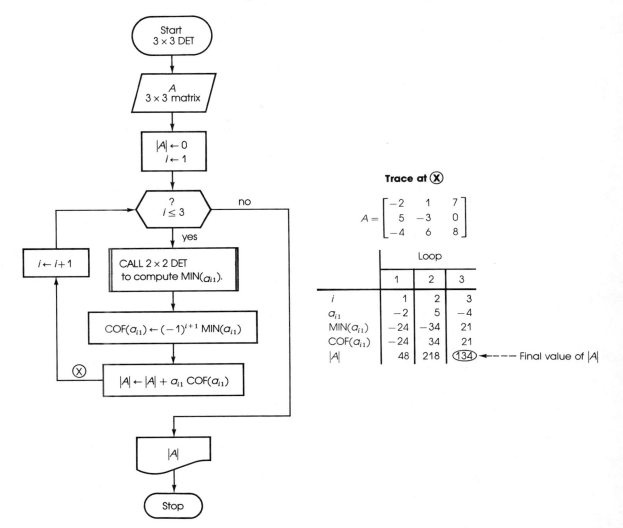

Figure 16

Now that we can calculate 3×3 determinants, we can apply an earlier definition to evaluate minors and cofactors of entries in a 4×4 matrix.

Example 5.4.4   **Minors and Cofactors**

Consider

$$A = \begin{bmatrix} -4 & 1 & 0 & 2 \\ 3 & 0 & -1 & 1 \\ 0 & 2 & 0 & 3 \\ 1 & 0 & 5 & 0 \end{bmatrix}$$

Then

$$\text{MIN}(a_{11}) = \det \begin{bmatrix} -4 & 1 & 0 & 2 \\ 3 & 0 & -1 & 1 \\ 0 & 2 & 0 & 3 \\ 1 & 0 & 5 & 0 \end{bmatrix} = \begin{vmatrix} 0 & -1 & 1 \\ 2 & 0 & 3 \\ 0 & 5 & 0 \end{vmatrix}$$

$$= 0(-1)^{1+1} \begin{vmatrix} 0 & 3 \\ 5 & 0 \end{vmatrix} + 2(-1)^{2+1} \begin{vmatrix} -1 & 1 \\ 5 & 0 \end{vmatrix} + 0(-1)^{3+1} \begin{vmatrix} -1 & 1 \\ 0 & 3 \end{vmatrix}$$

$$= \quad 0(0 - 15) \quad + \quad (-2)(0 - 5) \quad + \quad 0(-3 - 0)$$

$$= \boxed{10}$$

$$\text{COF}(a_{11}) = (-1)^{1+1} \text{MIN}(a_{11}) = (-1)^2(10) = \boxed{10}$$

$$\text{MIN}(a_{21}) = \det \begin{bmatrix} -4 & 1 & 0 & 2 \\ 3 & 0 & -1 & 1 \\ 0 & 2 & 0 & 3 \\ 1 & 0 & 5 & 0 \end{bmatrix} = \begin{vmatrix} 1 & 0 & 2 \\ 2 & 0 & 3 \\ 0 & 5 & 0 \end{vmatrix}$$

$$= 1(-1)^{1+1} \begin{vmatrix} 0 & 3 \\ 5 & 0 \end{vmatrix} + 2(-1)^{2+1} \begin{vmatrix} 0 & 2 \\ 5 & 0 \end{vmatrix} + 0(-1)^{3+1} \begin{vmatrix} 0 & 2 \\ 0 & 3 \end{vmatrix}$$

$$= \quad (1)(0 - 15) \quad + \quad (-2)(0 - 10) \quad + \quad 0(0 - 0)$$

$$= \boxed{5}$$

$$\text{COF}(a_{21}) = (-1)^{2+1} \text{MIN}(a_{21}) = (-1)^3(5) = \boxed{-5}$$

$$\text{MIN}(a_{31}) = \det \begin{bmatrix} -4 & 1 & 0 & 2 \\ 3 & 0 & -1 & 1 \\ 0 & 2 & 0 & 3 \\ 1 & 0 & 5 & 0 \end{bmatrix} = \begin{vmatrix} 1 & 0 & 2 \\ 0 & -1 & 1 \\ 0 & 5 & 0 \end{vmatrix}$$

$$= 1(-1)^{1+1} \begin{vmatrix} -1 & 1 \\ 5 & 0 \end{vmatrix} + 0(-1)^{2+1} \begin{vmatrix} 0 & 2 \\ 5 & 0 \end{vmatrix} + 0(-1)^{3+1} \begin{vmatrix} 0 & 2 \\ -1 & 1 \end{vmatrix}$$

$$= \quad (1)(0-5) \quad + \quad 0(0-10) \quad + \quad 0(0+2)$$

$$= \boxed{-5}$$

$$\text{COF}(a_{31}) = (-1)^{3+1} \text{ MIN}(a_{31}) = (-1)^4(-5) = \boxed{-5}$$

$$\text{MIN}(a_{41}) = \det \begin{bmatrix} -4 & 1 & 0 & 2 \\ 3 & 0 & -1 & 1 \\ 0 & 2 & 0 & 3 \\ 1 & 0 & 5 & 0 \end{bmatrix} = \begin{vmatrix} 1 & 0 & 2 \\ 0 & -1 & 1 \\ 2 & 0 & 3 \end{vmatrix}$$

$$= 1(-1)^{1+1} \begin{vmatrix} -1 & 1 \\ 0 & 3 \end{vmatrix} + 0(-1)^{2+1} \begin{vmatrix} 0 & 2 \\ 0 & 3 \end{vmatrix} + 2(-1)^{3+1} \begin{vmatrix} 0 & 2 \\ -1 & 1 \end{vmatrix}$$

$$= \quad (1)(-3-0) \quad + \quad 0(0-0) \quad + \quad (2)(0+2)$$

$$= \boxed{1}$$

$$\text{COF}(a_{41}) = (-1)^{4+1} \text{ MIN}(a_{41}) = (-1)^5(1) = \boxed{-1}$$

The minors and cofactors of the remaining 12 entries in A (namely, those in columns 2, 3, and 4) can be computed similarly. ∎

Now that we can calculate the cofactors of entries in a 4×4 matrix, we can extend the definition of determinant one additional step.

Definition

The ***determinant of a 4 × 4 matrix***

$$A = \begin{bmatrix} a_{11} & a_{12} & a_{13} & a_{14} \\ a_{21} & a_{22} & a_{23} & a_{24} \\ a_{31} & a_{32} & a_{33} & a_{34} \\ a_{41} & a_{42} & a_{43} & a_{44} \end{bmatrix}$$

is the number

$$a_{11} \text{ COF}(a_{11}) + a_{21} \text{ COF}(a_{21}) + a_{31} \text{ COF}(a_{31}) + a_{41} \text{ COF}(a_{41})$$

Notation: $|A|$ or $\det A$

As with the 3×3 determinant, equivalent definitions would result if we used the cofactors of all entries in any one row or column (see exercises 19 and 20).

Example 5.4.5

4 × 4 Determinant

Using the 4×4 matrix A and the results from the preceding example, we have

$$|A| = a_{11} \text{ COF}(a_{11}) + a_{21} \text{ COF}(a_{21}) + a_{31} \text{ COF}(a_{31}) + a_{41} \text{ COF}(a_{41})$$

$$= \quad (-4)(10) \quad + \quad (3)(-5) \quad + \quad 0(-5) \quad + \quad (1)(-1)$$

$$= \quad -40 \quad\quad\quad -15 \quad\quad\quad +0 \quad\quad\quad -1$$

$$= \boxed{-56}$$ ∎

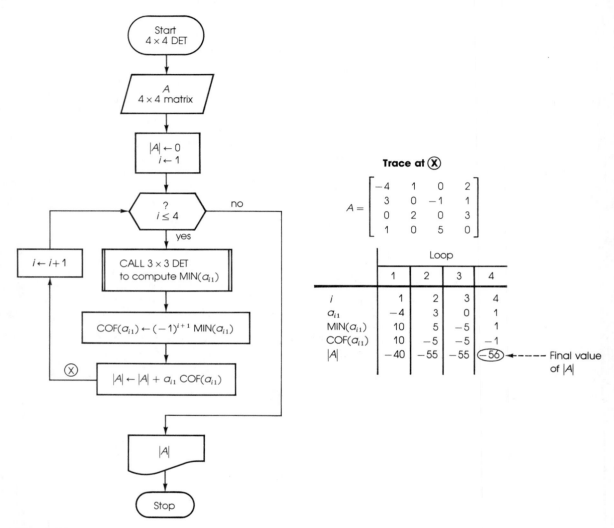

Figure 17

In Figure 17, we have drawn the flowchart for the procedure of calculating a 4×4 determinant. It involves four calls of the algorithm for computing a 3×3 determinant, each of which involves three calls of the 2×2 determinant algorithm. You are urged to compare the trace included in Figure 17 with the work of Example 5.4.5.

By now you may have discerned a pattern in our treatment of determinants. After the initial definition of a 2×2 determinant, a 3×3 determinant is defined in terms of the computation of three 2×2 determinants. A 4×4 determinant is defined in terms of the computation of four 3×3 determinants. *This process can be extended step-by-step indefinitely!* It is an example of what is called a **recursive** process.

We shall conclude this section with the appropriate generalized definition.

Definition

> The **determinant of an $n \times n$ matrix**
>
> $$A = (a_{ij})$$
>
> is the number
>
> $$a_{11} \, \mathrm{COF}(a_{11}) + a_{21} \, \mathrm{COF}(a_{21}) + \cdots + a_{n1} \, \mathrm{COF}(a_{n1})$$
>
> **Notation:** $|A|$ or $\det A$

Again, this definition uses cofactors of elements of the first column, but we could have used cofactors of any particular column or row. Consequently, it follows that *any matrix A with a column of all zeroes or row of all zeroes will have det A = 0.* The flowchart for the $n \times n$ determinant algorithm, a direct generalization of Figure 17, is drawn in Figure 18.

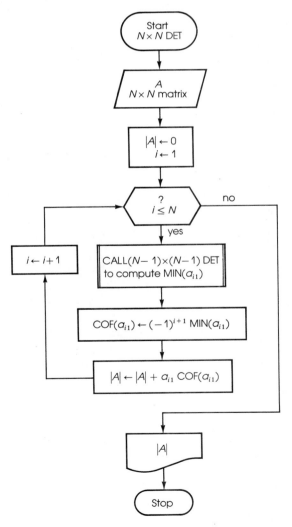

Figure 18

EXERCISES **5.4**

In the first twelve exercises, compute the determinant of the given matrix.

1. $\begin{bmatrix} -4 & 1 \\ 2 & 5 \end{bmatrix}$

2. $\begin{bmatrix} 6 & 4 \\ -1 & 5 \end{bmatrix}$

3. $\begin{bmatrix} 3 & 0 \\ 0 & -.5 \end{bmatrix}$

4. $\begin{bmatrix} -1 & 4 \\ 3 & -12 \end{bmatrix}$

5. $\begin{bmatrix} 5 & -1 & 2 \\ 0 & 3 & 4 \\ 1 & -6 & 2 \end{bmatrix}$

6. $\begin{bmatrix} 5 & -1 & 2 \\ 0 & -3 & 4 \\ 1 & 6 & 2 \end{bmatrix}$

7. $\begin{bmatrix} 0 & 0 & -2 \\ 0 & -4 & 0 \\ -1 & 0 & 0 \end{bmatrix}$

8. $\begin{bmatrix} 3 & -4 & 1 \\ 0 & 2 & -2 \\ 3 & 0 & -3 \end{bmatrix}$

9. $\begin{bmatrix} 1 & -1 & 0 & -2 \\ 2 & 0 & 4 & 0 \\ 3 & 5 & 1 & 0 \\ 4 & 2 & 0 & -1 \end{bmatrix}$

10. $\begin{bmatrix} 3 & 0 & -2 & 1 \\ 0 & 4 & 0 & 1 \\ -1 & 5 & -3 & -4 \\ 2 & 1 & 5 & 0 \end{bmatrix}$

11. $\begin{bmatrix} 1 & 0 & 3 & -2 & 0 \\ 2 & 4 & -2 & 0 & 3 \\ 3 & -1 & 5 & 0 & -1 \\ 4 & 2 & 0 & 3 & 0 \\ 5 & 0 & 1 & 2 & -1 \end{bmatrix}$

12. $\begin{bmatrix} 2 & 0 & 3 & -2 & 0 \\ 3 & 4 & -2 & 0 & 3 \\ 4 & -1 & 5 & 0 & -1 \\ 5 & 2 & 0 & 3 & 0 \\ 1 & 0 & 1 & 2 & -1 \end{bmatrix}$

13. Construct a trace at point $\textcircled{X}$ of the flowchart in Figure 16 for the 3×3 matrix given in exercise 5.

14. Construct a trace at point $\textcircled{X}$ of the flowchart in Figure 16 for the 3×3 matrix given in exercise 6.

15. Construct a trace at point $\textcircled{X}$ of the flowchart in Figure 17 for the 4×4 matrix given in exercise 9.

16. Construct a trace at point $\textcircled{X}$ of the flowchart in Figure 17 for the 4×4 matrix given in exercise 10.

17. For the 3×3 matrix A given in Example 5.4.3(a), compute the cofactor of each element in the second column and then verify that

$$|A| = a_{12} \, \text{COF}(a_{12}) + a_{22} \, \text{COF}(a_{22}) + a_{32} \, \text{COF}(a_{32})$$

18. For the 3×3 matrix A given in Example 5.4.3(a), compute the cofactor of each element in the third row and then verify that

$$|A| = a_{31} \, \text{COF}(a_{31}) + a_{32} \, \text{COF}(a_{32}) + a_{33} \, \text{COF}(a_{33})$$

19. For the 4×4 matrix A given in Example 5.4.4, compute the cofactor of each element in the third row and then verify that

$$|A| = a_{31} \, \text{COF}(a_{31}) + a_{32} \, \text{COF}(a_{32}) \\ + a_{33} \, \text{COF}(a_{33}) + a_{34} \, \text{COF}(a_{34})$$

20. For the 4×4 matrix A given in Example 5.4.4, compute the cofactor of each element in the fourth column and then verify that

$$|A| = a_{14} \, \text{COF}(a_{14}) + a_{24} \, \text{COF}(a_{24}) \\ + a_{34} \, \text{COF}(a_{34}) + a_{44} \, \text{COF}(a_{44})$$

21. Using Figure 17 as a model, construct a flowchart for the computation of a 5×5 determinant.

22. Using Figure 17 as a model, construct a flowchart for the computation of a 6×6 determinant.

23. In the computation of a 4×4 determinant, how many times is the algorithm for computing a 2×2 determinant called?

24. In the computation of a 5×5 determinant, how many times is the algorithm for computing a 2×2 determinant called?

25. The following algebraic properties are true in general:

1. $|AB| = |A|\,|B|$
2. $|AB| = |BA|$

Verify each of these equations for the matrices

$$A = \begin{bmatrix} -4 & 5 \\ 2 & 1 \end{bmatrix} \qquad B = \begin{bmatrix} 6 & 4 \\ -1 & 2 \end{bmatrix}$$

26. For square matrices A and B of the same dimension, is the following statement true?

$$|A + B| = |A| + |B|$$

27. If T is the transpose of square matrix A, then it is true that

$$|A| = |T|$$

Verify this property for the matrix

$$A = \begin{bmatrix} 2 & -1 & 5 \\ 3 & 0 & -4 \\ -6 & 2 & 1 \end{bmatrix}$$

(*Note:* The term ***transpose*** was defined in exercise 42 of Section 5.3.)

5.5

Systems of Linear Equations

Mathematical models of problems arising in management, engineering, and the physical sciences often entail *systems of linear equations*—that is, the ability to find numbers that *simultaneously* satisfy each of several linear equations. The topic of individual linear equations and their graphs is covered in beginning algebra, and so we shall assume familiarity with that material here.

Example 5.5.1

2 × 2 System

Consider a movie theater that charges \$4 per adult admission and \$2 per child admission. If one showing sold 390 tickets and grossed \$1330, then letting x_1 and x_2 be the numbers of adult and child tickets sold, respectively, we have

$$\begin{cases} x_1 + x_2 = 390 \\ 4x_1 + 2x_2 = 1330 \end{cases}$$

as a model for this particular showing. This is an example of two linear equations in the same two unknowns x_1 and x_2 or, more briefly, a *2 × 2 system* of linear equations.

Continuing with this system, let us denote the *matrix of coefficients* of the unknowns x_1 and x_2 by

$$A = \begin{bmatrix} 1 & 1 \\ 4 & 2 \end{bmatrix}$$

and the unknowns x_1 and x_2 and the constant terms 390 and 1330 by the column vectors (or 2×1 matrices)

$$X = \begin{bmatrix} x_1 \\ x_2 \end{bmatrix} \qquad C = \begin{bmatrix} 390 \\ 1330 \end{bmatrix}$$

respectively. Then using the definition of matrix multiplication, we have

$$AX = \begin{bmatrix} 1 & 1 \\ 4 & 2 \end{bmatrix}\begin{bmatrix} x_1 \\ x_2 \end{bmatrix} = \begin{bmatrix} x_1 + x_2 \\ 4x_1 + 2x_2 \end{bmatrix}$$

and using the definition of matrix equality, we get

$$\begin{bmatrix} x_1 + x_2 \\ 4x_1 + 2x_2 \end{bmatrix} = \begin{bmatrix} 390 \\ 1330 \end{bmatrix}$$

Thus, the 2×2 system of equations can be expressed as a single *matrix equation*

$$\begin{bmatrix} 1 & 1 \\ 4 & 2 \end{bmatrix} \begin{bmatrix} x_1 \\ x_2 \end{bmatrix} = \begin{bmatrix} 390 \\ 1330 \end{bmatrix}$$

that is,

$$AX = C$$

This is called the *matrix representation* of the given system.   ∎

Example 5.5.2

3×3 System

In a large manufacturing company, three particular products are handled by the assembly, testing, and shipping departments, in that order. The table below gives the number of labor hours expended on each unit by each department, along with the maximum number of labor-hours available in each department.

Product	Labor-Hours per Unit		
	Assembly	Testing	Shipping
I	5	2	2
II	3	1	1.5
III	4	1.5	1
Labor-hours available	295	110	100

Now if we let

x_1 = number of units of item I produced

x_2 = number of units of item II produced

x_3 = number of units of item III produced

then the total number of labor hours expended by the assembly department is

$$5x_1 + 3x_2 + 4x_3$$

Since the ideal would be to utilize *all* the labor hours available for assembly, we shall seek positive numbers x_1, x_2, and x_3 that satisfy

$$5x_1 + 3x_2 + 4x_3 = 295$$

Similar reasoning applied to the testing department yields the equation

$$2x_1 + x_2 + 1.5x_3 = 110$$

and for the shipping department

$$2x_1 + 1.5x_2 + x_3 = 100$$

So the collection of equations

$$\begin{cases} 5x_1 + 3x_2 + 4x_3 = 295 \\ 2x_1 + x_2 + 1.5x_3 = 110 \\ 2x_1 + 1.5x_2 + x_3 = 100 \end{cases}$$

is an example of three linear equations in the same three unknowns x_1, x_2, and x_3 or, briefly, a **3 × 3 system** of linear equations.

As with the 2×2 system of the preceding example, this 3×3 system has a **matrix representation.** In this case, the **matrix of coefficients** is

$$A = \begin{bmatrix} 5 & 3 & 4 \\ 2 & 1 & 1.5 \\ 2 & 1.5 & 1 \end{bmatrix}$$

and we let

$$X = \begin{bmatrix} x_1 \\ x_2 \\ x_3 \end{bmatrix} \qquad C = \begin{bmatrix} 295 \\ 110 \\ 100 \end{bmatrix}$$

Then the above 3×3 system can be represented by the **matrix equation**

$$\begin{bmatrix} 5 & 3 & 4 \\ 2 & 1 & 1.5 \\ 2 & 1.5 & 1 \end{bmatrix} \begin{bmatrix} x_1 \\ x_2 \\ x_3 \end{bmatrix} = \begin{bmatrix} 295 \\ 110 \\ 100 \end{bmatrix}$$

that is,

$$AX = C \qquad \blacksquare$$

Of course, with systems of equations in hand, we want to be able to find numbers that solve them.

Definitions

Let n be a positive integer, $n \geq 2$.

1. A collection of n linear equations each with unknowns

$$x_1, x_2, \ldots, x_n$$

is called an **n × n system of linear equations.**

2. A **solution to a linear equation** with unknowns

$$x_1, x_2, \ldots, x_n$$

is a vector of length n

$$(s_1, s_2, \ldots, s_n)$$

such that the linear equation is satisfied when

$$x_1 = s_1, x_2 = s_2, \ldots, x_n = s_n$$

3. A **solution to an n × n system** of linear equations is a vector of length n that is a solution to *each* of the linear equations in the system.

As suggested in the work for the two examples above, any $n \times n$ system of linear equations has a **matrix representation**

$$AX = C$$

where A is the **matrix of coefficients**, X is the column vector of the n-unknowns, and C is the column vector of constant terms taken from the right sides of the equations.

It is possible, and sometimes necessary, to deal with systems of equations in which the number of unknowns and the number of equations are not equal. This situation is generally studied in a course on linear algebra. In this text, we shall restrict ourselves to the "square" systems defined above, where the number of equations and the number of unknowns are equal.

Example 5.5.3

Solution to a System

a. Relative to the 2×2 system derived in Example 5.5.1, the vector $(275, 115)$ is a solution because when

$$x_1 = 275 \qquad x_2 = 115$$

then

$$\begin{cases} x_1 + x_2 = 275 + 115 = 390 \\ 4x_1 + 2x_2 = 1100 + 230 = 1330 \end{cases}$$

However, the vector $(250, 140)$ is not a solution to the system, since

$$4x_1 + 2x_2 = 1000 + 280 = 1280 \neq 1330$$

even though this vector is a solution to the equation

$$x_1 + x_2 = 390$$

b. In reference to the 3×3 system given in Example 5.5.2, the vector $(15, 20, 40)$ is a solution because when

$$x_1 = 15 \qquad x_2 = 20 \qquad x_3 = 40$$

we have

$$\begin{cases} 5x_1 + 3x_2 + 4x_3 = 75 + 60 + 160 = 295 \\ 2x_1 + x_2 + 1.5x_3 = 30 + 20 + 60 = 110 \\ 2x_1 + 1.5x_2 + x_3 = 30 + 30 + 40 = 100 \end{cases}$$

But the vector $(10, 24, 44)$ is not a solution to this system since

$$5x_1 + 3x_2 + 4x_3 = 50 + 72 + 176 = 298 \neq 295$$

even though this vector is a solution to each of the other two equations:

$$2x_1 + x_2 + 1.5x_3 = 20 + 24 + 66 = 110$$
$$2x_1 + 1.5x_2 + x_3 = 20 + 36 + 44 = 100 \qquad \blacksquare$$

In the next three examples, we shall illustrate different techniques for solving systems of linear equations as well as three types of possible outcomes. To simplify the notation, we shall use x and y for the unknowns instead of x_1 and x_2 when there are two variables and x, y, and z instead of x_1, x_2, and x_3 when there are three.

Example 5.5.4   **2 × 2 System: Unique Solution**

Again consider the 2×2 system presented at the beginning of this section:

$$E_1: \quad x + y = 390$$
$$E_2: \quad 4x + 2y = 1330$$

Notice that the coefficients 4 and 2 and the constant term 1330 are all even, so we can simplify E_2 as follows:

$$E_2: \quad 2x + y = 665$$

There are three methods we can use to determine the solution to the system E_1 and E_2.

Graphing: The graph of each of the equations E_1 and E_2 is a straight line, as shown in Figure 19. If we do this plotting very carefully on graph paper, we should be able to read off the coordinates (275, 115) of the point of intersection of these lines. Since this is the *only point* on both lines,

$$x = 275 \qquad y = 115$$

is the unique solution to the system.

Obviously, the accuracy of the result here depends on the accuracy with which we have graphed the given lines. Fortunately, there are other ways of obtaining an exact answer here without depending on the reading (estimating) of coordinates from a graph.

Substitution: First, solve either one of the two equations for either one of the two variables.

$$E_1: \quad x = 390 - y$$

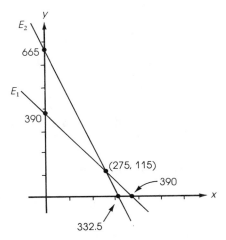

Figure 19

Second, substitute this expression into the *other* equation and solve for the one remaining variable.

$$E_2: \qquad 2x + y = 665$$
$$2(390 - y) + y = 665$$
$$780 - 2y + y = 665$$
$$-y = -115$$
$$y = 115$$

Third, substitute this numeric value into either of the given equations and solve for the numeric value of the other variable.

$$E_1: \qquad x + y = 390$$
$$x + 115 = 390$$
$$x = 275$$

So the unique solution is

$$x = 275 \qquad y = 115$$

Elimination: Is it possible to add or subtract respective sides of the given equations and eliminate either one of the variables? Sometimes it may be necessary to multiply one or both of the equations by appropriately chosen constants in order to make this possible. In this case, no such multiplication is necessary.

$$E_1: \qquad x + y = 390$$
$$E_2: \qquad 2x + y = 665$$
$$\overline{E_1 - E_2: \quad -x + 0 = -275}$$
$$x = 275$$

Now substitute this value into either E_1 or E_2 and solve for the other variable.

$$E_1: \qquad x + y = 390$$
$$275 + y = 390$$
$$y = 115$$

Again, the unique answer is

$$x = 275 \qquad y = 115$$

So using any one of the three methods above, we can conclude that 275 adults and 115 children attended the movie (see Example 5.5.1). ∎

Example 5.5.5   **2 × 2 System: No Solution**

Consider the system

$$E_1: \qquad 4x - 2y = 12$$
$$E_2: \quad -2x + y = 2$$

Again, we can simplify one of these equations.

$$E_1: \qquad 2x - y = 6$$

Graphing: In Figure 20 we have drawn the graphs of these two equations on the same coordinate axes. These lines are parallel since they have the same slope (namely, 2) but different y-intercepts. This means there is no point (x, y) common to the two graphs, so no single point (x, y) is a solution to each equation. Thus, the given system has *no solution*.

Substitution: First, solve either equation for either variable.

$$E_2: \quad y = 2 + 2x$$

Second, substitute this into the other equation and solve, if possible.

$$E_1: \qquad 2x - y = 6$$
$$2x - (2 + 2x) = 6$$
$$2x - 2 - 2x = 6$$
$$-2 = 6 \qquad \text{False!}$$

When all variables disappear and the resulting numeric statement is false, the given system has *no solution*.

Elimination: Is it possible to eliminate a variable by adding or subtracting the given equations? Yes.

$$\begin{array}{ll} E_1: & 2x - y = 6 \\ E_2: & -2x + y = 2 \\ \hline E_1 + E_2: & \qquad 0 = 8 \qquad \text{False!} \end{array}$$

Since *both* variables have canceled and the resulting equation is false, the system has *no solution*. ∎

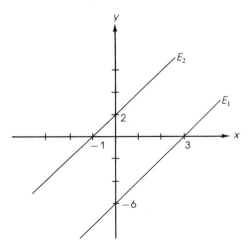

Figure 20

Example 5.5.6   **2 × 2 System: Infinitely Many Solutions**

Given the system

$$E_1:\quad 4x - 2y = 12$$
$$E_2:\quad -2x + y = -6$$

we can again simplify one of the equations.

$$E_1:\quad 2x - y = 6$$

Graphing: When we go to graph these equations on the same coordinate axes, we see that the graphs coincide (see Figure 21). Each is a line of slope 2 and y-intercept -6. Therefore, any of the infinitely many points on this line is a solution to the given system. For instance,

$$x = 0 \qquad y = -6$$
$$x = 1 \qquad y = -4$$
$$\text{and}\quad x = 5 \qquad y = 4$$

are three of the *infinitely many solutions.*

Substitution: First, solve one of the equations for one of the variables.

$$E_1:\quad y = 2x - 6$$

Second, substitute this into the other equation and solve, if possible.

$$E_2:\qquad -2x + y = -6$$
$$-2x + (2x - 6) = -6$$
$$-2x + 2x - 6 = -6$$
$$-6 = -6 \qquad \text{True!}$$

When all variables disappear and the resulting numeric statement is true, the system has *infinitely many solutions.*

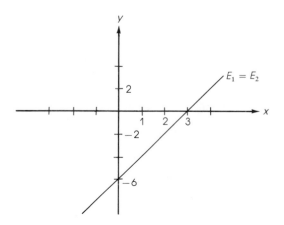

Figure 21

Elimination: Again, is it possible to eliminate a variable by adding or subtracting the equations? Yes.

$$
\begin{array}{rl}
E_1: & 2x - y = 6 \\
E_2: & -2x + y = -6 \\
\hline
& 0 = 0 \qquad \text{True!}
\end{array}
$$

Since *both* variables have canceled and the resulting equation is true, there are *infinitely many solutions* to the system.

Finally, our work here with the substitution method gives us an easy way of generating as many solutions as we'd like. Since

$$y = 2x - 6$$

then for any particular number t, the vector

$$(t, 2t - 6)$$

is one particular solution to the system. For instance, to specify four of the infinitely many solutions, when

$$
\begin{array}{lll}
t = 0 & \text{then } (0, -6) & \text{is a solution;} \\
t = 4 & \text{then } (4, 2) & \text{is a solution;} \\
t = -1 & \text{then } (-1, -8) & \text{is a solution; and} \\
t = -5.2 & \text{then } (-5.2, -16.4) & \text{is a solution.} \qquad \blacksquare
\end{array}
$$

The three possible outcomes illustrated in the preceding examples (namely, one solution, no solution, or infinitely many solutions) also apply to $n \times n$ linear systems where $n > 2$. In this context, the notion of determinant presented in Section 5.4 can be helpful.

Fact

> Let $AX = C$ be the matrix representation of an $n \times n$ system of linear equations. Then
>
> $$\left.\begin{array}{l} \text{This system has} \\ \text{a unique solution} \end{array}\right\} \quad \leftrightarrow \quad |A| \neq 0$$

Remember that the symbol $\leftrightarrow$ is read "if and only if" and means "is exactly the same as," and A is the matrix of coefficients of the given system.

Typically, you use this fact *before* starting to solve a given system in order to determine whether the system has a unique solution. For instance, at the beginning of Example 5.5.4, we could have computed the determinant of the matrix of coefficients:

$$|A| = \begin{vmatrix} 1 & 1 \\ 4 & 2 \end{vmatrix} = 2 - 4 = -2 \neq 0$$

The fact assures us that the system does have a unique solution, which we could then proceed to find by any of the methods described. In Examples

5.5.5 and 5.5.6, the matrix A is the same, and

$$|A| = \begin{vmatrix} 4 & -2 \\ -2 & 1 \end{vmatrix} = 4 - 4 = 0$$

The fact tells us that these systems *do not* have unique solutions, and we can proceed with our work knowing that each of these systems has either no solution or infinitely many.

As indicated previously, the graphing method is difficult to apply to two-variable systems. It becomes even more difficult for three-variable systems and impossible when four or more variables are involved. The technique presented in the remaining examples of this section combines the substitution and elimination methods. While we shall demonstrate this technique on 3×3 systems, it can also be applied to systems with four or more variables. It is a preliminary form of the algorithmic procedure to be presented in the next section.

Example 5.5.7   **3×3 System: Unique Solution**

Let's consider the system derived in Example 5.5.2:

$$5x + 3y + 4z = 295$$
$$2x + y + 1.5z = 110$$
$$2x + 1.5y + z = 100$$

First, we compute the determinant of the matrix of coefficients using cofactors, as discussed in Section 5.4,

$$\begin{vmatrix} 5 & 3 & 4 \\ 2 & 1 & 1.5 \\ 2 & 1.5 & 1 \end{vmatrix} = 5(-1)^{1+1}\begin{vmatrix} 1 & 1.5 \\ 1.5 & 1 \end{vmatrix} + 2(-1)^{2+1}\begin{vmatrix} 3 & 4 \\ 1.5 & 1 \end{vmatrix} + 2(-1)^{3+1}\begin{vmatrix} 3 & 4 \\ 1 & 1.5 \end{vmatrix}$$

$$= \quad 5(1 - 2.25) \quad + \quad (-2)(3 - 6) \quad + \quad 2(4.5 - 4)$$
$$= \quad\quad -6.25 \quad\quad + \quad\quad 6 \quad\quad + \quad\quad 1$$
$$= .75$$
$$\neq 0$$

Since this determinant does not equal 0, we know that the system has a unique solution.

To find this solution, we use a method that involves ***downward elimination*** and ***upward substitution***. In the elimination stage, you work with the main diagonal coefficients one at a time, from top left to lower right. For each of these, first make that coefficient equal to 1, and then use it to make all coefficients directly below equal to 0. A symbol like

$$(E_2 - 5E_1) \rightarrow E_2$$

means that *both sides* of the equation E_1 are multiplied by 5, and then *both sides* of this equation are subtracted from the respective sides of E_2; the resulting equation is designated as the new E_2.

$$E_1: 5x + 3y + 4z = 295$$
$$E_2: 2x + y + 1.5z = 110$$
$$E_3: 2x + 1.5y + z = 100$$

$$E_1/5 \rightarrow E_1: \quad x + .6y + .8z = 59$$
$$E_2: 2x + y + 1.5z = 110$$
$$E_3: 2x + 1.5y + z = 100$$

$$E_1: \quad x + .6y + .8z = 59$$
$$(E_2 - 2E_1) \rightarrow E_2: \quad -.2y - .1z = -8$$
$$E_3: 2x + 1.5y + z = 100$$

$$E_1: \quad x + .6y + .8z = 59$$
$$E_2: \quad -.2y - .1z = -8$$
$$(E_3 - 2E_1) \rightarrow E_3: \quad .3y - .6z = -18$$

$$E_1: \quad x + .6y + .8z = 59$$
$$E_2/(-.2) \rightarrow E_2: \quad y + .5z = 40$$
$$E_3: \quad .3y - .6z = -18$$

$$E_1: \quad x + .6y + .8z = 59$$
$$E_2: \quad y + .5z = 40$$
$$(E_3 - .3E_2) \rightarrow E_3: \quad -.75z = -30$$

$$E_1: \quad x + .6y + .8z = 59$$
$$E_2: \quad y + .5z = 40$$
$$E_3/(-.75) \rightarrow E_3: \quad z = 40$$

This completes the elimination stage. In the substitution stage, you work upward from the bottom to determine the numeric values of the three unknowns. From E_3 we have $z = 40$; substituting this into E_2 gives

$$y + .5z = 40$$
$$y + .5(40) = 40$$
$$y + 20 = 40$$
$$y = 20$$

and substituting these two values into E_1 yields

$$x + .6y + .8z = 59$$
$$x + .6(20) + .8(40) = 59$$
$$x + 12 + 32 = 59$$
$$x = 15$$

Recalling the context in which this system was derived in Example 5.5.2, we can conclude that in order to fully utilize the labor-hours available, the company should produce 15 units of item I, 20 units of item II, and 40 units of item III. ∎

Two comments should be made concerning the preceding example. First, the algebraic steps taken are by no means the only valid ones. In fact, you could judiciously alter the steps and get the same answer using simpler computations. The idea here is to present a general technique that can be applied to any system. Second, in the case when a diagonal coefficient *is* zero and thus cannot be made 1 by division, exchange the entire equation with one below so that the new diagonal coefficient is *not* zero. This exchange is always possible when the given system has a unique solution.

Example 5.5.8

3 × 3 System: No Solution

Given the system

$$x - 3y + 2 = -17$$
$$3x + y + 5z = -35$$
$$-2x + 6y - 4z = 30$$

we first compute the determinant of the matrix of coefficients:

$$\begin{vmatrix} 1 & -3 & 2 \\ 3 & 1 & 5 \\ -2 & 6 & -4 \end{vmatrix} = 1(-1)^{1+1}\begin{vmatrix} 1 & 5 \\ 6 & -4 \end{vmatrix} + 3(-1)^{2+1}\begin{vmatrix} -3 & 2 \\ 6 & -4 \end{vmatrix} + (-2)(-1)^{3+1}\begin{vmatrix} -3 & 2 \\ 1 & 5 \end{vmatrix}$$

$$\begin{array}{cccccc} = & 1(-4 - 30) & + & (-3)(12 - 12) & + & (-2)(-15 - 2) \\ = & -34 & + & 0 & + & 34 \end{array}$$

$$= 0$$

Thus we know that this system has either no solution or infinitely many. Let's try to apply the downward elimination–upward substitution method described above.

$$\begin{array}{lrl} E_1: & x - 3y + 2z = -17 \\ E_2: & 3x + y + 5z = -35 \\ E_3: & -2x + 6y - 4z = 30 \end{array} \qquad \begin{array}{lrl} E_1: & x - 3y + 2z = -17 \\ (E_2 - 3E_1) \to E_2: & 10y - z = 16 \\ E_3: & -2x + 6y - 4z = 30 \end{array}$$

$$\begin{array}{lrl} E_1: & x - 3y + 2z = -17 \\ E_2: & 10y - z = 16 \\ (E_3 + 2E_1) \to E_3: & 0 = -4 \quad \text{False!} \end{array}$$

When a false statement arises in the elimination stage, the system has *no solution*. ∎

Example 5.5.9

3 × 3 System: Infinitely Many Solutions

Consider the system

$$x - 3y + 2z = -17$$
$$3x + y + 5z = -35$$
$$-2x + 6y - 4z = 34$$

The matrix of coefficients here is the same as in the preceding example, so its determinant equals 0 and the system has either no solution or infinitely many.

$$\begin{array}{lrl} E_1: & x - 3y + 2z = -17 \\ E_2: & 3x + y + 5z = -35 \\ E_3: & -2x + 6y - 4z = 34 \end{array} \qquad \begin{array}{lrl} E_1: & x - 3y + 2z = -17 \\ (E_2 - 3E_1) \to E_2: & 10y - z = 16 \\ E_3: & -2x + 6y - 4z = 34 \end{array}$$

$$E_1: \quad x - 3y + 2z = -17 \qquad\qquad E_1: \quad x - 3y + 2z = -17$$
$$E_2: \quad\quad 10y - z = 16 \qquad E_2/10 \to E_2: \quad\quad y - .1z = 1.6$$
$$(E_3 + 2E_1) \to E_3: \quad\quad\quad 0 = 0 \qquad E_3: \quad\quad\quad 0 = 0$$

This completes the downward elimination; now for the upward substitution. From E_2,

$$y = 1.6 + .1z$$

Substituting this into E_1 gives

$$
\begin{aligned}
x - 3y + 2z &= -17 \\
x - 3(1.6 + .1z) + 2z &= -17 \\
x - 4.8 - .3z + 2z &= -17 \\
x + 1.7z &= -12.2 \\
x &= -1.7z - 12.2
\end{aligned}
$$

Now let t be any number and let $z = t$. Then the vector

$$(-1.7t - 12.2, \; 1.6 + .1t, \; t)$$

is a solution of the given system. For instance, when $t = 0$

$$(-12.2, \; 1.6, \; 0)$$

is one particular solution, and when $t = 10$

$$(-29.2, \; 2.6, \; 10)$$

is another.   ■

In theory we could proceed with the downward elimination–upward substitution method to handle systems of larger dimension—that is, $n \times n$ systems where $n > 3$. But the notation becomes simpler if we use matrices, as we shall in the next section.

EXERCISES **5.5**

In each of the first ten exercises:

a. *Write the matrix representation of the given system of equations.*
b. *Without solving, determine if the system has a unique solution.*
c. *Solve the system by graphing.*
d. *Solve the system by substitution.*
e. *Solve the system by elimination.*

1. $2x - y = 0$
      $x + 3y = 7$

2. $2x - y = 4$
      $-12x + 6y = -24$

3. $-x + 4y = 1$
      $3x - 12y = 6$

4. $-6x + y = 20$
      $x - .5y = -4$

5. $3x - 2y = 11$
      $5x + y = 14$

6. $x + y = -9$
      $2x - 4y = 6$

7. $-x + 4y = 1$
      $3x - 12y = -3$

8. $x - y = 25$
      $2x + 3y = 475$

9. $x - y = 80$
      $2x + y = 610$

10. $2x - y = 4$
       $x - .5y = 1$

In each of the next six exercises:

a. *Write the matrix representation of the given system of equations.*
b. *Without solving, determine if the system has a unique solution.*

c. *Solve the system using downward elimination and upward substitution.*

11.
$$
\begin{aligned}
2x - 3y + z &= 2 \\
x - y + 4z &= 1 \\
-3x + 5y + 2z &= 3
\end{aligned}
$$

12.
$$
\begin{aligned}
2x - 3y + z &= 2 \\
x - y + 4z &= 1 \\
-3x + 5y + 2z &= -3
\end{aligned}
$$

13.
$$
\begin{aligned}
x + 4y - 2z &= -4 \\
-3x \phantom{{}+ 4y} + z &= -7 \\
-x + 2y + 5z &= -11
\end{aligned}
$$

14.
$$
\begin{aligned}
-3x \phantom{{}+ 4y} + 2z &= 1 \\
x + 4y + 5z &= -1 \\
-x + 8y + 12z &= 1
\end{aligned}
$$

15.
$$
\begin{aligned}
-3x \phantom{{}+ 4y} + 2z &= 1 \\
x + 4y + 5z &= -1 \\
-x + 8y + 12z &= -1
\end{aligned}
$$

16.
$$
\begin{aligned}
5x - y + 2z &= 1 \\
-x + 2y - 2z &= 3 \\
2x + 3y + 4z &= -6
\end{aligned}
$$

17. If possible, find a number t so that the system

$$
\begin{cases}
6x - y = 2 \\
3x + ty = 1
\end{cases}
$$

has (a) a unique solution, (b) no solution, and (c) infinitely many solutions.

18. If possible, find a number t so that the system

$$
\begin{cases}
x + 6y \phantom{{}- z} = 3 \\
2y - z = 2 \\
-x \phantom{{}+ 2y} + tz = 1
\end{cases}
$$

has (a) a unique solution, (b) no solution, and (c) infinitely many solutions.

19. A person has a pocketful of dimes and quarters totaling $2.30. If there are 14 coins altogether, how many dimes and how many quarters are there?

20. A person has two investments earning annual interest rates of 8% and 10%. After one year, the interest earned is $230 and the total investment value is $2730. What were the amounts of the two initial investments?

21. A certain blend of snack food contains peanuts, raisins, and cashews in such proportions that the weight of the peanuts equals the combined weight of the raisins and cashews. A certain batch of this snack weighed 500 pounds, and the cost of the three components totaled $362.50, where peanuts, raisins, and cashews cost $.50, $.75, and $1.25 per pound, respectively. Determine the weight of each component in this batch.

22. A person has a pocketful of nickels, dimes, and quarters totaling $1.75. If there are 13 coins altogether and twice as many dimes as nickels, how many of each coin are there?

23. The sum of two numbers is 72 and the difference is 12. Find the numbers.

24. The length of a rectangle is double its width, and the rectangle's perimeter is 54 inches. Find the dimensions of the rectangle.

5.6

Gaussian Elimination

In working through the computations in the last three examples, you probably noticed that the variables x, y, and z and the equal sign served no role other than to keep the coefficients and constant terms in their correct positions. Consequently, we can eliminate the need to rewrite these symbols by keeping the coefficients of a particular variable vertically aligned in the same column and placing the constant terms adjacent to these in a column on the right side. The result is a matrix of coefficients augmented on the right by a column of constant terms, or the ***augmented matrix of coefficients***.

For Examples 5.5.7, 5.5.8, 5.5.9, the augmented matrices of coefficients would be written as

$$\begin{bmatrix} 5 & 3 & 4 & \vdots & 295 \\ 2 & 1 & 1.5 & \vdots & 110 \\ 2 & 1.5 & 1 & \vdots & 100 \end{bmatrix} \quad \begin{bmatrix} 1 & -3 & 2 & \vdots & -17 \\ 3 & 1 & 5 & \vdots & -35 \\ -2 & 6 & -4 & \vdots & 30 \end{bmatrix} \quad \begin{bmatrix} 1 & -3 & 2 & \vdots & -17 \\ 3 & 1 & 5 & \vdots & -35 \\ -2 & 6 & -4 & \vdots & 34 \end{bmatrix}$$

respectively. Ordinarily we separate the matrix of coefficients from the column of constant terms with a dotted vertical line, as done here. Remember that the unwritten column headings are x, y, z, and constant, in that order. The same algebraic maneuvers performed on the equations before can now be performed on the rows of the augmented matrix, but the clutter of symbols is greatly reduced.

Elementary Row Operations

1. Any two rows can be exchanged.
2. A row can be multiplied by or divided by a nonzero constant.
3. Nonzero multiples of one row can be added to or subtracted from another row.

Similar to our work with equations, we shall use a symbol like

$$R_3 - 2R_2 \rightarrow R_3$$

to mean that row R_2 is multiplied by 2 and subtracted from row R_3 to obtain a new R_3. Applying these elementary row operations produces matrices whose associated systems of equations have exactly the same solution(s), if any, as the original system.

With this augmented matrix notation, we can now introduce a technique that is related to the downward elimination–upward substitution method already discussed. This new method is aptly called the downward elimination–upward elimination process, or **Gaussian elimination** (named after Karl Friedrich Gauss [1777–1855], generally considered to be one of the greatest mathematicians of all time). First, downward elimination is performed on the augmented matrix so as to produce all 1's on the main diagonal of the coefficient portion and all 0's below.

$$\begin{bmatrix} 1 & & & & \vdots & b_1 \\ & 1 & \text{Numbers} & & \vdots & b_2 \\ & & 1 & & \vdots & \\ \text{All 0's} & & & \ddots & \vdots & \vdots \\ & & & 1 & \vdots & b_n \end{bmatrix}$$

Then upward elimination is performed by using the diagonal elements so as to produce all 0's above the diagonal.

$$\begin{bmatrix} 1 & & & & & & d_1 \\ & 1 & & \text{All 0's} & & & d_2 \\ & & 1 & & & & \\ & \text{All 0's} & & \ddots & & & \vdots \\ & & & & 1 & & d_n \end{bmatrix}$$

If these two steps are successfully accomplished, the coefficient portion of the augmented matrix is transformed into the identity matrix. In this case, the numbers $d_1, d_2, \ldots, d_n$ in the right column provide the unique solution for the given system. The next example illustrates the process.

Example 5.6.1 **3 × 3 System: Unique Solution**

Consider the system

$$2x + y - 3z = 7$$
$$4x - 2y + z = 0$$
$$x - 8y - z = 2.5$$

Then

$$\begin{vmatrix} 2 & 1 & -3 \\ 4 & -2 & 1 \\ 1 & -8 & -1 \end{vmatrix} = 2(-1)^{1+1}\begin{vmatrix} -2 & 1 \\ -8 & -1 \end{vmatrix} + 4(-1)^{2+1}\begin{vmatrix} 1 & -3 \\ -8 & -1 \end{vmatrix} + 1(-1)^{3+1}\begin{vmatrix} 1 & -3 \\ -2 & 1 \end{vmatrix}$$

$$= \quad 2(2+8) \quad + \quad (-4)(-1-24) \quad + \quad (1)(1-6)$$
$$= \quad\quad 20 \quad\quad + \quad\quad 100 \quad\quad - \quad\quad 5$$
$$= 115 \neq 0$$

so the given system has a unique solution. Let's find it by using Gaussian elimination.

$$\begin{bmatrix} 2 & 1 & -3 & | & 7 \\ 4 & -2 & 1 & | & 0 \\ 1 & -8 & -1 & | & 2.5 \end{bmatrix} \begin{array}{c} R_1/2 \to R_1 \end{array} \begin{bmatrix} 1 & .5 & -1.5 & | & 3.5 \\ 4 & -2 & 1 & | & 0 \\ 1 & -8 & -1 & | & 2.5 \end{bmatrix} (R_2 - 4R_1) \to R_2 \begin{bmatrix} 1 & .5 & -1.5 & | & 3.5 \\ 0 & -4 & 7 & | & -14 \\ 1 & -8 & -1 & | & 2.5 \end{bmatrix}$$

$$(R_3 - R_1) \to R_3 \begin{bmatrix} 1 & .5 & -1.5 & | & 3.5 \\ 0 & -4 & 7 & | & -14 \\ 0 & -8.5 & .5 & | & -1 \end{bmatrix} \begin{array}{c} R_2/(-4) \to R_2 \end{array} \begin{bmatrix} 1 & .5 & -1.5 & | & 3.5 \\ 0 & 1 & -1.75 & | & 3.5 \\ 0 & -8.5 & .5 & | & -1 \end{bmatrix}$$

$$(R_3 + 8.5R_2) \to R_3 \begin{bmatrix} 1 & .5 & -1.5 & | & 3.5 \\ 0 & 1 & -1.75 & | & 3.5 \\ 0 & 0 & -14.375 & | & 28.75 \end{bmatrix} \begin{array}{c} R_3/(-14.375) \to R_3 \end{array} \begin{bmatrix} 1 & .5 & -1.5 & | & 3.5 \\ 0 & 1 & -1.75 & | & 3.5 \\ 0 & 0 & 1 & | & -2 \end{bmatrix}$$

This completes the downward elimination; the upward elimination proceeds bottom to top along the 1's on the main diagonal so as to make 0 all entries *above* the diagonal.

$$(R_2 + 1.75R_3) \to R_2 \begin{bmatrix} 1 & .5 & -1.5 & \vdots & 3.5 \\ 0 & 1 & 0 & \vdots & 0 \\ 0 & 0 & 1 & \vdots & -2 \end{bmatrix} \quad (R_1 + 1.5R_3) \to R_1 \begin{bmatrix} 1 & .5 & 0 & \vdots & .5 \\ 0 & 1 & 0 & \vdots & 0 \\ 0 & 0 & 1 & \vdots & -2 \end{bmatrix}$$

$$(R_1 - .5R_2) \to R_1 \begin{bmatrix} 1 & 0 & 0 & \vdots & .5 \\ 0 & 1 & 0 & \vdots & 0 \\ 0 & 0 & 1 & \vdots & -2 \end{bmatrix}$$
$$\underbrace{\phantom{\begin{matrix}1&0&0\end{matrix}}}_{I_3}$$

The upward elimination is now finished, because we have obtained the identity matrix I_3 in the coefficient portion of the matrix. Recalling the unwritten column headings, this last matrix represents the system of equations

$$x = .5 \qquad y = 0 \qquad z = -2$$

which is the unique solution of the given system.   ∎

The technique of the above example yields the unique solution to the system

$$AX = C$$

whenever $\det A \neq 0$. A summary of the major steps is given in Figure 22. Let's now describe the specific steps in the two elimination phases.

In downward elimination, we start with a particular diagonal element a_{ii} and divide each element of row R_i by a_{ii}, calling the resulting row R_i.

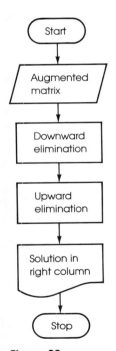

$$\begin{bmatrix} 1 & & & & \\ 0 & 1 & & \text{Numbers} & \\ 0 & 0 & \ddots & & \\ \vdots & \vdots & & a_{ii} & \\ 0 & 0 & & & \ddots \end{bmatrix} \quad R_i/a_{ii} \to R_i \begin{bmatrix} 1 & & & & \\ 0 & 1 & & \text{Numbers} & \\ 0 & 0 & \ddots & & \\ \vdots & \vdots & & 1_{ii} & \\ 0 & 0 & & & \ddots \end{bmatrix}$$

This produces a 1 in the (i, i) position of the new matrix, denoted by 1_{ii}. The flowchart symbol for this step is

$$\boxed{R_i \leftarrow R_i/a_{ii}}$$

Two comments here. First, if $a_{ii} = 0$, thus making division by a_{ii} undefined, exchange the entire row R_i with one below it so that the new a_{ii} entry is *not* zero. Then divide the new row R_i by the new, nonzero a_{ii}. This exchange is always possible if the given system has a unique solution.

The second comment is an important point on notation. As always, the symbol a_{ij} denotes the matrix entry in the ith row and jth column. After performing a row operation, in the new matrix we shall still denote the (i, j) entry by a_{ij} even though its value may have changed. For instance, in the first matrix written in Example 5.6.1,

$$a_{11} = 2 \qquad a_{12} = 1 \qquad a_{13} = -3 \qquad a_{14} = 7$$

Figure 22

whereas in the second matrix, after the first row operation has been performed,

$$a_{11} = 1 \qquad a_{12} = .5 \qquad a_{13} = -1.5 \qquad a_{14} = 3.5$$

Continuing with downward elimination, we now use the entry 1_{ii} to make each entry below it equal to 0. That is, for $i+1 \le j \le n$,

$$\begin{bmatrix} 1 & & & \\ 0 & 1 & & \text{Numbers} \\ 0 & 0 & \ddots & \\ & & & 1_{ii} \\ \vdots & \vdots & & \textcircled{a_{ji}} & \ddots \\ & & & \vdots \\ 0 & 0 & & a_{ni} \end{bmatrix} \quad (R_j - a_{ji}R_i) \to R_j \quad \begin{bmatrix} 1 & & & \\ 0 & 1 & & \text{Numbers} \\ 0 & 0 & \ddots & \\ & & & 1_{ii} \\ \vdots & \vdots & & \textcircled{0_{j1}} & \ddots \\ & & & \vdots \\ 0 & 0 & & a_{ni} \end{bmatrix}$$

The result here is that the entry a_{ji} is made 0. In flowchart notation

$$\boxed{R_j \leftarrow (R_j - a_{ji}R_i)}$$

So the loop that will make *each* entry *below* 1_{ii} equal to 0 is given in Figure 23. To complete the downward elimination, this loop must be placed in a larger one, as drawn in Figure 24.

Once the downward phase has been completed, the coefficient portion of the matrix has all 1's on the main diagonal and all 0's below. In the upward elimination phase, you move up the main diagonal starting at the bottom entry 1_{nn} and make into 0 each entry above. That is, for $i-1 \ge j \ge 1$,

$$\begin{bmatrix} 1 & & & a_{1i} & \\ & 1 & & \vdots & \\ & & \ddots & \textcircled{a_{ji}} & \\ & & & 1_{ii} & \\ & \text{All 0's} & & & \ddots \\ & & & & & 1 \end{bmatrix} \quad (R_j - a_{ji}R_i) \to R_j \quad \begin{bmatrix} 1 & & & a_{1i} & \\ & 1 & & \vdots & \\ & & \ddots & \textcircled{0_{ji}} & \\ & & & 1_{ii} & \\ & \text{All 0's} & & & \ddots \\ & & & & & 1 \end{bmatrix}$$

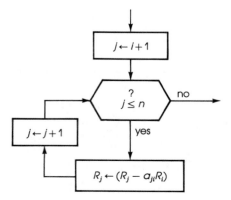

Figure 23

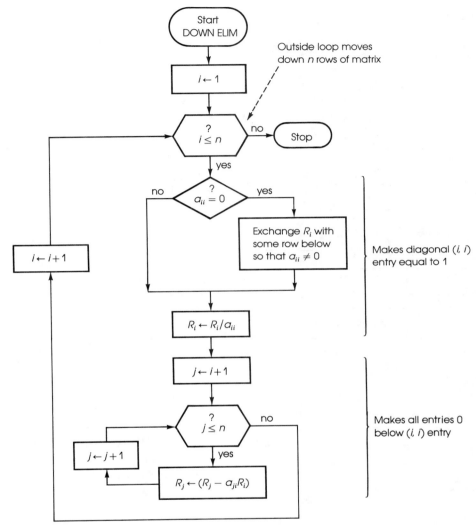

Figure 24

Here the result is that entry a_{ji} is made 0. In flowchart notation,

$$R_j \leftarrow (R_j - a_{ji}R_i)$$

The loop that will make equal to 0 each entry *above* 1_{ii} is drawn in Figure 25. To complete the upward elimination, this loop must be placed inside a larger one, as shown in Figure 26.

Recall that at the end of Section 5.4 we constructed an algorithm $N \times N$ DET to evaluate any $n \times n$ determinant. We can combine this algorithm with

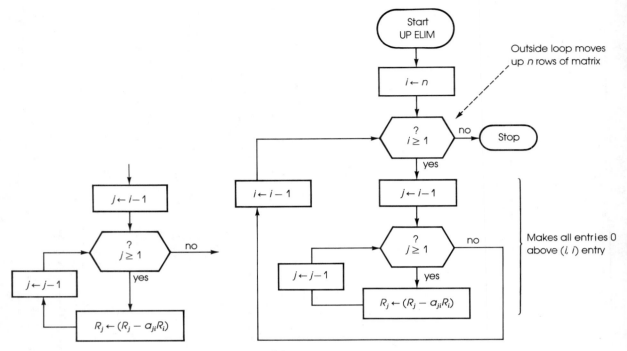

Figure 25

Figure 26

the two just presented, DOWN ELIM and UP ELIM, to form a flowchart that will process any $(n \times n)$ system of linear equations. If the system has a unique solution, the algorithm will find it by Gaussian elimination; otherwise, it merely indicates that the system has no solution or infinitely many. This flowchart is given in Figure 27. In it we use

$$A = [a_{ij}]$$

for the matrix of coefficients, and

$$A_{\text{aug}} = [a_{ij} \mid c_i]$$

for the augmented matrix of coefficients that is the input to DOWN ELIM. The matrix produced as output from DOWN ELIM is taken as input by UP ELIM, and the output of UP ELIM has constants

$$d_1, d_2, \ldots, d_n$$

in the right column. This process for evolving the flowchart of Figure 27 is another example of *modular algorithm development*.

We shall conclude this section with illustrations of how the Gaussian elimination method can be adapted to handle systems that do *not* have a unique solution.

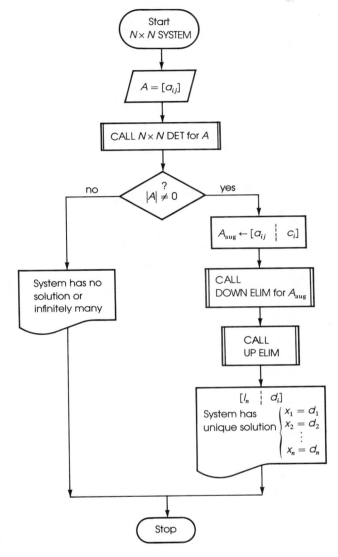

Figure 27

Example 5.6.2   4 × 4 System: Infinitely Many Solutions

Let us solve the system

$$\begin{cases} x - 5y + 2z - v = 0 \\ 7x - y + v = -20 \\ 3x + 2y - z + v = -10 \\ -2x - 7y + 3z - 2v = 10 \end{cases}$$

The first thing we should do is compute the determinant of the matrix of coefficients; this turns out to equal 0 (see exercise 15), so we know that the system has either no solution or many. Nevertheless, we begin downward elimination as usual.

$$
\begin{bmatrix}
1 & -5 & 2 & -1 & \vdots & 0 \\
7 & -1 & 0 & 1 & \vdots & -20 \\
3 & 2 & -1 & 1 & \vdots & -10 \\
-2 & -7 & 3 & -2 & \vdots & 10
\end{bmatrix}
$$

$$
(R_2 - 7R_1) \rightarrow R_2 \begin{bmatrix}
1 & -5 & 2 & -1 & \vdots & 0 \\
0 & 34 & -14 & 8 & \vdots & -20 \\
3 & 2 & -1 & 1 & \vdots & -10 \\
-2 & -7 & 3 & -2 & \vdots & 10
\end{bmatrix}
$$

$$
(R_3 - 3R_1) \rightarrow R_3 \begin{bmatrix}
1 & -5 & 2 & -1 & \vdots & 0 \\
0 & 34 & -14 & 8 & \vdots & -20 \\
0 & 17 & -7 & 4 & \vdots & -10 \\
-2 & -7 & 3 & -2 & \vdots & 10
\end{bmatrix}
$$

$$
(R_4 + 2R_1) \rightarrow R_4 \begin{bmatrix}
1 & -5 & 2 & -1 & \vdots & 0 \\
0 & 34 & -14 & 8 & \vdots & -20 \\
0 & 17 & -7 & 4 & \vdots & -10 \\
0 & -17 & 7 & -4 & \vdots & 10
\end{bmatrix}
$$

$$
R_2/34 \rightarrow R_2 \begin{bmatrix}
1 & -5 & 2 & -1 & \vdots & 0 \\
0 & 1 & -\frac{7}{17} & \frac{4}{17} & \vdots & -\frac{10}{17} \\
0 & 17 & -7 & 4 & \vdots & -10 \\
0 & -17 & 7 & -4 & \vdots & 10
\end{bmatrix}
$$

$$
(R_3 - 17R_2) \rightarrow R_3 \begin{bmatrix}
1 & -5 & 2 & -1 & \vdots & 0 \\
0 & 1 & -\frac{7}{17} & \frac{4}{17} & \vdots & -\frac{10}{17} \\
0 & 0 & 0 & 0 & \vdots & 0 \\
0 & -17 & 7 & -4 & \vdots & 10
\end{bmatrix}
$$

$$
(R_4 + 17R_2) \rightarrow R_4 \begin{bmatrix}
1 & -5 & 2 & -1 & \vdots & 0 \\
0 & 1 & -\frac{7}{17} & \frac{4}{17} & \vdots & -\frac{10}{17} \\
0 & 0 & 0 & 0 & \vdots & 0 \\
0 & 0 & 0 & 0 & \vdots & 0
\end{bmatrix}
$$

This is as far as one can go with downward elimination. Now we perform the upward elimination using only the 1's on the diagonal and starting with the lowest. In this case, we start (and end) upward elimination with the 1 in the (2, 1) position.

$$\begin{bmatrix} 1 & -5 & 2 & -1 & | & 0 \\ 0 & 1 & -\frac{7}{17} & \frac{4}{17} & | & -\frac{10}{17} \\ 0 & 0 & 0 & 0 & | & 0 \\ 0 & 0 & 0 & 0 & | & 0 \end{bmatrix} \begin{matrix} (R_1 + 5R_2) \to R_1 \\ \\ \\ \end{matrix} \begin{bmatrix} 1 & 0 & -\frac{1}{17} & \frac{3}{17} & | & -\frac{50}{17} \\ 0 & 1 & -\frac{7}{17} & \frac{4}{17} & | & -\frac{10}{17} \\ 0 & 0 & 0 & 0 & | & 0 \\ 0 & 0 & 0 & 0 & | & 0 \end{bmatrix}$$

This is the end of upward elimination. The first two rows of the last matrix represent the equations

$$\begin{cases} x & -\frac{1}{17}z + \frac{3}{17}v = -\frac{50}{17} \\ y - \frac{7}{17}z + \frac{4}{17}v = -\frac{10}{17} \end{cases}$$

or

$$\begin{cases} x = \frac{1}{17}z - \frac{3}{17}v - \frac{50}{17} \\ y = \frac{7}{17}z - \frac{4}{17}v - \frac{10}{17} \end{cases}$$

or

$$\begin{cases} x = (z - 3v - 50)/17 \\ y = (7z - 4v - 10)/17 \end{cases}$$

Now let s and t be any numbers and let $z = s$ and $v = t$. Then the vector

$$\left(\frac{s - 3t - 50}{17}, \frac{7s - 4t - 10}{17}, s, t \right)$$

is a solution to the given system. For instance, when

$$s = 2 \qquad t = 1$$

then

$$x = \frac{s - 3t - 50}{17} = \frac{2 - 3 - 50}{17} = -\frac{51}{17} = -3$$

and

$$y = \frac{7s - 4t - 10}{17} = \frac{14 - 4 - 10}{17} = \frac{0}{17} = 0$$

So $x = -3$, $y = 0$, $z = 2$, and $v = 1$ is one of the infinitely many solutions to the system. ∎

Example 5.6.3   **5 × 5 System: No Solution**

Consider the system

$$\begin{cases} x + 3y - 2z + 5v - w = -1 \\ 6y - z + 12v - 2w = -6 \\ -2x + 3z + 2v = -4 \\ 4x - y - v + 2w = 9 \\ 4x + 5y - z + 11v = 5 \end{cases}$$

Computing the determinant of the matrix of coefficients here gives 0 (see exercise 16), so this system has no solution or infinitely many. Again, we proceed with downward elimination.

$$\begin{bmatrix} 1 & 3 & -2 & 5 & -1 & | & -1 \\ 0 & 6 & -1 & 12 & -2 & | & -6 \\ -2 & 0 & 3 & 2 & 0 & | & -4 \\ 4 & -1 & 0 & -1 & 2 & | & 9 \\ 4 & 5 & -1 & 11 & 0 & | & 5 \end{bmatrix}$$

$$(R_3 + 2R_1) \rightarrow R_3 \begin{bmatrix} 1 & 3 & -2 & 5 & -1 & | & -1 \\ 0 & 6 & -1 & 12 & -2 & | & -6 \\ 0 & 6 & -1 & 12 & -2 & | & -6 \\ 4 & -1 & 0 & -1 & 2 & | & 9 \\ 4 & 5 & -1 & 11 & 0 & | & 5 \end{bmatrix}$$

$$(R_4 - 4R_1) \rightarrow R_4 \begin{bmatrix} 1 & 3 & -2 & 5 & -1 & | & -1 \\ 0 & 6 & -1 & 12 & -2 & | & -6 \\ 0 & 6 & -1 & 12 & -2 & | & -6 \\ 0 & -13 & 8 & -21 & 6 & | & 13 \\ 4 & 5 & -1 & 11 & 0 & | & 5 \end{bmatrix}$$

$$(R_5 - 4R_1) \rightarrow R_5 \begin{bmatrix} 1 & 3 & -2 & 5 & -1 & | & -1 \\ 0 & 6 & -1 & 12 & -2 & | & -6 \\ 0 & 6 & -1 & 12 & -2 & | & -6 \\ 0 & -13 & 8 & -21 & 6 & | & 13 \\ 0 & -7 & 7 & -9 & 4 & | & 9 \end{bmatrix}$$

$$R_2/6 \rightarrow R_2 \begin{bmatrix} 1 & 3 & -2 & 5 & -1 & | & -1 \\ 0 & 1 & -\frac{1}{6} & 2 & -\frac{1}{3} & | & -1 \\ 0 & 6 & -1 & 12 & -2 & | & -6 \\ 0 & -13 & 8 & -21 & 6 & | & 13 \\ 0 & -7 & 7 & -9 & 4 & | & 9 \end{bmatrix}$$

$$(R_3 - 6R_2) \rightarrow R_3 \begin{bmatrix} 1 & 3 & -2 & 5 & -1 & | & -1 \\ 0 & 1 & -\frac{1}{6} & 2 & -\frac{1}{3} & | & -1 \\ 0 & 0 & 0 & 0 & 0 & | & 0 \\ 0 & -13 & 8 & -21 & 6 & | & 13 \\ 0 & -7 & 7 & -9 & 4 & | & 9 \end{bmatrix}$$

$$(R_4 + 13R_2) \rightarrow R_4 \begin{bmatrix} 1 & 3 & -2 & 5 & -1 & | & -1 \\ 0 & 1 & -\frac{1}{6} & 2 & -\frac{1}{3} & | & -1 \\ 0 & 0 & 0 & 0 & 0 & | & 0 \\ 0 & 0 & \frac{35}{6} & 5 & \frac{5}{3} & | & 0 \\ 0 & -7 & 7 & -9 & 4 & | & 9 \end{bmatrix}$$

$$\begin{bmatrix} 1 & 3 & -2 & 5 & -1 & | & -1 \\ 0 & 1 & -\frac{1}{6} & 2 & -\frac{1}{3} & | & -1 \\ 0 & 0 & 0 & 0 & 0 & | & 0 \\ 0 & 0 & \frac{35}{6} & 5 & \frac{5}{3} & | & 0 \\ 0 & 0 & \frac{35}{6} & 5 & \frac{5}{3} & | & 2 \end{bmatrix}$$

$(R_5 + 7R_2) \to R_5$

At this point we exchange R_3 and R_5, since $a_{33} = 0$.

Exchange R_3 and R_5
$$\begin{bmatrix} 1 & 3 & -2 & 5 & -1 & | & -1 \\ 0 & 1 & -\frac{1}{6} & 2 & -\frac{1}{3} & | & -1 \\ 0 & 0 & \frac{35}{6} & 5 & \frac{5}{3} & | & 2 \\ 0 & 0 & \frac{35}{6} & 5 & \frac{5}{3} & | & 0 \\ 0 & 0 & 0 & 0 & 0 & | & 0 \end{bmatrix}$$

$R_3/(\frac{35}{6}) \to R_3$
$$\begin{bmatrix} 1 & 3 & -2 & 5 & -1 & | & -1 \\ 0 & 1 & -\frac{1}{6} & 2 & -\frac{1}{3} & | & -1 \\ 0 & 0 & 1 & \frac{6}{7} & \frac{2}{7} & | & \frac{12}{35} \\ 0 & 0 & \frac{35}{6} & 5 & \frac{5}{3} & | & 0 \\ 0 & 0 & 0 & 0 & 0 & | & 0 \end{bmatrix}$$

$(R_4 - \frac{35}{6}R_3) \to R_4$
$$\begin{bmatrix} 1 & 3 & -2 & 5 & -1 & | & -1 \\ 0 & 1 & -\frac{1}{6} & 2 & -\frac{1}{3} & | & -1 \\ 0 & 0 & 1 & \frac{6}{7} & \frac{2}{7} & | & \frac{12}{35} \\ 0 & 0 & 0 & 0 & 0 & | & -2 \\ 0 & 0 & 0 & 0 & 0 & | & 0 \end{bmatrix}$$

At this stage the fourth row of the matrix represents the equation

$$0x + 0y + 0z + 0v + 0w = -2$$

which has no solution. Thus, we stop here and conclude that the given system has no solution. ∎

Two final comments are in order here. First, the row operations in all the examples were performed according to the established patterns for downward and upward elimination. Certainly there are other valid row operations, some that would simplify the manual computations, which lead to the same results.

Second, in the last two examples, we worked with fractions rather than decimals when performing the row operations. This was done to avoid rounding error and to keep our answers exact. The Gaussian elimination procedure can be readily translated into a computer program. However, as we have seen in earlier chapters, the computer works with fixed or floating-point numeric forms, not fractional forms, and can store only finitely many digits of a given number. Thus, rounding takes place often, particularly when the number has infinitely many digits (for example, recall that the binary form of $.6_{dec}$ contains infinitely many digits). As a result, row and equation reduction techniques of the type illustrated in the last two sections can lead to significant cumulative error when implemented on a computer. In the study of **error analysis**, techniques are discussed that minimize such errors.

EXERCISES **5.6**

In each of the first fourteen exercises:

a. *Write the matrix equation for the given system of equations.*

b. *Without solving, determine if the system has a unique solution.*

c. *Use the methods of this section to find solution(s), if any.*

1. $\begin{aligned} x - y &= 25 \\ 2x + 3y &= 475 \end{aligned}$

2. $\begin{aligned} -x + 4y &= 1 \\ 3x - 12y &= 6 \end{aligned}$

3. $\begin{aligned} 2x - y &= 4 \\ -12x + 6y &= -24 \end{aligned}$

4. $\begin{aligned} 3x - 2y &= 11 \\ 5x + y &= 14 \end{aligned}$

5. $\begin{aligned} 2x - 3y + z &= 2 \\ x - y + 4z &= 1 \\ -3x + 5y + 2z &= -3 \end{aligned}$

6. $\begin{aligned} 2x - 3y + z &= 2 \\ x - y + 4z &= 1 \\ -3x + 5y + 2z &= 3 \end{aligned}$

7. $\begin{aligned} -3x \quad\;\; + 2z &= 1 \\ x + 4y + 5z &= -1 \\ -x + 8y + 12z &= 1 \end{aligned}$

8. $\begin{aligned} x + 4y - 2z &= -4 \\ -3x \quad\;\; + z &= -7 \\ -x + 2y + 5z &= -11 \end{aligned}$

9. $\begin{aligned} 5x - y + 2z &= 1 \\ -x + 2y - 2z &= 3 \\ 2x + 3y + 4z &= -6 \end{aligned}$

10. $\begin{aligned} -3x \quad\;\; + 2z &= 1 \\ x + 4y + 5z &= -1 \\ -x + 8y + 12z &= -1 \end{aligned}$

11. $\begin{aligned} -2x + y \qquad\;\; + 3v &= 1 \\ x \qquad + 2z - v &= 4 \\ -3y \qquad\;\; + 2v &= -2 \\ -x - 2y + 2z + 4v &= 0 \end{aligned}$

12. $\begin{aligned} 3x \qquad - z - v &= -4 \\ -2x + y \qquad\;\; + 3v &= 7 \\ 4y - z - 2v &= -15 \\ -x \qquad + 2z + v &= 5 \end{aligned}$

13. $\begin{aligned} x + y \qquad - 3v + 2w &= 5 \\ -2y + z \qquad + w &= 1 \\ -3x \qquad - z + v \qquad &= -2 \\ 4x + 2y + 3z \qquad - 2w &= -3 \\ -y + z - 2v + 3w &= 5 \end{aligned}$

14. $\begin{aligned} x + y \qquad - 3v + 2w &= 5 \\ -2y + z \qquad + w &= 1 \\ -3x \qquad - z + v \qquad &= -2 \\ -x \qquad - 5v + 5w &= 9 \\ -y + z - 2v + 3w &= 5 \end{aligned}$

15. Verify that the determinant of the matrix of coefficients in Example 5.6.2 equals 0.

16. Verify that the determinant of the matrix of coefficients in Example 5.6.3 equals 0.

5.7

Inverses

Up to now in our work with systems of linear equations, we have used the matrix representation of a system

$$AX = C$$

merely as an abbreviated form of the system. However, in certain instances this equation can also be treated as an algebraic equation and solved for the unknown X, so as to provide an alternate method for solving the system. But before we can do this, we must introduce another algebraic tool that is related to the notion of matrix; namely, the idea of an **inverse of a matrix**.

Recall the $n \times n$ identity matrix I_n presented in Section 5.3, which has all 1's on the main diagonal and all 0's elsewhere.

Definitions

> Let A be an $n \times n$ matrix. If there exists an $n \times n$ matrix, denoted A^{-1}, with the property that
>
> $$AA^{-1} = A^{-1}A = I_n$$
>
> then A^{-1} is called the **inverse of A**. In this case, we say that **A has an inverse**, or is **nonsingular**, or is **invertible**. A matrix that has *no* inverse is called **singular**.

Notice the use of the definite article *the* in "the inverse of A." This is because it can be shown that if A has an inverse, this inverse is unique (see exercise 33).

Example **5.7.1**   Singular Matrix

Consider the *2 × 2* matrix $A = \begin{bmatrix} 1 & -2 \\ -2 & 4 \end{bmatrix}$

Suppose that A is nonsingular, so that there is a 2×2 matrix

$$A^{-1} = \begin{bmatrix} a & b \\ c & d \end{bmatrix}$$

with the property that

$$AA^{-1} = A^{-1}A = I_2$$

In particular,

$$AA^{-1} = I_2$$

hence

$$\begin{bmatrix} 1 & -2 \\ -2 & 4 \end{bmatrix} \begin{bmatrix} a & b \\ c & d \end{bmatrix} = \begin{bmatrix} 1 & 0 \\ 0 & 1 \end{bmatrix}$$

and

$$\begin{bmatrix} a-2c & b-2d \\ -2a+4c & -2b+4d \end{bmatrix} = \begin{bmatrix} 1 & 0 \\ 0 & 1 \end{bmatrix}$$

Now equating just the first-column elements gives

$$E_1: \quad a - 2c = 1 \qquad\qquad E_1: \quad a - 2c = 1$$
$$E_2: \quad -2a + 4c = 0 \qquad E_2/(-2) \to E_2: \quad a - 2c = 0$$

However, no numbers a and c satisfy the system on the right. So the supposition that A is nonsingular is false, and the given matrix A is singular, or has no inverse. ∎

Example 5.7.2 **Nonsingular Matrix**

For the matrix $A = \begin{bmatrix} 3 & 1 \\ 5 & 2 \end{bmatrix}$ the matrix $B = \begin{bmatrix} 2 & -1 \\ -5 & 3 \end{bmatrix}$ has the property that

$$AB = \begin{bmatrix} 3 & 1 \\ 5 & 2 \end{bmatrix} \begin{bmatrix} 2 & -1 \\ -5 & 3 \end{bmatrix} = \begin{bmatrix} 3(2)+1(-5) & 3(-1)+1(3) \\ 5(2)+2(-5) & 5(-1)+2(3) \end{bmatrix}$$

$$= \begin{bmatrix} 1 & 0 \\ 0 & 1 \end{bmatrix} = I_2$$

$$BA = \begin{bmatrix} 2 & -1 \\ -5 & 3 \end{bmatrix} \begin{bmatrix} 3 & 1 \\ 5 & 2 \end{bmatrix} = \begin{bmatrix} 2(3)+(-1)(5) & 2(1)+(-1)(2) \\ -5(3)+3(5) & -5(1)+3(2) \end{bmatrix}$$

$$= \begin{bmatrix} 1 & 0 \\ 0 & 1 \end{bmatrix} = I_2$$

So A is nonsingular and $A^{-1} = B$. In addition, we can say that B is non-singular and $B^{-1} = A$. ∎

At this stage you may justifiably be asking: Given A in the preceding example, where did $B = A^{-1}$ come from? We'll get to that answer shortly, but first we present a statement that will enable us to determine whether a given matrix has an inverse.

Fact

$$\left. \begin{array}{c} A \text{ has} \\ \text{an inverse} \end{array} \right\} \quad \longleftrightarrow \quad |A| \neq 0$$

For instance, in the first example of this section

$$|A| = \begin{vmatrix} 1 & -2 \\ -2 & 4 \end{vmatrix} = 4 - 4 = 0$$

Thus, the above fact tells us that A has no inverse (and with considerably less effort than we exerted previously!). However, from Example 5.7.2,

$$|A| = \begin{vmatrix} 3 & 1 \\ 5 & 2 \end{vmatrix} = 6 - 5 = 1 \neq 0$$

The fact assures us that A is nonsingular, but it does not tell us how to compute the inverse of A.

One method for computing the inverse of a nonsingular matrix is closely related to Gaussian elimination.

Example **5.7.3**   Inverse of a Matrix

Let's take the matrix

$$A = \begin{bmatrix} 3 & 1 \\ 5 & 2 \end{bmatrix}$$

which we have just verified as being nonsingular; we can find its inverse as follows. First, augment the given matrix with the identity matrix of the same dimension:

$$[A \mid I_2] = \begin{bmatrix} 3 & 1 & \vdots & 1 & 0 \\ 5 & 2 & \vdots & 0 & 1 \end{bmatrix}$$

Second, using the same downward–upward elimination process as before, reduce the A portion of this matrix to I_2 so that the entire augmented matrix is reduced to the form

$$[I_2 \mid B]$$

The reduction of A to I_2 is always possible when A is nonsingular.

$$\underbrace{\begin{bmatrix} 3 & 1 \\ 5 & 2 \end{bmatrix}}_{A} \underbrace{\begin{bmatrix} \vdots & 1 & 0 \\ \vdots & 0 & 1 \end{bmatrix}}_{I_2} \begin{matrix} R_1/3 \to R_1 \end{matrix} \begin{bmatrix} 1 & \frac{1}{3} & \vdots & \frac{1}{3} & 0 \\ 5 & 2 & \vdots & 0 & 1 \end{bmatrix}$$

$$(R_2 - 5R_1) \to R_2 \begin{bmatrix} 1 & \frac{1}{3} & \vdots & \frac{1}{3} & 0 \\ 0 & \frac{1}{3} & \vdots & -\frac{5}{3} & 1 \end{bmatrix} (R_2/\tfrac{1}{3}) \to R_2 \begin{bmatrix} 1 & \frac{1}{3} & \vdots & \frac{1}{3} & 0 \\ 0 & 1 & \vdots & -5 & 3 \end{bmatrix}$$

$$(R_1 - \tfrac{1}{3}R_2) \to R_1 \underbrace{\begin{bmatrix} 1 & 0 \\ 0 & 1 \end{bmatrix}}_{I_2} \underbrace{\begin{bmatrix} 2 & -1 \\ -5 & 3 \end{bmatrix}}_{B}$$

Third, the matrix B in the augmented portion of this matrix is the inverse of A. That is,

$$A^{-1} = B = \begin{bmatrix} 2 & -1 \\ -5 & 3 \end{bmatrix} \qquad \blacksquare$$

One of the nice things about the (often long and tedious) work involved in computing a matrix inverse is that the answer is easily checked. Whenever you claim to have found an inverse A^{-1}, you should immediately verify that each of the products AA^{-1} and $A^{-1}A$ equals the identity matrix. For the 2×2 matrices here, these computations have already been done in Example 5.7.2.

The general algorithm for the method of this last example is given in Figure 28, which uses the $N \times N$ DET, DOWN ELIM, and UP ELIM algorithms previously developed. In this flowchart

$$A = [a_{ij}]$$

is the given $n \times n$ matrix, and

$$A_{\text{aug}} = \begin{bmatrix} A & | & I_n \end{bmatrix}$$

is the augmented matrix that is the input to the DOWN ELIM algorithm. The matrix produced as output from DOWN ELIM is used as input for UP

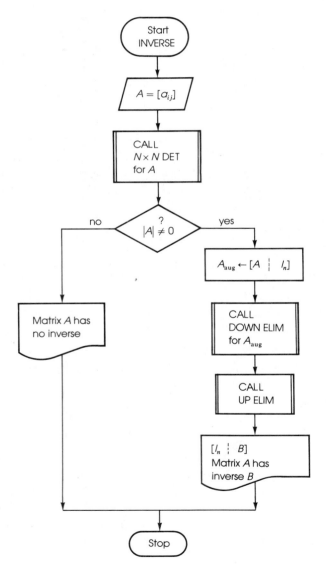

Figure 28

ELIM, and the output of UP ELIM is the augmented matrix

$$[I_n \mid B]$$

The next example is an application of Figure 28 to a 3×3 matrix.

Example 5.7.4   **Inverse Matrix**

Consider the 3×3 matrix $A = \begin{bmatrix} 2 & -3 & -2 \\ -3 & 1 & 2 \\ 7 & -2 & -6 \end{bmatrix}$

To begin with

$$|A| = \begin{vmatrix} 2 & -3 & -2 \\ -3 & 1 & 2 \\ 7 & -2 & -6 \end{vmatrix}$$

$$= 2(-1)^2 \begin{vmatrix} 1 & 2 \\ -2 & -6 \end{vmatrix} + (-3)(-1)^3 \begin{vmatrix} -3 & -2 \\ -2 & -6 \end{vmatrix} + 7(-1)^4 \begin{vmatrix} -3 & -2 \\ 1 & 2 \end{vmatrix}$$

$$= 2(-6 + 4) \qquad\qquad + 3(18 - 4) \qquad\qquad + 7(-6 + 2)$$

$$= 10 \neq 0$$

so A does have an inverse. Now we can proceed with the downward–upward elimination phases, knowing that if the arithmetic is done correctly, A^{-1} will be produced.

$$[A \mid I_3] = \begin{bmatrix} 2 & -3 & -2 & \mid & 1 & 0 & 0 \\ -3 & 1 & 2 & \mid & 0 & 1 & 0 \\ 7 & -2 & -6 & \mid & 0 & 0 & 1 \end{bmatrix} \xrightarrow{R_1/2 \to R_1} \begin{bmatrix} 1 & -\frac{3}{2} & -1 & \mid & \frac{1}{2} & 0 & 0 \\ -3 & 1 & 2 & \mid & 0 & 1 & 0 \\ 7 & -2 & -6 & \mid & 0 & 0 & 1 \end{bmatrix}$$

$$(R_2 + 3R_1) \to R_2 \begin{bmatrix} 1 & -\frac{3}{2} & -1 & \mid & \frac{1}{2} & 0 & 0 \\ 0 & -\frac{7}{2} & -1 & \mid & \frac{3}{2} & 1 & 0 \\ 7 & -2 & -6 & \mid & 0 & 0 & 1 \end{bmatrix} \xrightarrow{(R_3 - 7R_1) \to R_3} \begin{bmatrix} 1 & -\frac{3}{2} & -1 & \mid & \frac{1}{2} & 0 & 0 \\ 0 & -\frac{7}{2} & -1 & \mid & \frac{3}{2} & 1 & 0 \\ 0 & \frac{17}{2} & 1 & \mid & -\frac{7}{2} & 0 & 1 \end{bmatrix}$$

$$R_2/(-\frac{7}{2}) \to R_2 \begin{bmatrix} 1 & -\frac{3}{2} & -1 & \mid & \frac{1}{2} & 0 & 0 \\ 0 & 1 & \frac{2}{7} & \mid & -\frac{3}{7} & -\frac{2}{7} & 0 \\ 0 & \frac{17}{2} & 1 & \mid & -\frac{7}{2} & 0 & 1 \end{bmatrix} \xrightarrow{(R_3 - \frac{17}{2}R_2) \to R_3} \begin{bmatrix} 1 & -\frac{3}{2} & -1 & \mid & \frac{1}{2} & 0 & 0 \\ 0 & 1 & \frac{2}{7} & \mid & -\frac{3}{7} & -\frac{2}{7} & 0 \\ 0 & 0 & -\frac{10}{7} & \mid & \frac{1}{7} & \frac{17}{7} & 1 \end{bmatrix}$$

$$R_3/(-\frac{10}{7}) \to R_3 \begin{bmatrix} 1 & -\frac{3}{2} & -1 & \mid & \frac{1}{2} & 0 & 0 \\ 0 & 1 & \frac{2}{7} & \mid & -\frac{3}{7} & -\frac{2}{7} & 0 \\ 0 & 0 & 1 & \mid & -\frac{1}{10} & -\frac{17}{10} & -\frac{7}{10} \end{bmatrix} \xrightarrow{(R_2 - \frac{2}{7}R_3) \to R_2} \begin{bmatrix} 1 & -\frac{3}{2} & -1 & \mid & \frac{1}{2} & 0 & 0 \\ 0 & 1 & 0 & \mid & -\frac{2}{5} & \frac{1}{5} & \frac{1}{5} \\ 0 & 0 & 1 & \mid & -\frac{1}{10} & -\frac{17}{10} & -\frac{7}{10} \end{bmatrix}$$

$$(R_1 + R_3) \to R_1 \begin{bmatrix} 1 & -\frac{3}{2} & 0 & \mid & \frac{2}{5} & -\frac{17}{10} & -\frac{7}{10} \\ 0 & 1 & 0 & \mid & -\frac{2}{5} & \frac{1}{5} & \frac{1}{5} \\ 0 & 0 & 1 & \mid & -\frac{1}{10} & -\frac{17}{10} & -\frac{7}{10} \end{bmatrix} \xrightarrow{(R_1 + \frac{3}{2}R_2) \to R_1} \begin{bmatrix} 1 & 0 & 0 & \mid & -\frac{1}{5} & -\frac{7}{5} & -\frac{2}{5} \\ 0 & 1 & 0 & \mid & -\frac{2}{5} & \frac{1}{5} & \frac{1}{5} \\ 0 & 0 & 1 & \mid & -\frac{1}{10} & -\frac{17}{10} & -\frac{7}{10} \end{bmatrix}$$

$$\underbrace{\phantom{\begin{matrix} 1 & 0 & 0 \\ 0 & 1 & 0 \\ 0 & 0 & 1 \end{matrix}}}_{I_3} \underbrace{\phantom{\begin{matrix} -\frac{1}{5} & -\frac{7}{5} & -\frac{2}{5} \\ -\frac{2}{5} & \frac{1}{5} & \frac{1}{5} \\ -\frac{1}{10} & -\frac{17}{10} & -\frac{7}{10} \end{matrix}}}_{B}$$

The result is that

$$A^{-1} = \begin{bmatrix} -\frac{1}{5} & -\frac{7}{5} & -\frac{2}{5} \\ -\frac{2}{5} & \frac{1}{5} & \frac{1}{5} \\ -\frac{1}{10} & -\frac{17}{10} & -\frac{7}{10} \end{bmatrix}$$

which the reader should verify by multiplying with A. ■

Example 5.7.5 No Inverse Matrix

We know from the preceding work in Example 5.7.1 that the matrix

$$A = \begin{bmatrix} 1 & -2 \\ -2 & 4 \end{bmatrix}$$

is singular and so has no inverse. If we apply the elimination technique anyhow, we do not get very far.

$$\begin{bmatrix} 1 & -2 & | & 1 & 0 \\ -2 & 4 & | & 0 & 1 \end{bmatrix} (R_2 + 2R_1) \to R_2 \begin{bmatrix} 1 & -2 & | & 1 & 0 \\ \boxed{0} & \boxed{0} & | & 2 & 1 \end{bmatrix}$$

The row of zeroes on the left side of this augmented matrix means that it is impossible to reduce the left side to I_2. This is another way of showing that A has no inverse. ■

Now that we have a method of computing A^{-1} when it exists, let us return to the matrix equation $AX = C$ mentioned at the beginning of this section. If A is nonsingular so that A^{-1} exists, we can perform some "matrix algebra":

$$AX = C$$
$$A^{-1}(AX) = A^{-1}C \qquad \text{Multiplication on left by } A^{-1}$$
$$(A^{-1}A)X = A^{-1}C \qquad \text{Associativity of matrix multiplication}$$
$$I_nX = A^{-1}C \qquad \text{Definition of } A^{-1}$$
$$X = A^{-1}C \qquad \text{Property of } I_n$$

Therefore, the matrix product

$$A^{-1}C$$

provides the unique solution to the given system represented by the equation

$$AX = C$$

Fact

If $AX = C$ is a matrix equation that represents a system of linear equations and if $|A| \neq 0$, then

$$X = A^{-1}C$$

is the unique solution to the system.

Take note that A^{-1} multiplies C *on the left*; the order is important here, since matrix multiplication is *not* commutative.

Example 5.7.6

2 × 2 System Solution Using Inverse

The system

$$\begin{cases} x + y = 390 \\ 4x + 2y = 1330 \end{cases}$$

was first presented and solved in Section 5.5. In matrix form, the system is

$$AX = C$$

where

$$A = \begin{bmatrix} 1 & 1 \\ 4 & 2 \end{bmatrix} \qquad X = \begin{bmatrix} x \\ y \end{bmatrix} \qquad C = \begin{bmatrix} 390 \\ 1330 \end{bmatrix}$$

Since

$$|A| = \begin{vmatrix} 1 & 1 \\ 4 & 2 \end{vmatrix} = 2 - 4 = -2 \neq 0$$

the system has a unique solution. Let's find A^{-1}.

$$\begin{bmatrix} 1 & 1 & \vdots & 1 & 0 \\ 4 & 2 & \vdots & 0 & 1 \end{bmatrix} (R_2 - 4R_1) \rightarrow R_2 \begin{bmatrix} 1 & 1 & \vdots & 1 & 0 \\ 0 & -2 & \vdots & -4 & 1 \end{bmatrix}$$

$$R_2/(-2) \rightarrow R_2 \begin{bmatrix} 1 & 1 & \vdots & 1 & 0 \\ 0 & 1 & \vdots & 2 & -\frac{1}{2} \end{bmatrix} (R_1 - R_2) \rightarrow R_1 \begin{bmatrix} 1 & 0 & \vdots & -1 & \frac{1}{2} \\ 0 & 1 & \vdots & \underbrace{2 \quad -\frac{1}{2}} \end{bmatrix}$$

$$A^{-1}$$

By the preceding fact, the unique solution to the system is

$$X = A^{-1}C$$

or

$$\begin{bmatrix} x \\ y \end{bmatrix} = \begin{bmatrix} -1 & \frac{1}{2} \\ 2 & -\frac{1}{2} \end{bmatrix} \begin{bmatrix} 390 \\ 1330 \end{bmatrix} = \begin{bmatrix} (-1)(390) + \frac{1}{2}(1330) \\ 2(390) - \frac{1}{2}(1330) \end{bmatrix} = \begin{bmatrix} 275 \\ 115 \end{bmatrix}$$

That is,

$$x = 275 \qquad y = 115$$

One of the advantages of this method is that *other* solutions can be computed quickly when matrix A remains the same but matrix C changes. For instance, in the context of the original problem (see Example 5.5.1), suppose another showing of the movie grossed $1440 and 430 people attended. Then the associated system is

$$\begin{cases} x + y = 430 \\ 4x + 2y = 1440 \end{cases}$$

whose unique solution is

$$X = A^{-1}C$$

or

$$\begin{bmatrix} x \\ y \end{bmatrix} = \begin{bmatrix} -1 & \frac{1}{2} \\ 2 & -\frac{1}{2} \end{bmatrix} \begin{bmatrix} 430 \\ 1440 \end{bmatrix} = \begin{bmatrix} (-1)(430) + \frac{1}{2}(1440) \\ 2(430) - \frac{1}{2}(1440) \end{bmatrix} = \begin{bmatrix} 290 \\ 140 \end{bmatrix}$$

So

$$x = 290 \qquad y = 140$$

that is, 290 adults and 140 children attended that particular showing.   ∎

Example 5.7.7   **3 × 3 System Solution Using Inverse**

The system

$$\begin{cases} 5x + 3y + 4z = 295 \\ 2x + y + 1.5z = 110 \\ 2x + 1.5y + z = 100 \end{cases}$$

was derived in Example 5.5.2 and solved later in that section. It relates to a production situation in which certain numbers of labor-hours are available in each of three departments. The matrix form of this system is

$$AX = C$$

where

$$A = \begin{bmatrix} 5 & 3 & 4 \\ 2 & 1 & 1.5 \\ 2 & 1.5 & 1 \end{bmatrix} \qquad X = \begin{bmatrix} x \\ y \\ z \end{bmatrix} \qquad C = \begin{bmatrix} 295 \\ 110 \\ 100 \end{bmatrix}$$

First, we should check to see that

$$|A| = .75 \neq 0$$

This work was done in Example 5.5.7. Here we shall find the unique solution by using A^{-1}.

$$\begin{bmatrix} 5 & 3 & 4 & | & 1 & 0 & 0 \\ 2 & 1 & 1.5 & | & 0 & 1 & 0 \\ 2 & 1.5 & 1 & | & 0 & 0 & 1 \end{bmatrix} \xrightarrow{R_1/5 \to R_1} \begin{bmatrix} 1 & .6 & .8 & | & .2 & 0 & 0 \\ 2 & 1 & 1.5 & | & 0 & 1 & 0 \\ 2 & 1.5 & 1 & | & 0 & 0 & 1 \end{bmatrix}$$

$$(R_2 - 2R_1) \to R_2 \begin{bmatrix} 1 & .6 & .8 & | & .2 & 0 & 0 \\ 0 & -.2 & -.1 & | & -.4 & 1 & 0 \\ 2 & 1.5 & 1 & | & 0 & 0 & 1 \end{bmatrix}$$

$$(R_3 - 2R_1) \to R_3 \begin{bmatrix} 1 & .6 & .8 & | & .2 & 0 & 0 \\ 0 & -.2 & -.1 & | & -.4 & 1 & 0 \\ 0 & .3 & -.6 & | & -.4 & 0 & 1 \end{bmatrix}$$

$$R_2/(-.2) \to R_2 \begin{bmatrix} 1 & .6 & .8 & | & .2 & 0 & 0 \\ 0 & 1 & .5 & | & 2 & -5 & 0 \\ 0 & .3 & -.6 & | & -.4 & 0 & 1 \end{bmatrix}$$

$$(R_3 - .3R_2) \to R_3 \begin{bmatrix} 1 & .6 & .8 & | & .2 & 0 & 0 \\ 0 & 1 & .5 & | & 2 & -5 & 0 \\ 0 & 0 & -.75 & | & -1 & 1.5 & 1 \end{bmatrix}$$

In the next step, we convert two entries in the last row into fractional form to avoid rounding:

$$\frac{1}{.75} = \tfrac{4}{3} = 1.333\ldots$$

In the succeeding step, we convert all remaining decimals to fractions.

$$R_3/(-.75) \to R_3 \begin{bmatrix} 1 & .6 & .8 & | & .2 & 0 & 0 \\ 0 & 1 & .5 & | & 2 & -5 & 0 \\ 0 & 0 & 1 & | & \tfrac{4}{3} & -2 & -\tfrac{4}{3} \end{bmatrix} \xrightarrow{\text{Convert}} \begin{bmatrix} 1 & \tfrac{3}{5} & \tfrac{4}{5} & | & \tfrac{1}{5} & 0 & 0 \\ 0 & 1 & \tfrac{1}{2} & | & 2 & -5 & 0 \\ 0 & 0 & 1 & | & \tfrac{4}{3} & -2 & -\tfrac{4}{3} \end{bmatrix}$$

$$(R_2 - \tfrac{1}{2}R_3) \to R_2 \begin{bmatrix} 1 & \tfrac{3}{5} & \tfrac{4}{5} & | & \tfrac{1}{5} & 0 & 0 \\ 0 & 1 & 0 & | & \tfrac{4}{3} & -4 & \tfrac{2}{3} \\ 0 & 0 & 1 & | & \tfrac{4}{3} & -2 & -\tfrac{4}{3} \end{bmatrix}$$

$$(R_1 - \tfrac{4}{5}R_3) \to R_1 \begin{bmatrix} 1 & \tfrac{3}{5} & 0 & | & -\tfrac{13}{15} & \tfrac{8}{5} & \tfrac{16}{15} \\ 0 & 1 & 0 & | & \tfrac{4}{3} & -4 & \tfrac{2}{3} \\ 0 & 0 & 1 & | & \tfrac{4}{3} & -2 & -\tfrac{4}{3} \end{bmatrix}$$

$$(R_1 - \tfrac{3}{5}R_2) \to R_1 \begin{bmatrix} 1 & 0 & 0 & | & -\tfrac{5}{3} & 4 & \tfrac{2}{3} \\ 0 & 1 & 0 & | & \tfrac{4}{3} & -4 & \tfrac{2}{3} \\ 0 & 0 & 1 & | & \underbrace{\tfrac{4}{3} \quad -2 \quad -\tfrac{4}{3}}_{A^{-1}} \end{bmatrix}$$

Therefore, the desired solution is $X = A^{-1}C$ or

$$\begin{bmatrix} x \\ y \\ z \end{bmatrix} = \begin{bmatrix} -\tfrac{5}{3} & 4 & \tfrac{2}{3} \\ \tfrac{4}{3} & -4 & \tfrac{2}{3} \\ \tfrac{4}{3} & -2 & -\tfrac{4}{3} \end{bmatrix} \begin{bmatrix} 295 \\ 110 \\ 100 \end{bmatrix} = \begin{bmatrix} 15 \\ 20 \\ 40 \end{bmatrix}$$

So

    $x = 15$ units of product I

    $y = 20$ units of product II

    $z = 40$ units of product III

will fully utilize all available labor.   ■

Continuing with the setting of the last example, suppose the hours available in the assembly, testing, and shipping departments changed to 305, 115, and 107, respectively. Then the solution to the related system

$$\begin{cases} 5x + 3y + 4z = 305 \\ 2x + y + 1.5z = 115 \\ 2x + 1.5y + z = 107 \end{cases}$$

is

$$X = A^{-1}C$$

or

$$\begin{bmatrix} x \\ y \\ z \end{bmatrix} = \begin{bmatrix} -\frac{5}{3} & 4 & \frac{2}{3} \\ \frac{4}{3} & -4 & \frac{2}{3} \\ \frac{4}{3} & -2 & -\frac{4}{3} \end{bmatrix} \begin{bmatrix} 305 \\ 115 \\ 107 \end{bmatrix} = \begin{bmatrix} 23 \\ 18 \\ 34 \end{bmatrix}$$

Hence, in this case producing 23, 18, and 34 units of products I, II, and III, respectively, will utilize the available labor-hours.

E X E R C I S E S **5.7**

In the first 24 exercises:
  a. *Determine whether the given matrix has an inverse.*
  b. *For each matrix that is nonsingular, use the method of this section to find the inverse.*

1. $\begin{bmatrix} 1 & -1 \\ 2 & -3 \end{bmatrix}$ **2.** $\begin{bmatrix} -1 & 2 \\ 4 & -7 \end{bmatrix}$ **3.** $\begin{bmatrix} 1 & -1 \\ 2 & -2 \end{bmatrix}$

4. $\begin{bmatrix} -1 & 2 \\ 4 & -8 \end{bmatrix}$ **5.** $\begin{bmatrix} 0 & 2 \\ 3 & 4 \end{bmatrix}$ **6.** $\begin{bmatrix} 1 & 0 \\ 3 & 4 \end{bmatrix}$

7. $\begin{bmatrix} 1 & 2 \\ 0 & 4 \end{bmatrix}$ **8.** $\begin{bmatrix} 1 & 2 \\ 3 & 0 \end{bmatrix}$ **9.** $\begin{bmatrix} 1 & 0 \\ 0 & 1 \end{bmatrix}$

10. $\begin{bmatrix} 0 & 1 \\ 1 & 0 \end{bmatrix}$ **11.** $\begin{bmatrix} 0 & 5 \\ -1 & 0 \end{bmatrix}$ **12.** $\begin{bmatrix} 2 & 0 \\ 0 & -3 \end{bmatrix}$

13. $\begin{bmatrix} 3 & 0 \\ 0 & 0 \end{bmatrix}$ **14.** $\begin{bmatrix} 0 & -.5 \\ 0 & 0 \end{bmatrix}$

15. $\begin{bmatrix} 5 & -1 & 3 \\ 2 & 0 & 1 \\ 3 & 2 & 1 \end{bmatrix}$ **16.** $\begin{bmatrix} 3 & 5 & -2 \\ 7 & 0 & 1 \\ 0 & 2 & -1 \end{bmatrix}$

17. $\begin{bmatrix} 5 & -3 & 1 \\ 0 & 1 & 4 \\ 0 & 0 & -5 \end{bmatrix}$ **18.** $\begin{bmatrix} -3 & 0 & 2 \\ 1 & 7 & 6 \\ 7 & 7 & 2 \end{bmatrix}$

19. $\begin{bmatrix} 5 & -3 & 1 \\ 0 & 0 & 4 \\ 0 & 0 & -5 \end{bmatrix}$ **20.** $\begin{bmatrix} 0 & 3 & 1 \\ 1 & 1 & 0 \\ 2 & 3 & 3 \end{bmatrix}$

21. $\begin{bmatrix} 2 & 0 & 0 \\ 0 & -1 & 0 \\ 0 & 0 & 3 \end{bmatrix}$ **22.** $\begin{bmatrix} 1 & 0 & 0 \\ 0 & 1 & 0 \\ 0 & 0 & 1 \end{bmatrix}$

23. $\begin{bmatrix} 1 & 0 & 1 & 0 \\ 1 & 3 & 1 & -1 \\ 1 & 1 & 5 & 2 \\ -3 & 9 & 2 & 0 \end{bmatrix}$

24. $\begin{bmatrix} 0 & -2 & -3 & 1 \\ 4 & 1 & 0 & -3 \\ 1 & 5 & 8 & 0 \\ -3 & 0 & 2 & 5 \end{bmatrix}$

In the next four exercises, find value(s) of t so that the given matrix is (a) singular and (b) nonsingular.

25. $\begin{bmatrix} 2 & 4 \\ 5 & t \end{bmatrix}$ **26.** $\begin{bmatrix} t & 3 \\ 12 & t \end{bmatrix}$

27. $\begin{bmatrix} 0 & t & -1 \\ 1 & -1 & 2 \\ t & 6 & -5 \end{bmatrix}$ **28.** $\begin{bmatrix} 1 & 0 & 2 \\ 3 & t & -1 \\ -1 & t & 0 \end{bmatrix}$

29. a. Solve the system
$$\begin{cases} 5x + 3y = 2 \\ 7x + 4y = 1 \end{cases}$$
  b. Solve the system in (a) where constants 2 and 1 are replaced by 3 and 4, respectively.

30. a. Solve the system
$$\begin{cases} 9x - 2y = 0 \\ 5x - y = 1 \end{cases}$$
  b. Solve the system in (a) where constants 0 and 1 are replaced by -2 and 2, respectively.

31. a. Solve the system
$$\begin{cases} -x + 3y + 2z = -2 \\ x \quad\quad - z = 0 \\ x + 2y \quad\quad = 1 \end{cases}$$

b. Solve the system in (a) where constants $-2, 0$, and 1 are replaced by $3, -4$, and -1, respectively.

32. a. Solve the system

$$\begin{cases} 5x \quad\quad - z = -3 \\ x - 2y + 4z = \quad 1 \\ \quad - y + 2z = \quad 0 \end{cases}$$

b. Solve the system in (a) where constants $-3, 1$, and 0 are replaced by $4, -2$, and 1, respectively.

33. Show that if a matrix has an inverse, that inverse is unique. (*Hint:* Suppose matrix A has inverses B and D. Use identity matrix I_n and the associative property of matrix multiplication to show that $B = D$.)

34. If matrices A and B are nonsingular, then each of the following statements is true:

a. AB is nonsingular and $(AB)^{-1} = B^{-1}A^{-1}$.
b. A^{-1} is nonsingular and $(A^{-1})^{-1} = A$.
c. For any number $t \neq 0$, tA is nonsingular and

$$(tA)^{-1} = \frac{1}{t} A^{-1}$$

Verify the above three properties for the matrices

$$A = \begin{bmatrix} 2 & 5 \\ 3 & 7 \end{bmatrix} \quad B = \begin{bmatrix} 7 & 2 \\ 4 & 1 \end{bmatrix}$$

and scalar $t = 4$.

35. Let A and B be nonsingular matrices. Is it true that matrix $A + B$ is nonsingular and

$$(A + B)^{-1} = A^{-1} + B^{-1} \quad ?$$

CHAPTER **6**

Selected Topics

The varied topics of this chapter were chosen for one of two reasons. Either they are connected to work done in preceding sections of this text, or they provide background in certain areas of computer-related applied math, or both. Because of this, not all the sections are related. The four major topics are:

Math of Finance	Sections 6.1 and 6.2
Statistics	Sections 6.3 and 6.4
Functions and Graphs	Section 6.5
Induction	Section 6.6

This material has been written so that each of these four topics is developed independently from the others. Therefore, you may choose to cover only certain ones, with the assurance that no reference will be made to those topics you may have skipped.

Two of the most common reasons for purchasing a computer system or a piece of computer software are to process financial data (such as payroll and bookkeeping), and to analyze statistically large collections of numerical data (such as sales records and population shifts). The first two sections of this chapter present formulas pertaining to single- and multiple-payment financial transactions involving interest. Sections 6.3 and 6.4 contain introductory information on the presentation and interpretation of large collections of data (an area known as *descriptive statistics*) and the important normal probability distribution.

The notions of function and graph, presented in Section 6.5, are useful in many areas of applied mathematics, and computer-related math is no exception. The chapter concludes with the topic of mathematical induction, a device that can be used to prove important applied formulas as well as to confirm certain algorithmic procedures.

6.1

Mathematics of Finance I: Single Payments

At some time in our lives, most of us will need to handle specific computations involving numeric financial data. Arranging for a home mortgage or new car loan, investing in a savings plan, or purchasing a retirement annuity are typical examples. In this section, we present formulas that apply to such single-payment, interest-bearing situations.

If you borrow $P = \$1500$ from your Uncle Harry to buy a car and promise to pay him $R = .01 = 1\%$ interest each month until you have the $1500 to pay back, then each month you will pay him interest equaling

$$PR = \$1500(.01) = \$15$$

Over $N = 6$ months, you would pay him a total of

$$\$15N = \$15(6) = \$90$$

Let us generalize on this example.

Fact

Simple Interest

If

> P = principal
>
> R = rate of interest per period
>
> N = number of periods

then

> $I = PRN$

is the total **simple interest** earned over the N periods.

Example 6.1.1

Simple Interest

A $5000 bond earns 5.6% interest every 6 months. Determine the total interest earned over a 2-year period.

In this case

> $P = \$5000$
>
> $R = .056$ per half-year
>
> $N = 2$ years $= 4$ half-years

So the total interest earned over 2 years is

$$I = PRN$$
$$= \$5000(.056)(4)$$
$$= \boxed{\$1120} \qquad \blacksquare$$

Observe in this example that *interest is not earned on interest*. For instance, over the first half-year

$$\$5000(.056) = \$280$$

is the interest earned, but this amount is *not* added to the principal of $5000 when the interest earned during the second half-year period is computed. Rather, an additional $280 is earned during the second half-year and each succeeding half-year. This is quite different from investing in a typical savings account, in which interest is earned on interest, a procedure called the *compounding of interest*.

Example 6.1.2   **Compound Interest**

You invest $5000 in a bank that pays an annual rate of 11.2% compounded semiannually. The following table gives information for each of the 4 half-year periods over 2 years:

Period	Principal at beginning	Interest earned	Principal at end
1	5000	$5000 * \dfrac{.112}{2} = 280$	5280
2	5280	$5280 * \dfrac{.112}{2} = 295.68$	5575.68
3	5575.68	$5575.68 * \dfrac{.112}{2} = 312.24$	5887.92
4	5887.92	$5887.92 * \dfrac{.112}{2} = 329.72$	6217.64

And

$$
\begin{array}{ll}
6217.64 & = \text{principal after 2 years} \\
-\,5000.00 & = \text{original principal} \\
\hline
\boxed{1217.64} & = \text{total interest earned over 2 years}
\end{array}
$$

As expected, the amount of interest earned here is larger than the simple interest earned in Example 6.1.1. ∎

We should make several remarks about the preceding example. First, in the third column of the table, the product of

Principal ∗ annual rate

is always divided by 2 because the bank computes interest semiannually, or twice per year. The factor

$$\frac{.112}{2} = .056$$

is the *rate per period* (in this example, rate per half-year). Second, the principal at the end of a period (fourth column) is always the principal at the beginning of the next period (second column). Third, each computation of interest after the first one involves a principal larger than $5000, so here we're earning interest on interest.

The technique just described is an algorithmic one, readily adapted to flowcharting and programming for computer execution.

Example 6.1.3

Flowchart: Compound Interest

If principal P is invested for N periods in a bank that pays interest rate R per period, then the flowchart in Figure 1 outputs the final principal value after N periods. You should compare the trace in Figure 1 with the table given in Example 6.1.2. ∎

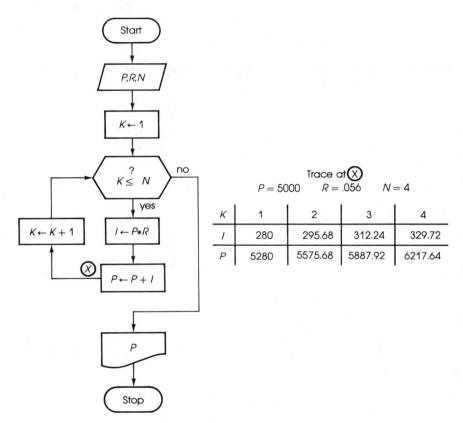

Trace at Ⓧ

$P = 5000$ $R = .056$ $N = 4$

K	1	2	3	4
I	280	295.68	312.24	329.72
P	5280	5575.68	5887.92	6217.64

Figure 1

While the process used in the two preceding examples is correct, a little bit of algebra suggests an easier method, particularly for large values of N. Given P, R, and N, the value of the principal after the first period is

$$P + PR = P(1 + R)$$

At the end of the second period it is

$$P(1 + R) + [P(1 + R)]R = [P(1 + R)]1 + [P(1 + R)]R$$

principal   interest

$$= [P(1 + R)](1 + R)$$
$$= P(1 + R)^2$$

and after the third period it is

$$P(1 + R)^2 + [P(1 + R)^2]R = [P(1 + R)^2]1 + [P(1 + R)^2]R$$

principal   interest

$$= [P(1 + R)^2](1 + R)$$
$$= P(1 + R)^3$$

The principle of mathematical induction can be used to prove the following generalization of this work (see Example 6.6.4).

Fact

> **Compound Amount Formula**
>
> If
>
> P = principal
>
> R = rate of interest per period
>
> N = number of periods
>
> then
>
> $$S = P(1 + R)^N$$
>
> is the value of the principal after N periods.

It is sometimes helpful to associate this formula with the diagram

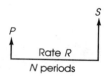

where the principal, or initial deposit value P, is smaller than the value of the principal after N periods, or compound amount S. When using this formula, remember that

R = rate per period

$$= \frac{\text{annual rate of interest}}{\text{number of times per year interest computed}}$$

In order to obtain numeric answers using the compound amount formula (CAF), you need either a calculator with a power button, like y^x, or a table of values for the *compound amount factor*

$$(1 + R)^N$$

Table 1 in the back of this text contains values of this factor for certain values of R and N. For other values of R and N, you must use more complete tables found elsewhere or a calculator. For instance, we can verify the results of Example 6.1.2 as follows:

When $P = \$5000$

$$R = \frac{.112}{2} = .056 \text{ per half-year}$$

$$N = 2 \text{ years} = 4 \text{ half-years}$$

then using the CAF we have

$$S = P(1 + R)^N$$
$$= 5000(1 + .056)^4$$
$$= 5000(1.2435283)$$
$$= \boxed{\$6217.64}$$

Here the value of the factor $(1 + .056)^4$ was obtained from a calculator, since $R = .056$ does not appear in Table 1.

Example 6.1.4

Compound Amount Formula

Suppose $10000 is invested in savings certificates that yield 12% annual interest compounded quarterly. Then

$$P = 10000$$

$$R = \frac{.12}{4} = .03 \text{ per quarter}$$

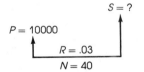

Figure 2

a. After 10 years = 40 quarters (see Figure 2), the initial deposit value P will grow to

$$S = P(1 + R)^N$$
$$= 10000(1 + .03)^{40}$$
$$= 10000(3.26204)$$
$$= \boxed{\$32,620.40}$$

Figure 3

b. After 20 years = 80 quarters (see Figure 3), the value of the deposit will equal

$$S = P(1 + R)^N$$
$$= 10000(1 + .03)^{80}$$
$$= 10000(10.64089)$$
$$= \boxed{\$106,408.90}$$

In (a) the compound amount factor was read from Table 1, but in (b) the value of N was too large for this table, so a calculator was used.   ∎

Above we calculated S given values for variables P, R, and N. However, we can, of course, use the same compound amount formula

$$S = P(1 + R)^N$$

to solve for *any one* of the four variables S, P, R, or N if we have values for the other three. This is illustrated in the next three examples.

Example 6.1.5

Present Value

If a bank offers a savings plan that earns interest at an annual rate of 11.4% compounded monthly, what amount P must be invested now so that the amount will grow to $15000 in 5 years?
Here

$$R = \frac{.114}{12} = .0095 \text{ per month}$$

$$N = 5 \text{ years} = 60 \text{ months}$$

$$S = 15000$$

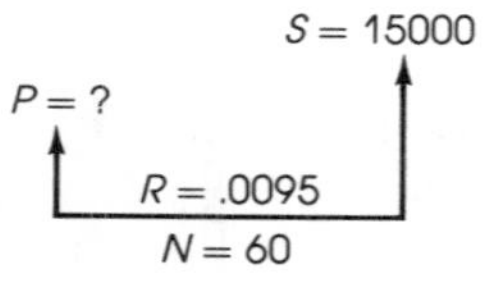

(see Figure 4). Using the CAF we get:

$$S = P(1 + R)^N$$
$$15000 = P(1 + .0095)^{60}$$
$$15000 = P(1.763516)$$

So

$$P = \frac{15000}{1.763516}$$
$$= \boxed{\$8505.74}$$

is the amount that should be invested now. ∎

$S = 15000$

$P = ?$

$R = .0095$

$N = 60$

Figure 4

Example 6.1.6

Doubling Time

How long will it take for an investment to double in value if it earns 8% interest per year compounded quarterly?
We can begin with an initial deposit of any value, so for simplicity let's take

$$P = \$1$$

then

$$S = \$2$$

and

$$R = \frac{.08}{4} = .02 \text{ per quarter}$$

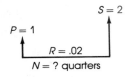

Figure 5

(see Figure 5). Using the CAF we get

$$S = P(1 + R)^N$$
$$2 = (1 + .02)^N$$

There are two ways we can proceed from this point. Using the compound amount factor table, we can *read down the .02 column* to find the first entry that exceeds 2. In our Table 1, this is

2.03989

for $N = 36$. This means that after

36 quarters = 9 years

our investment will double; actually, the value will be $2.04, which is more than double.

Alternately, using logarithms and their properties, where ln is the natural logarithm function,

$$(1.02)^N = 2$$
$$\log_{1.02}(1.02)^N = \log_{1.02} 2$$
$$N = \log_{1.02} 2$$
$$N = \frac{\ln 2}{\ln 1.02}$$
$$N = \frac{.69314718}{.01980263}$$
$$N = 35.002789$$

Since N is a number of quarters and must have an integer value, it will take 36 quarters or 9 years for the deposit to double in value.   ■

Example 6.1.7

Minimum Rate

If a person wants a $6000 investment to grow to at least $10000 in 4 years, what minimum annual rate of interest is necessary, assuming interest is compounded semiannually?

Here we have

$$P = 6000$$
$$S = 10000$$
$$N = 4 \text{ years} = 8 \text{ half-years}$$

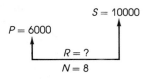

Figure 6

(see Figure 6). Using the CAF:

$$S = P(1 + R)^N$$
$$10000 = 6000(1 + R)^8$$
$$1.66667 = (1 + R)^8$$

Again there are two ways to proceed here. *Reading across the N = 8 row in Table 1, we find*

1.71819   at $R = .07$

as the first value exceeding 1.66667. Therefore, an approximate answer is

$R = .07$ per half-year

or

$.14 = 14\%$ per year

To get a more precise answer, use a calculator to take the eighth root of both sides of

$$1.66667 = (1 + R)^8$$

getting

$$1.0659362 = 1 + R$$

or

$R = .0659362$ per half-year

This translates to

.1318724

or approximately 13.19% per year.   ∎

Occasionally in newspaper ads for savings plans you see the phrases "nominal rate" and "effective annual rate." The ***nominal rate*** is the annual rate used in the computation of compound interest. For example, if the annual rate is 10% compounded quarterly, then the nominal rate is 10%. However, the ***effective annual rate*** (EAR) is the annual rate that, if applied just once per year, would produce the same interest yield per year as the given nominal rate.

Example 6.1.8   Effective Annual Rate

What is the EAR of the nominal rate of 10% per year compounded quarterly?

The idea here is to determine the interest earned in 1 year on an initial deposit of, say, $100. (The $100 figure was chosen to facilitate the computations below.) The numeric value, excluding the dollar sign, of this interest amount is precisely the EAR, in percent.

Step 1: Apply the CAF where

$$P = 100$$

$$R = \frac{.10}{4} = .025 \text{ per quarter}$$

$$N = 1 \text{ year} = 4 \text{ quarters,}$$

getting

$$S = P(1 + R)^N$$
$$= 100(1 + .025)^4$$
$$= 100(1.1038)$$
$$= 110.38$$

Step 2: Compute the interest earned:

$110.38 =$ value after 1 year
$- 100.00 =$ initial deposit value
$\overline{\$\ 10.38} =$ interest earned

Step 3: Interpret the answer in step 2. An annual rate of 10.38% applied just once a year will yield the same interest, $10.38.

Hence $\boxed{10.38\%}$ is the desired EAR.   ■

You also may have noticed in the advertising of certain investment plans a reference to **_continuous compounding_** of interest. This means that the plan pays interest equivalent to what would be earned if the nominal rate were compounded infinitely many times per year. Of course, from experience we know that this does not mean that the interest earned grows wildly without bound (see exercise 23). To understand this concept fully requires an understanding of the notion of _limit_, which is a central topic in the study of calculus. The underlying formula relating to continuous compounding is the following.

Fact

Continuous Compounding Formula

If

$P =$ principal
$R =$ annual interest rate compounded continuously
$N =$ number of years

then

$$S = Pe^{RN}$$

is the value of the principal after N years.

Recall that the irrational number e has the approximate value 2.71828.

Example 6.1.9   **Continuous Compounding**

Let $P = \$10000$ be invested in an account that earns 12% annual interest compounded continuously; thus, $R = .12$.

a. After 10 years, the deposit value is

$$S = Pe^{RN}$$
$$= 10000e^{.12(10)}$$
$$= 10000e^{1.2}$$
$$= 10000(3.320117)$$
$$= \boxed{\$33,201.17}$$

b. After 20 years, the principal will grow to

$$S = Pe^{RN}$$
$$= 10000e^{.12(20)}$$
$$= 10000e^{2.4}$$
$$= 10000(11.023176)$$
$$= \boxed{\$110,231.76}$$ ∎

Example 6.1.10

Effective Annual Rate

What is the EAR of the nominal rate of 10% per year compounded continuously?

Let's use the technique illustrated in Example 6.1.8 and find the interest earned in 1 year on a deposit of $100.

Step 1: Using the continuous compounding formula, we are given

$$P = 1000$$
$$R = .10$$
$$N = 1$$

Thus, we get:

$$S = Pe^{RN}$$
$$= 100e^{.10(1)}$$
$$= 100e^{.10}$$
$$= 100(1.10517)$$
$$= 110.52$$

Step 2: Compute the interest earned:

$$\begin{array}{rl} \$110.52 = & \text{value after 1 year} \\ -100.00 = & \text{initial deposit value} \\ \hline \$\ 10.52 = & \text{interest earned} \end{array}$$

Step 3: Interpret the result of step 2: An annual rate of 10.52% applied once a year will yield the same interest amount, $10.52.

Therefore, $\boxed{10.52\%}$ is the EAR. ∎

You might find it interesting to compare the results of the last two examples with those of Examples 6.1.4 and 6.1.8, respectively.

EXERCISES **6.1**

1. A bond earns 7% annual interest, which is paid semi-annually to the bondholder. How much interest will the bondholder receive each 6 months on a bond of face value $2000?

2. In reference to exercise 1, how many years will it take for the total interest received to equal or exceed the face value?

3. A person negotiates a short-term, 3-month loan of $1500 from a bank that charges 18% annual interest. How much must the person repay the bank at the end of 3 months?

4. A friend will lend you $100 if you pay him $105 at the end of one month. What annual rate of interest are you paying?

5. The sum of $12,500 is invested in an account that earns 7% per year compounded quarterly.

   a. Construct a table of the type shown in Example 6.1.2 for 4 periods.
   b. After 1 year, what will the deposit value be and how much interest will have been earned?

6. The sum of $67,000 is invested in an account that earns 8% per year compounded semiannually.

   a. Construct a table of the type shown in Example 6.1.2 but for 6 periods.
   b. After 3 years, what is the deposit value and the total interest earned?

7. Using the initial data of exercise 5, construct a trace at point Ⓧ of the flowchart in Figure 1 for the variables K, I, and P.

8. Using the initial data of exercise 6, construct a trace at point Ⓧ of the flowchart in Figure 1 for the variables K, I, and P.

In the remaining exercises, use the compound amount formula or continuous compounding formula, as appropriate.

9. Suppose $12,500 is invested in an account that earns 7% per year compounded quarterly. After 1 year, what is the deposit value and total interest earned?

10. Suppose $67,000 is invested in an account that earns 8% per year compounded semiannually. After 3 years, what is the deposit value and total interest earned?

11. If you had a choice between a bank that pays 8% per year compounded quarterly and another bank that pays 8.2% per year compounded annually, where would you invest your money (assuming all other considerations were equal)?

12. If you had a choice between a bank that pays 8% per year compounded semiannually and another bank that pays 8.2% per year compounded annually, where would you invest your money (assuming all other considerations were equal)?

13. Two parents wish to build a college tuition fund of $20,000 to be available 18 years from the birthdate of their child. If they invest in a savings plan that pays 10% per year compounded quarterly, what amount must be invested at the child's birth to assure that the goal will be attained?

14. A family wants to have $2500 at the time of a trip that is planned for 3 years from now. If a local bank offers a 3-year savings certificate that earns 12% per year compounded monthly, what amount must be invested now to guarantee that the desired amount will be available?

15. How long will it take for a deposit of $3500 to grow to $10,000 if the interest rate is 7.6% per year compounded quarterly?

16. What is the tripling time (that is, the time it takes for an initial deposit to triple in value) if interest is compounded semiannually at an annual rate of 14%?

17. If $1000 is to grow to at least $2000 over an 8-year period, what minimum annual rate of interest is required if the interest is compounded semiannually?

18. If $12,000 is to grow to at least $18,000 over a 10-year period, what minimum annual rate of interest is required if the interest is compounded quarterly?

19. The sum of $12,500 is invested in an account that earns 7% per year compounded continuously. What is the deposit value after (a) 1 year? (b) 5 years?

20. The sum of $67,000 is invested in an account earning 8% per year compounded continuously. What is the deposit value after (a) 3 years? (b) 10 years?

21. What is the EAR of the nominal rate of 8% per year compounded (a) semiannually? (b) quarterly? (c) continuously?

22. What is the EAR of the nominal rate of 12% per year compounded (a) monthly? (b) quarterly? (c) continuously?

23. Suppose $1000 is deposited in a bank that pays an interest rate of 6% per year. Determine the deposit value at the end of one year if interest is compounded

   a. annually   **b.** semiannually
   c. quarterly   **d.** monthly

   e. daily   **f.** hourly
   g. continuously

24. Solve for P in the compound amount formula. (The factor $(1 + R)^{-N}$ is called the **present value factor**.)

6.2

Mathematics of Finance II: Annuities

The formulas of the preceding section apply to situations where a single deposit or single payment is made. However, many important cases involve a series of deposits or payments, such as monthly auto, mortgage, or insurance payments and regular savings account deposits. In this section, certain formulas are presented that pertain to these multiple-payment plans, or *annuities*.

Example **6.2.1**

Future Value: Annuity

In order to create a contingency fund, a small business invests $2000 at the end of each quarter into an account earning 8% interest per year compounded quarterly. We wish to find the value of the fund after 1 year, that is, at the time of the fourth payment.

    In one method of solution, the Compound Amount Formula is applied to each of the individual $2000 payments, as shown in Figure 7. Here the interest per period is $.08/4 = .02$, and the value of the contingency fund after 1 year is $8243.22. ∎

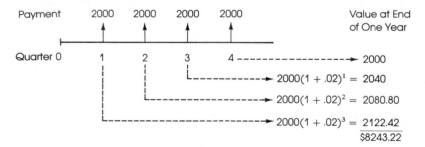

Figure 7

    The technique of Figure 7 can still be applied, although tediously, to a large number of regular payments. But again a little bit of mathematics comes to the rescue!

    If we use the symbol S_4 to represent the fund value at the fourth payment, then

$$S_4 = 2000 + 2000(1 + .02)^1 + 2000(1 + .02)^2 + 2000(1 + .02)^3$$

We can use the principle of mathematical induction (see Section 6.6) to prove that for any positive integer N,

$$S_N = 2000 + 2000(1 + .02)^1 + \cdots + 2000(1 + .02)^{N-1}$$

is the value S_N at the time of the Nth payment. More generally, if A is the periodic payment and R is the interest rate per period, it can be shown that (see Example 6.6.5)

$$S_N = A + A(1 + R)^1 + A(1 + R)^2 + \cdots + A(1 + R)^{N-1}$$

Now for a helpful algebra move: Multiply both sides of the last equation by $(1 + R)$, getting

$$S_N(1 + R) = A(1 + R) + A(1 + R)^2 + A(1 + R)^3 + \cdots + A(1 + R)^N$$

If we vertically align like powers of $(1 + R)$ in the last two equations and subtract, we get

$$S_N(1 + R) = \qquad A(1 + R) + A(1 + R)^2 + \cdots + A(1 + R)^{N-1} + A(1 + R)^N$$
$$- \qquad S_N = A + A(1 + R) + A(1 + R)^2 + \cdots + A(1 + R)^{N-1}$$
$$\overline{S_N(1 + R) - S_N = -A \qquad\qquad\qquad\qquad\qquad\qquad + A(1 + R)^N}$$

or

$$S_N = A\left[\frac{(1 + R)^N - 1}{R}\right]$$

Fact

Series Compound Amount Formula

If

A = payment at end of each period

R = rate of interest per period

N = number of periods

then

$$S_N = A\left[\frac{(1 + R)^N - 1}{R}\right]$$

is the total payment value at the time of the Nth payment.

The term in brackets is called the **_series compound amount factor_**. It is important to remember that the period (such as month, quarter, or year) must be the same for A, R, and N. Hence, the computation of interest coincides with the time of each payment. Finally, this formula can be associated with the diagram

As an illustration, let's apply the series compound amount formula (SCAF) to the data given in the example above, where

A = $2000 per quarter

$$R = \frac{.08}{4} = .02 \text{ per quarter}$$

$$N = 1 \text{ year} = 4 \text{ quarters}$$

(see Figure 8). Then

$$S_4 = A\left[\frac{(1+R)^4 - 1}{R}\right]$$

$$= 2000\left[\frac{(1+.02)^4 - 1}{.02}\right]$$

$$= 2000(4.1216079)$$

$$= \$8243.22$$

Here the series compound amount factor 4.1216079 was determined by using a calculator, but it could have been read (to fewer decimal places) from Table 2 at the back of this book.

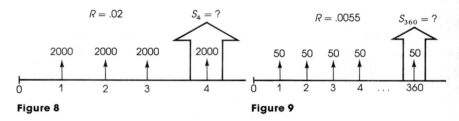

Figure 8   **Figure 9**

Example 6.2.2 **Compound Amount: Annuity**

To save for retirement, a person plans to invest $50 at the end of each month in a bank that compounds interest monthly at a rate of 6.6% per year. What is the value of the retirement fund after 30 years?

Here we have

$$A = \$50 \text{ per month}$$

$$R = \frac{.066}{12} = .0055 \text{ per month}$$

$$N = 30 \text{ years} = 360 \text{ months}$$

(see Figure 9); so

$$S_{360} = A\left[\frac{(1+R)^{360} - 1}{R}\right]$$

$$= 50\left[\frac{(1+.0055)^{360} - 1}{.0055}\right]$$

$$= 50[1127.9196]$$

$$= \boxed{\$56,395.98}$$

Also, since the total of all payments is

$$\$50 * 360 = \$18,000$$

then the difference

$$\$56,395.98 - \$18,000 = \boxed{\$38,395.98}$$

is the total interest earned.   ■

Of course, this same formula can be used to solve for the value of a periodic payment A if all of the other variables are known.

Example 6.2.3

Payment Value: Annuity

A particular investment institution pays 11.4% per year compounded semi-annually. What semiannual payment must be made in order to attain a total investment value of $15000 in five years?

In this case,

$$N = 5 \text{ years} = 10 \text{ half-years}$$

$$R = \frac{.114}{2} = .057 \text{ per half-year}$$

$$S_{10} = \$15000$$

(see Figure 10); using the SCAF, we get

$$S_{10} = A \left[\frac{(1 + R)^{10} - 1}{R} \right]$$

$$15000 = A \left[\frac{(1 + .057)^{10} - 1}{.057} \right]$$

$$= A(12.996561)$$

So

$$A = \frac{15000}{12.996561} = \boxed{\$1154.15}$$

is the semiannual payment that will assure the desired goal. In addition, the payment total is

$$\$1154.15 * 10 = \$11541.50$$

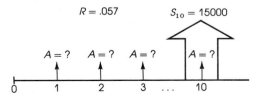

$R = .057$ $S_{10} = 15000$

$A = ?$ $A = ?$ $A = ?$ $A = ?$

Figure 10   0   1   2   3   ...   10

so

$$\$15000 - \$11541.50 = \boxed{\$3458.50}$$

is the total interest earned.   ■

  These examples have related a series of payments A to their future value S_N. However, in some cases it is important to relate the series of payments A with the present value P (see Figure 11). From the SCAF,

$$S_N = A\left[\frac{(1 + R)^N - 1}{R}\right]$$

and from the CAF,

$$S_N = P(1 + R)^N$$

Therefore, equating the right sides of these two equations gives

$$P(1 + R)^N = A\left[\frac{(1 + R)^N - 1}{R}\right]$$

or

$$P = A\left[\frac{(1 + R)^N - 1}{R(1 + R)^N}\right]$$

Figure 11

Fact

> **Series Present Value Formula**
>
> If
>
> A = payment at end of each period
>
> R = rate of interest per period
>
> N = number of periods
>
> then
>
> $$P = A\left[\frac{(1 + R)^N - 1}{R(1 + R)^N}\right]$$
>
> is the present value of these N payments.

The expression in brackets is called the ***series present value factor***, some values of which are given in Table 3 at the back of the book. As before, the period used must be the same for A, R, and N.

Example 6.2.4

Present Value: Annuity

Assuming that you can earn interest at the rate of 8% per year compounded annually, what amount of money must be invested now in order to assure yearly payments of $3500 at the ends of each of the next 4 years?

Here we are given

$A = \$3500$ per year

$R = .08$ per year

$N = 4$ years

(see Figure 12), so using the series present value formula,

$$P = A\left[\frac{(1 + R)^N - 1}{R(1 + R)^N}\right]$$

$$= 3500\left[\frac{(1 + .08)^4 - 1}{.08(1 + .08)^4}\right]$$

$$= 3500[3.3121268]$$

$$= \boxed{\$11,592.44}$$

Also,

$$
\begin{array}{rll}
4 * \$3500 = & \$14000.00 & = \text{total of payments} \\
& -11592.44 & = \text{present value} \\
\hline
& \boxed{\$\ 2407.56} & = \text{total interest earned}
\end{array}
$$

∎

In this example we used a calculator to get the series present value factor of 3.3121268, but a less accurate value can be read from Table 3.

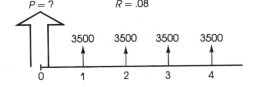

Figure 12

Example 6.2.5

Present Value: Car Payments

You can arrange a 4-year automobile loan at 15% per year compounded monthly. If $200 is all you can afford for the monthly payment, find the maximum amount you can borrow at this rate.

In this situation

$A = \$200$ per month

$R = \dfrac{.15}{12} = .0125$ per month

$N = 4$ years $= 48$ months

(see Figure 13). So

$$P = A\left[\frac{(1 + R)^N - 1}{R(1 + R)^N}\right]$$

$$= 200\left[\frac{(1 + .0125)^{48} - 1}{.0125(1 + .0125)^{48}}\right]$$

$$= 200(35.93148)$$

$$= \boxed{\$7186.30}$$

is the largest loan you can afford.   ■

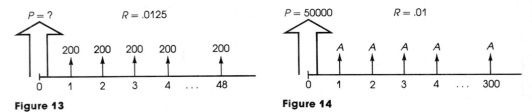

Figure 13   **Figure 14**

As you may have guessed, this series present value formula can be applied to find the regular payment A when all the other variables are known.

Example 6.2.6 **Monthly Payment: Home Mortgage**

A person wishes to borrow $50000 to purchase a home, and the mortgage rate is 12% per year compounded monthly. What is the monthly payment if the term of the mortgage is 25 years?

We are given that

$$R = \frac{.12}{12} = .01 \text{ per month}$$

$$N = 25 \text{ years} = 300 \text{ months}$$

$$P = \$50000$$

(see Figure 14). Substituting these values into the formula yields

$$50000 = A\left[\frac{(1 + .01)^{300} - 1}{.01(1 + .01)^{300}}\right]$$

$$= A[94.946551]$$

so

$$A = \frac{50000}{94.946551}$$

$$= \boxed{\$526.61}$$

Thus, $526.61 is the monthly payment.   ■

In this section we have dealt exclusively with multiple-payment situations in which payments and the compounding of interest coincide. To handle other cases, such as monthly payments with quarterly compounding, other formulas are required.

EXERCISES **6.2**

In the first two exercises, use only the CAF of Section 6.1.

1. A person invests $5000 at the end of each 6-month period into a savings account that yields 9% per year compounded semiannually.

    a. At the end of 3 years, what is the deposit value and the total interest earned?

    b. What is the present value of the first six payments?

2. At the end of each of 4 successive years, a person deposits $2500 into a fund that earns 10% per year compounded annually.

    a. At the time of the fourth deposit, what is the fund value and total interest earned?

    b. What is the present value of the first four payments?

In the remaining exercises use the series compound amount formula or series present value formula, as appropriate.

3. Redo exercise 1(a).

4. Redo exercise 2(a).

5. A parent pays $50 per month into a whole-life insurance plan, which earns $7\frac{1}{2}\%$ interest per year compounded monthly. What is the cash value of this plan after 15 years? How much interest has been earned?

6. As a tax payment fund, a small business deposits $3000 quarterly into a bank that pays $8\frac{1}{2}\%$ annual interest compounded quarterly. Determine the value of the fund and the total interest earned at the end of 1 year.

7. Parents wish to build a college tuition fund of $20000 to be available 18 years from the birthdate of their child. If they invest at the child's birth in a savings plan that pays 10% per year compounded quarterly, what quarterly payment is needed to assure the goal will be attained? (Compare exercise 13 of Section 6.1.)

8. A family wants to have $2500 at the time of a trip that is planned for 3 years from now. If a local bank offers 12% per year interest compounded monthly, what monthly payment will guarantee that the desired amount will be available? (Compare exercise 14 of Section 6.1.)

9. Redo exercise 1(b).

10. Redo exercise 2(b).

11. In a certain retirement plan, the retiree receives a payment at the end of each month for 20 years. If each monthly check to the retiree equals $500, and it can be invested at an annual rate of 6% compounded monthly, what is the present value of the plan?

12. Assuming that you can earn interest at a rate of 8.5% per year compounded semiannually, what is the present value of $1000 payments made at the end of each 6-month period over 10 years?

13. You can arrange a home mortgage at a rate of 11% per year compounded monthly for 20 years. If $42900 is the amount you wish to borrow, what is your monthly mortgage payment?

14. If you wish to borrow $7850 to purchase an auto at an annual percentage rate (APR) of 9.9% compounded monthly for a term of 4 years, what is your monthly car payment?

15. Lucky you have won the state lottery and are given a choice between receiving a lump sum of $60000 now or receiving $500 at the end of each month for 20 years. If you can invest in an account that pays 9% per year compounded monthly, which option would yield the greater monthly income over 20 years? (*Note*: Ignore the obviously important tax considerations here!)

16. Solve for A the series compound amount formula. (The factor

$$\frac{R}{(1 + R)^N - 1}$$

is called the **sinking fund factor**.)

17. Solve for A the series present value formula. (The factor

$$\frac{R(1 + R)^N}{(1 + R)^N - 1}$$

is called the **capital recovery factor**.)

6.3

Statistics I: Descriptive Statistics

Statistics is the branch of mathematics that deals with the collection, classification, presentation, and, most importantly, the interpretation of numerical data. Its impact and significance in our modern world are enormous. Whenever sample data are gathered in studies of sociological, psychological, educational, scientific, or organizational problems, statistical methods are generally used in drawing conclusions about the population from which the sample was taken. As a result, the study of elementary statistics is a book (and a course!) unto itself. However, since software is available for many computer systems, even microcomputers, to perform statistical analyses, we have included in this section and the next a brief introduction to statistics. Fundamental terms and techniques are presented by way of example, rather than by general definition and explanation.

The first two examples concern the classification and presentation of data by means of a *frequency distribution*.

Example 6.3.1

Frequency Distribution: Discrete Variable

A computer processing service has collected figures on the number of batch jobs completed between 9 A.M. and noon on each of 75 randomly selected workdays. The ordered list of those 75 numbers, called *ranked data*, is given in Figure 15.

27	36	41	45	50	53	57	61
28	36	41	45	51	54	57	61
28	37	42	46	51	54	57	61
30	38	43	46	51	55	59	62
31	38	44	48	51	55	59	62
31	39	44	49	52	55	59	
33	40	45	49	52	56	60	
34	40	45	49	52	56	60	
35	40	45	50	52	57	60	
36	40	45	50	52	57	60	

Figure 15

In this ranked form, it is difficult to perceive what information these sample data hold concerning the batch job counts for all days of operation, called the *population*. Therefore, we'll construct a tabular representation of this data known as a *frequency distribution*.

The *range* of this data is the highest score (62) minus the lowest score (27), or 35. We shall "cover" this range of 35 with eight *classes*, each having a *class width* of 5, although there are several other possibilities. (The class width should be chosen so as to produce 5 to 13 classes. Generally, there is no single, best choice.) This division into classes is summarized in the two leftmost columns of Table 1.

Now we can proceed item by item through the ranked data to determine the total number of entries in each class, called the *frequency* of the class; the results are given in Table 1. A *class mark* is the numerical value that is exactly in the middle of the class limits. It is calculated by adding

27   Range 35   62

$25 \to 29,\ 30 \to 34,\ 35 \to 39,\ 40 \to 44,\ 45 \to 49,\ 50 \to 54,\ 55 \to 59,\ 60 \to 64$

Table 1

Class	Class Limits	Frequency	Class Mark
1	25–29	3	27
2	30–34	5	32
3	35–39	8	37
4	40–44	10	42
5	45–49	12	47
6	50–54	15	52
7	55–59	13	57
8	60–64	9	62
		75	

the class limits and dividing by 2; in class 3, for instance, the class mark is

$$\frac{35 + 39}{2} = 37$$

Table 1 is an illustration of a *grouped frequency distribution*. Notice that some information is lost in the grouping of the given ranked data; for example, Table 1 tells us that the 35–39 class has eight entries without specifying their values.

In Figure 16 we have drawn a geometric representation, called a *histogram*, of the frequency distribution in Table 1. The frequencies are plotted on the vertical axis, and the horizontal axis (*x*-axis) is divided into eight intervals by the *class boundaries*. A class boundary is the number halfway between the upper limit of one class and the lower limit of the next. So, for instance, the boundary between class 1 and class 2 is the point 29.5. Observe that in this example the class boundaries cannot be among the sample data since *x* is a *discrete variable*—that is, *x* can assume only integer values.

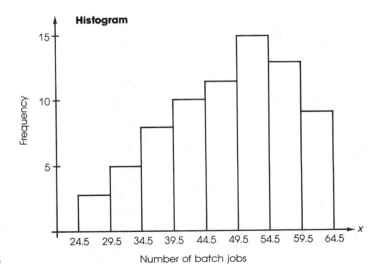

Histogram

Figure 16

Number of batch jobs

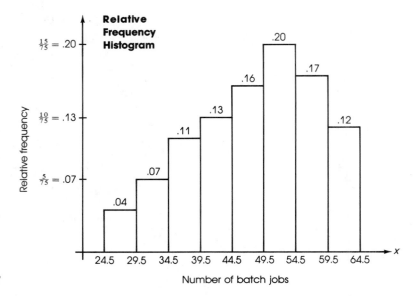

Figure 17

The ***relative frequency histogram*** in Figure 17 is obtained from the histogram of Figure 16 by dividing the vertical scale by 75, the total number of observations in our sample. Hence, the relative frequency

$$\text{of class 1 is } \frac{3}{75} = .04 \qquad \text{of class 2 is } \frac{5}{75} = .07$$

and so on. The sum of all the relative frequencies equals 1. If our sample of 75 was drawn to be representative of the population, then the relative frequencies can be interpreted as approximate proportions of the population. For instance, roughly

$$.16 = 16\%$$

of the population of all batch counts lies in the class 45–49.

Finally, a histogram that has the shape of those in Figures 16 and 17 is said to be ***skewed to the left***. ∎

Example 6.3.2

Frequency Distribution: Continuous Variable

A manufacturer has collected data on a particular model of dot-matrix printer, the variable x of interest being the time (in hours) until the first failure. Of the 2000 units randomly sampled, Table 2 summarizes the results.

In this table a ***class mark*** is that number exactly in the middle of the class boundaries, and it is computed in the same fashion as in the preceding example. The frequency and relative frequency histograms are drawn in Figures 18 and 19, respectively.

Observe that in this example the class boundary points (.5, 75.5, 150.5, and so on) *can* lie among the sample data, since hours can be measured in noninteger numbers. Should a piece of data fall exactly on a boundary point between two classes, we shall count it in the class to the left, that is, in the

Table 2

Class	Class Boundaries	Frequency	Class Mark
1	.5– 75.5	9	38
2	75.5–150.5	32	113
3	150.5–225.5	182	188
4	225.5–300.5	295	263
5	300.5–375.5	452	338
6	375.5–450.5	437	413
7	450.5–525.5	319	488
8	525.5–600.5	201	563
9	600.5–675.5	55	638
10	675.5–750.5	18	713
		2000	

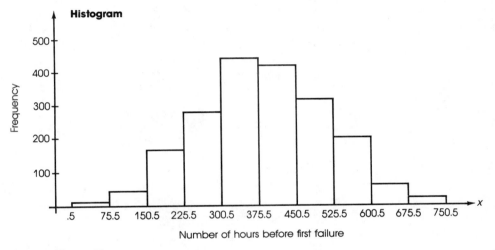

Figure 18

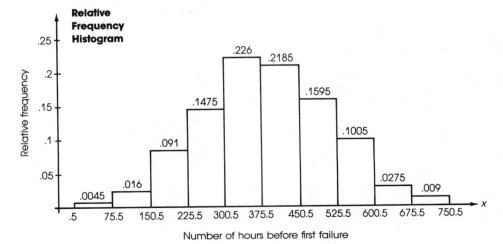

Figure 19

class with the smaller boundaries. Since the variable x can assume any of the infinitely many number values between .5 and 750.5, we call x a ***continuous variable***, as opposed to a discrete variable.

The relative frequency histogram is obtained from the frequency histogram by dividing all frequencies by 2000, the total number of cases sampled. Once again, if this sample is representative of the population of all such printers, then the relative frequencies

.0045   .016   .091   .1475   and so on

can be thought of as approximate proportions of the entire population. To illustrate, the relative frequency of .1475 for class 4 means that about

$.1475 = 14.75\%$

of the population of all printers fail for the first time after 225.5 to 300.5 hours of operation.

Also, the histograms in Figures 18 and 19 approximate what is called a ***normal histogram*** because they are almost symmetric about a high point in the center. Such "bell-shaped" distributions are the topic of Section 6.4. ■

In statistics the three most commonly used ***measures of central tendency***, or measures of average value, are the mode, median, and mean.

Mode

1. In a collection of ungrouped data, the ***mode*** is the number that occurs with the greatest frequency.
2. In a frequency distribution, the ***modal class*** is the class that occurs with the greatest frequency.

If there is no single number satisfying (1) or no single class satisfying (2), then there is no mode or no modal class, respectively.

Example 6.3.3   **Mode and Modal Class**

a. In Example 6.3.1, where 75 pieces of data were given, 45 occurs more often, namely six times, than any other number. So 45 is the mode. However, when these data are grouped as shown in Table 1, the modal class is 50–54, since its frequency of 15 is the largest.

b. In Example 6.3.2, the ungrouped data are not available, so there is no mode. But in the given frequency distribution of Table 2, the greatest frequency of 452 is associated with the class 300.5–375.5, so this is the modal class.   ■

Median

In a collection of ranked data or in a frequency distribution, the ***median*** is the middle value.

Admittedly, the phrase "middle value" needs clarification; the next example will help.

Example 6.3.4 **Median**

a. Whenever a collection of ranked data has an odd number m of entries, then the middle value is the

$$\frac{m + 1}{2}$$

piece of data. For instance, Figure 15 contains $m = 75$ ordered numbers, so the middle value is the

$$\frac{m + 1}{2} = \frac{75 + 1}{2} = 38\text{th}$$

piece of data, leaving 37 pieces on either side. Counting from the top of that list gives the 38th entry to be 49, which is the median.

b. Whenever a collection of ranked data contains an even number n of entries, life is more difficult. The middle value is that halfway between the

$$\frac{n}{2} \quad \text{and} \quad \frac{n}{2} + 1$$

pieces of data, leaving $n/2$ pieces of data on either side.

For example, if we delete the last five entries from the list in Figure 15, there are $n = 70$ pieces of data, an even number. So the middle value is that number halfway between the

$$\frac{n}{2} = \frac{70}{2} = 35\text{th} \quad \text{and} \quad \frac{n}{2} + 1 = 36\text{th}$$

entries. Since the 35th entry is 48 and the 36th is 49, then the median is 48.5, which does not even appear in the list.

c. For a frequency distribution you usually must use *interpolation* to approximate the median. For the distribution given in Table 2 (Example 6.3.2), we seek to estimate the 1000th score from among the total of 2000 tallied. By adding frequencies starting at the top, we see that the sum of the frequencies of classes 1, 2, 3, 4, and 5 equals 970. So the 1000th score is the 30th one into class 6, which has frequency 437.

Now we assume that these 437 scores are evenly distributed throughout class 6, so that the 1000th score is $\frac{30}{437}$ of the way into class 6; that is,

$$\tfrac{30}{437} * 75 = 5.2$$

units into class 6. Hence the 1000th score, or median, is approximately

$$375.5 + 5.2 = 380.7$$

(see Figure 20). ■

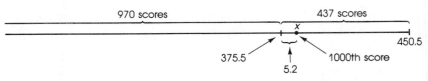

Figure 20

However, of the three measures of central tendency presented here, by far the most important is the mean.

Mean

1. If
$$x_1, x_2, \ldots, x_N$$
is a collection of numerical data, then the *sample mean* $\bar{x}$ is given by
$$\bar{x} = \frac{\sum_{i=1}^{N} x_i}{N}$$

2. If a frequency distribution has class marks
$$X_1, X_2, \ldots, X_M$$
and respective class frequencies
$$F_1, F_2, \ldots, F_M$$
whose sum is N, then the *sample mean* $\bar{x}$ is given by
$$\bar{x} = \frac{\sum_{i=1}^{M} X_i F_i}{N}$$

The effect of statement 2 is that each piece of data is treated as though it fell at the class mark of the class to which it belongs.

Example 6.3.5   **Sample Mean**

a. In Example 6.3.1, for the ranked data of Figure 15, the sample mean is

$$\frac{\text{Sum of 75 given numbers}}{75} = \frac{3570}{75} = \boxed{47.6}$$

For the associated frequency distribution of Table 1, the mean is

$$\bar{x} = \frac{\sum_{i=1}^{8} X_i F_i}{75}$$

$$= \frac{27(3) + 32(5) + 37(8) + 42(10) + 47(12) + 52(15) + 57(13) + 62(9)}{75}$$

$$= \frac{3600}{75} = \boxed{48}$$

It's interesting to compare the last two answers. When each piece of ranked data (Figure 15) is considered to fall at its class mark (Table 1), the error introduced in the value of the mean is only

$$48 - 47.6 = .4$$

b. The mean of the frequency distribution given in Table 2 of Example 6.3.2 is

$$\bar{x} = \frac{\sum\limits_{i=1}^{10} X_i F_i}{2000}$$

$$= \frac{\begin{array}{c} 38(9) + 113(32) + 188(182) + 263(295) + 338(452) \\ + 413(437) + 488(319) + 563(201) + 638(55) + 713(18) \end{array}}{2000}$$

$$= \boxed{382.9}$$ ∎

While the mean is the most important and most widely used statistic for measuring central tendency, the **standard deviation** is the most important statistic for measuring the dispersion of data about the mean.

Standard Deviation

   1. If

$$x_1, x_2, \ldots, x_N$$

is a collection of numerical data with mean $\bar{x}$, then the **sample variance** s^2 is given by

$$s^2 = \frac{\sum\limits_{i=1}^{N} (x_i - \bar{x})^2}{N - 1}$$

The quantity $s = \sqrt{s^2}$ is called the **sample standard deviation**.

   2. If a frequency distribution has class marks

$$X_1, X_2, \ldots, X_M$$

respective class frequencies

$$F_1, F_2, \ldots, F_M$$

whose sum is N, and mean $\bar{x}$, then the **sample variance** s^2 is given by

$$s^2 = \frac{\sum\limits_{i=1}^{M} (X_i - \bar{x})^2 F_i}{N - 1}$$

And $s = \sqrt{s^2}$ is called the **sample standard deviation**.

As with the definition of *mean* given earlier, statement 2 treats each piece of data as though it fell on a class mark. Obviously, statement 1 is preferred *if* the raw data are available. Also, if you're wondering why the denominators above are $N - 1$ rather than N, it's because statisticians can prove that $N - 1$ yields a value for s^2 that better approximates the variance of the entire population.

Example 6.3.6

Sample Variance and Standard Deviation

a. For the ranked data given in Figure 15, we found in the preceding example that $\bar{x} = 47.6$, so the sample variance is

$$\frac{(27 - 47.6)^2 + (28 - 47.6)^2 + (28 - 47.6)^2 + \cdots + (62 - 47.6)^2}{75 - 1} = \frac{6810}{74} = \boxed{92.0}$$

The standard deviation equals

$$\sqrt{92.0} = \boxed{9.6}$$

In the preceding example we also calculated the mean $\bar{x}$ of the frequency distribution in Table 1 to be 48. So the variance of this distribution is

$$s^2 = \frac{\sum_{i=1}^{8} (X_i - 48)^2 F_i}{75 - 1}$$

$$= \frac{\begin{array}{c}(27 - 48)^2 3 + (32 - 48)^2 5 + (37 - 48)^2 8 + (42 - 48)^2 10 \\ + (47 - 48)^2 12 + (52 - 48)^2 15 + (57 - 48)^2 13 + (62 - 48)^2 9\end{array}}{74}$$

$$= \frac{7000}{74} = \boxed{94.6}$$

and the standard deviation is

$$s = \sqrt{94.6} = \boxed{9.7}$$

Once again, we notice the small difference between the variances of the ranked data and grouped data (namely, $94.6 - 92.0 = 2.6$), and the even slighter difference between the standard deviations of the ranked data and grouped data (namely, $9.7 - 9.6 = .1$).

b. The frequency distribution of Table 2 was found to have mean $\bar{x} = 382.9$, therefore the variance is

$$s^2 = \frac{\sum_{i=1}^{10} (X_i - 382.9)^2 F_i}{2000 - 1}$$

$$= \frac{[(38 - 382.9)^2 9 + (113 - 382.9)^2 32 + \cdots + (713 - 382.9)^2 18}{1999}$$

$$= \frac{31,447,100}{1999} = \boxed{15731.4}$$

and the standard deviation is

$$s = \sqrt{15731.4} = \boxed{125.4}$$ ■

Two concluding comments are in order here. First, the means calculated in Example 6.3.5 should be taken as average values of their respective samples, and the standard deviations computed in the preceding example give some measure of the spreads of the data about their respective means.

More specifically, there is a result called Chebyshëv's theorem,[†] which states that in any distribution of data with mean $\bar{x}$ and standard deviation s, at least 75% of the data lie between

$$\bar{x} - 2s \quad \text{and} \quad \bar{x} + 2s$$

and at least 89% of the data lie between

$$\bar{x} - 3s \quad \text{and} \quad \bar{x} + 3s$$

Thus, a small value for s indicates a high concentration of data about $\bar{x}$, while a larger s indicates the data are more dispersed about $\bar{x}$ (see Figure 21).

	Small s	Larger s

Figure 21   Data concentrated about $\bar{x}$ Data more dispersed about $\bar{x}$

Second, in practice $\bar{x}$ and s are calculated only when the samples are considered representative of the populations. (In our examples, the populations are *all* batch job counts and first-failure times for *all* printers of the given model.) In this way $\bar{x}$ and s provide estimates for the population mean and population standard deviation, and it is the background population about which we wish to make inferences or draw conclusions.

We shall finish this section with flowcharts in Figures 22 and 23 for computing the mean and standard deviation of N pieces of numerical data. Notice that a one-dimensional array $x(I)$ is utilized in calculating the standard deviation but not the mean. Actually, there are alternate, yet equivalent, formulas for standard deviation that do *not* require computation of the mean first (see exercises 15 through 18).

[†] After Russian mathematician P. L. Chebyshëv (1821–1894).

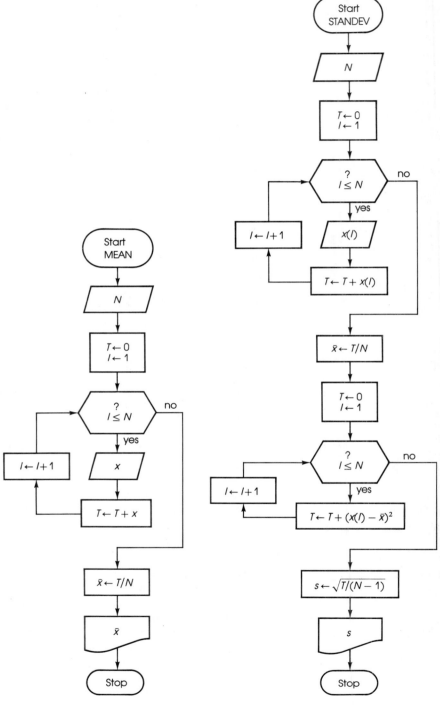

Figure 22

Figure 23

EXERCISES **6.3**

The two collections of ranked data in Figures 24 and 25 are used in many of the exercises below.

26	28	30	31	32	33	35	38	41	43
26	28	30	31	32	33	36	38	41	44
27	28	30	31	32	34	36	39	42	45
27	28	30	31	32	34	36	39	42	45
27	29	31	32	32	35	38	39	43	47
28	29	31	32	33	35	38	41	43	48

Figure 24 Numbers of students enrolled in 60 sections of computer science courses

16.2	22.9	28.1	31.5	33.8	36.0	39.3	44.7	49.4
18.7	23.6	28.9	31.9	34.1	37.2	40.0	45.1	50.5
20.0	24.4	29.2	32.2	35.4	37.8	41.4	45.8	51.9
21.8	26.0	30.1	33.0	35.8	38.1	42.0	47.4	53.6
22.5	27.3	30.8	33.6	35.9	38.5	43.1	48.2	55.1

Figure 25 Speeds of 45 autos passing through a particular intersection

1. For the data set of Figure 24, calculate the following:

    a. Mode   **b.** Median

    c. Mean   **d.** Standard deviation

2. Same as exercise 1, but for the data set in Figure 25.

3. Construct a frequency distribution for the data set in Figure 24 using the five classes 25–29, 30–34, 35–39, 40–44, and 45–49 with class width 5.

4. Construct a frequency distribution for the data set in Figure 25 using the six classes 14.5–21.5, 21.5–28.5, 28.5–35.5, 35.5–42.5, 42.5–49.5, and 49.5–56.5 with class width 7.

5. Construct a frequency histogram and relative frequency histogram for the distribution of exercise 3.

6. Construct a frequency histogram and relative frequency histogram for the distribution of exercise 4.

7. For the frequency distribution of exercise 3, calculate the following:

    a. Modal class   **b.** Median

    c. Mean   **d.** Standard deviation

8. For the frequency distribution of exercise 4, calculate the following:

    a. Modal class   **b.** Median

    c. Mean   **d.** Standard deviation

9. Construct a frequency distribution for the data set in Figure 24 using the nine classes 24–26, 27–29, 30–32, 33–35, 36–38, 39–41, 42–44, 45–47, and 48–50 with class width 3.

10. Construct a frequency distribution for the data set in Figure 25 using the eight classes 15.5–20.5, 20.5–25.5, 25.5–30.5, 30.5–35.5, 35.5–40.5, 40.5–45.5, 45.5–50.5, and 50.5–55.5 with the class width 5.

11. Construct a frequency histogram and relative frequency histogram for the distribution of exercise 9.

12. Construct a frequency histogram and relative frequency histogram for the distribution of exercise 10.

13. For the frequency distribution of exercise 9, calculate the following:

    a. Modal class   **b.** Median

    c. Mean   **d.** Standard deviation

14. For the frequency distribution of exercise 10, calculate the following:

    a. Modal class   **b.** Median

    c. Mean   **d.** Standard deviation

15. An equivalent formula for computing the variance of the collection of numerical data $x_1, x_2, \ldots, x_N$ is given by

$$s^2 = \frac{N\left(\sum_{i=1}^{N} x_i^2\right) - \left(\sum_{i=1}^{N} x_i\right)^2}{N(N-1)}$$

Use the formula to find the variance and standard deviation of the data set in Figure 24. Does the standard deviation coincide with the answer to exercise 1(d)?

16. Use the formula given in the preceding exercise to find the variance and standard deviation of the data set in Figure 25. Does the standard deviation coincide with the answer to exercise 2(d)?

17. Consider the grouped data with class marks

$$X_1, X_2, \ldots, X_M$$

and respective class frequencies

$$F_1, F_2, \ldots, F_M$$

whose sum is N. An equivalent formula for computing the variance is given by

$$s^2 = \frac{N\left(\sum_{i=1}^{M} X_i^2 F_i\right) - \left(\sum_{i=1}^{M} X_i F_i\right)^2}{N(N-1)}$$

Use this formula to find the variance and standard deviation of the data with the frequency distribution found in exercise 3. Does the standard deviation coincide with the answer to exercise 7(d)?

18. Use the formula given in the preceding exercise to find the variance and standard deviation of the data with the frequency distribution found in exercise 4. Does the

standard deviation coincide with the answer to exercise 8(d)?

19. Draw a flowchart for the computation of the standard deviation using the formula given in exercise 15. Is a one-dimensional array necessary here to represent the data? Compare this flowchart with the one in Figure 23.

6.4

Statistics II: The Normal Distribution

Take a moment to review the relative frequency histogram of Figure 17, which concerns the number of batch jobs completed in a given time interval. By changing the scale of the vertical axis (specifically, by dividing by class width 5), we can produce another histogram of the same shape but where the relative frequencies are exactly the areas of the eight vertical bars, as shown in Figure 26. Remember from Example 6.3.1 that these relative frequencies (that is, the areas) can be interpreted as proportions of the population. So, for instance, the observation that the fourth bar from the left in Figure 26 has area .13 means that approximately 13% of the population of all batch counts fall between 39.5 and 44.5. Differently stated, the probability is .13 that a randomly chosen value of x falls between 39.5 and 44.5. Notationally, these last two sentences can be expressed as

$$\Pr(39.5 \leq x \leq 44.5) = .13$$

Therefore Figure 26 can be considered as a ***probability distribution*** for the population.

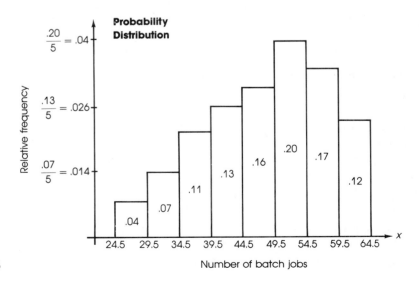

Figure 26

The variable x just discussed is a ***discrete variable*** insofar as its values arise from a count (of batch jobs completed in a given time interval); so, in this example, x can take on only integer values. A variable x is called ***continuous*** when it represents quantities that are measured rather than counted. For instance, time, length, weight, temperature, volume, and speed are quantities that can assume an infinity of values between any two numbers.

The population distribution of a continuous random variable is often represented by a smooth curve rather than by a histogram. As above, the statement

$$Pr(a \leq x \leq b)$$

can be interpreted in two equivalent ways: as a proportion of the population between fixed numbers a and b or as the probability that one randomly selected value of x will fall between a and b.

Continuous Probability Distribution

A ***probability distribution of a continuous variable x*** is represented by a smooth curve

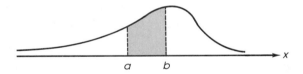

and has the following properties:

  1. $Pr(a \leq x \leq b)$ = area under curve and above axis between a and b

  2. For any fixed number a,
$$Pr(a \leq x \leq a) = P(x = a) = 0$$

  3. The total area under the curve and above the x-axis equals 1; that is,
$$Pr(-\infty < x < \infty) = 1$$

Such a distribution is used to represent a population rather than a sample. Property (2) says that $Pr(x = a) = Pr(x = b) = 0$, so as a consequence we have

$$
\begin{aligned}
Pr(a \leq x \leq b) &= Pr(a < x \leq b) \\
&= Pr(a \leq x < b) \\
&= Pr(a < x < b)
\end{aligned}
$$

which we shall use whenever necessary or appropriate.

Of the many types of continuous probability distributions that are studied in statistics, ***normal probability distributions*** are certainly the most noteworthy. The reason is that many important variables—like incomes, IQs, test scores, heights, and weights—have distributions that closely approximate a normal distribution.

Normal Probability Distribution

A ***normal probability distribution*** is a continuous probability distribution whose graph is a bell-shaped, smooth curve symmetric about a center point μ (called the ***population mean***) and possessing a spread parameter σ (called the ***population standard deviation***).

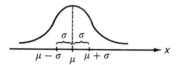

This distribution exhibits the following properties:

1. Within one standard deviation of the mean lies approximately 68% of the population.

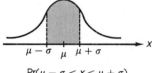

$$\Pr(\mu - \sigma < x < \mu + \sigma)$$
$$= \text{shaded area} \approx .68$$

2. Within two standard deviations of the mean lies approximately 95% of the population.

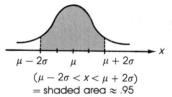

$$(\mu - 2\sigma < x < \mu + 2\sigma)$$
$$= \text{shaded area} \approx .95$$

3. Within three standard deviations of the mean lies approximately 99.7% of the population.

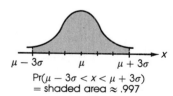

$$\Pr(\mu - 3\sigma < x < \mu + 3\sigma)$$
$$= \text{shaded area} \approx .997$$

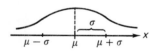

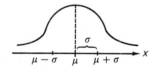

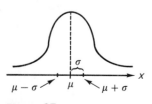

Figure 27

Several comments are in order here. First, a normal distribution is uniquely determined by its mean μ and standard deviation σ. Three different normal distributions are drawn in Figure 27; note how σ measures the spread of the population about μ. Second, knowing μ and σ for a particular distribution, the equation of the bell-shaped curve is the graph of the function

$$f(x) = \frac{1}{\sigma\sqrt{2\pi}}\, e^{(-1/2)((x-\mu)/\sigma)^2}$$

and the probability (proportion, or area) can be expressed using a definite integral:

$$\Pr(a < x < b) = \int_a^b f(x)\, dx$$

These last two expressions, which are studied in calculus and succeeding probability and statistics courses, will not be pursued here.

Example 6.4.1

Normal Probability Distribution

Recall Example 6.3.2 and the associated frequency distribution (Table 2), which concern the number of hours before the first printer failure. In the preceding section we computed the sample mean and standard deviation to be

$$\bar{x} = 382.9 \quad \text{and} \quad s = 125.4$$

respectively. Therefore, we have

$$\bar{x} - s = 257.5 \quad \text{and} \quad \bar{x} + s = 508.3$$
$$\bar{x} - 2s = 132.1 \quad \text{and} \quad \bar{x} + 2s = 633.7$$
$$\bar{x} - 3s = 6.7 \quad \text{and} \quad \bar{x} + 3s = 759.1$$

Using Figure 19 and taking into account, when necessary, fractional portions of the rectangular sections of this histogram, the following values can be calculated:

$$\Pr(257.5 < x < 508.3) = .0846 + .226 + .2185 + .1229$$
$$= .652$$

$$\Pr(132.1 < x < 633.7) = .0039 + .091 + .1475 + .226 + .2185$$
$$+ .1595 + .1005 + .0122$$
$$= .9591$$

$$\Pr(6.7 < x < 759.1) \;\; = .0041 + .016 + .091 + .1475 + .226$$
$$+ .2185 + .1595 + .1005 + .0275 + .009$$
$$= .9996$$

Observe that these three numbers are very close to the values

$$.68 \quad .95 \quad .997$$

contained in the definition of a normal probability distribution. If this sample is representative of the population of all printers, then the population is likely to have a normal distribution, and we can estimate the population mean μ and standard deviation σ by using the above values of $\bar{x}$ and s. We said *estimate* here because it is very unlikely that the values of $\bar{x}$ and s of a

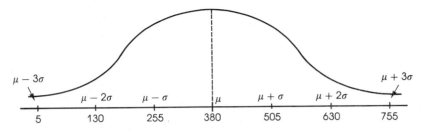

Figure 28

Number of hours before first failure

sample, even a representative sample, coincide exactly with μ and σ, respectively. To simplify any arithmetic involved, we'll take the approximations

$$\mu = 380 \quad \text{and} \quad \sigma = 125$$

The normal distribution for this population is shown in Figure 28. ∎

Example 6.4.2 **Standard Normal Distribution**

The *standard normal distribution* is that particular normal distribution with mean $\mu = 0$ and standard deviation $\sigma = 1$. For this distribution, it is customary to use z rather than x to denote the random variable and to call a particular value of z a *standard score* or *z-score* (see Figure 29).

The entries in Table 4 (entitled "Standard Normal Distribution") in the back of this text can be used as follows: For a particular number $z' > 0$ that is accurate to two decimal places, the table entry gives the area beneath the standard normal curve, above the z-axis, and between the mean 0 and z'. That is,

$$\Pr(0 < z < z') = \text{area } R$$
$$= \text{table entry for } z'$$

(see Figure 30).

a. $z' = 1$

$$\Pr(0 < z < 1) = \text{area } R$$
$$= \text{table entry for } z = 1$$
$$= \boxed{.3413}$$

(see Figure 31).

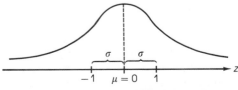

Standard Normal Probability Distribution

Figure 29

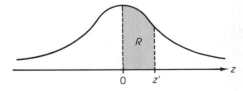

Figure 30

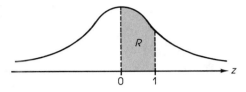

Figure 31   **Figure 32**

Because of the symmetry about the mean of a normal curve, Table 4 can also be used to find areas bounded by negative z-scores, as the next two examples show.

b. $z' = -2$

$$\Pr(-2 < z < 0) = \text{area } R$$
$$= \text{area } R'$$
$$= \text{table entry for } z = 2$$
$$= \boxed{.4772}$$

(see Figure 32).

c. $\Pr(-1 < z < .52) = \quad\quad \text{area } R_1 \quad\quad + \quad\quad \text{area } R_2$
$$= (\text{table entry for } z = 1) + (\text{table entry for } z = .52)$$
$$= \quad\quad\quad .3413 \quad\quad\quad + \quad\quad\quad .1985$$
$$= \boxed{.5398}$$

(see Figure 33).

d. $\Pr(.97 < z < 2) = \quad\quad \text{area } R_2$
$$= \quad \text{area } (R_1 \cup R_2) \quad\quad - \quad\quad \text{area } R_1$$
$$= (\text{table entry for } z = 2) - (\text{table entry for } z = .97)$$
$$= \quad\quad\quad .4772 \quad\quad\quad - \quad\quad\quad .3340$$
$$= \boxed{.1432}$$

(see Figure 34). ■

Table 4 can be used to evaluate *any* normal variable x if we first convert x-values into corresponding z-scores by means of the formula

$$\boxed{z = \dfrac{x - \mu}{\sigma}}$$

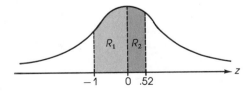

Figure 33

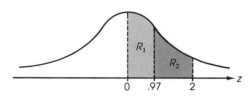

Figure 34

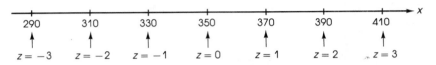

Figure 35

where μ and σ are the mean and standard deviation, respectively, of the population of x-values. In effect, the z-score as computed by this formula is just the distance of the x-value from the mean μ, measured in units of standard deviation σ. Note that the mean μ is always associated with $z = 0$ and $x = \mu + \sigma$ corresponds to $z = 1$ (see Figure 35).

Example 6.4.3 | **Normal Distribution: z-values**

Consider the normal distribution with

$$\mu = 350 \quad \text{and} \quad \sigma = 20$$

Several corresponding z-values are given in Figure 36.

Figure 36

a. For $x = 400$,

$$z = \frac{400 - 350}{20} = 2.5$$

So

$$\Pr(350 < x < 400) = P(0 < z < 2.5)$$
$$= \text{area } R$$
$$= \text{table entry for } z = 2.5$$
$$= \boxed{.4938}$$

(see Figure 37).

b. For $x = 311$,   For $x = 360$,

$$z = \frac{311 - 350}{20} = -1.95 \qquad z = \frac{360 - 350}{20} = .5$$

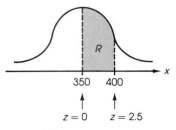

Figure 37

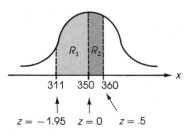

Figure 38

So

$$\begin{aligned}
\Pr(311 < x < 360) &= \Pr(-1.95 < z < .5) \\
&= \quad \Pr(-1.95 < z < 0) \quad + \quad \Pr(0 < z < .5) \\
&= \quad\quad \text{area } R_1 \quad\quad\quad + \quad\quad \text{area } R_2 \\
&= (\text{table entry for } z = 1.95) + (\text{table entry for } z = .5) \\
&= \quad\quad\quad\quad .4744 \quad\quad\quad + \quad\quad\quad .1915 \\
&= \boxed{.6659}
\end{aligned}$$

(see Figure 38). ■

The concluding three examples of this section illustrate practical applications of normal distributions.

Example 6.4.4

Normal Distribution: Failure Time

Recall that in Example 6.4.1 we estimated the mean and standard deviation of the population of x-values (number of hours before first failure) to be

$$\mu = 380 \quad \text{and} \quad \sigma = 125$$

We'd like to find the proportion of all printers that failed within the first 500 hours of use. That is, we want

$$\Pr(x < 500)$$

(see Figure 39). For $x = 500$,

$$z = \frac{500 - 380}{125} = .96$$

$$\begin{aligned}
\Pr(x < 500) &= \text{area } R_1 + \quad\quad \text{area } R_2 \\
&= \quad .5 \quad + \Pr(380 < x < 500) \\
&= \quad .5 \quad + \quad \Pr(0 < z < .96) \\
&= \quad .5 \quad + (\text{table entry for .96}) \\
&= \quad .5 \quad + \quad\quad\quad .3315 \\
&= \boxed{.8315}
\end{aligned}$$

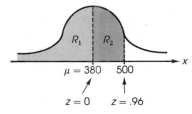

Figure 39

This result could also be interpreted to mean that if one printer is selected at random, the probability is about .8 that it will fail within the first 500 hours of operation. ■

Example 6.4.5

Normal Distribution: Salaries

Suppose that the yearly salaries of workers in a particular industry are normally distributed with mean $\mu = \$15,670$ and standard deviation $\sigma = \$1430$.

What proportion of the workers in this industry earn more than $18,000 per year (see Figure 40)?

For $x = 18000$

$$z = \frac{18000 - 15670}{1430} = 1.63$$

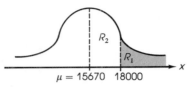

Figure 40

$$
\begin{aligned}
\Pr(x > 18000) = \text{area } R_1 &= \text{area } (R_1 \cup R_2) - \quad \text{area } R_2 \\
&= \quad .5 \quad\quad - P(15670 < x < 18000) \\
&= \quad .5 \quad\quad - \quad P(0 < z < 1.63) \\
&= \quad .5 \quad\quad - \text{(table entry for 1.63)} \\
&= \quad .5 \quad\quad - \quad\quad .4484 \\
&= \boxed{.0516}
\end{aligned}
$$

Again, we could also interpret this result as saying that if one worker is randomly chosen, the probability is about .05 that this person earns more than $18,000 per year. ∎

Example 6.4.6   **Normal Distribution: Exam Scores**

Assume that the scores on a particular exam are normally distributed about a mean $\mu = 73$ with standard deviation $\sigma = 12$. If the instructor states that the top 10% of the scores will receive an A, what is the lowest exam score that will earn an A grade? (see Figure 41).

Using Table 4 in reverse, we locate the entry closest to .40. (Interpolation is possible here, but we shall bypass it.) The closest entry is .3997, with corresponding $z' = 1.28$. Thus, substituting into the formula that relates x-values and z-scores, we have

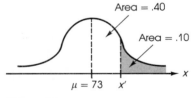

Figure 41

$$z' = \frac{x' - \mu}{\sigma}$$

$$1.28 = \frac{x' - 73}{12}$$

$$15.36 = x' - 73$$

$$\boxed{x' = 88.36}$$

So 88 approximates the lowest score that will receive an A. ∎

EXERCISES 6.4

1. In the standard normal distribution, find the proportion of scores that satisfy the given inequality(ies) and interpret each answer as an area.

a. $0 < z < 1.23$ **b.** $-.6 < z < 0$
c. $-1.05 < z < .82$ **d.** $.2 < z < 2.75$
e. $-.45 < z$

2. Same as for exercise 1, but for the following inequality(ies):

a. $0 < z < .5$ **b.** $-1.83 < z < 0$
c. $-.41 < z < 1.6$ **d.** $-2.1 < z < -1.92$
e. $z < .1$

3. For each of the five parts of exercise 1, what proportion of x-values taken from a normal population will have z-scores in the given interval?

4. For each of the five parts of exercise 2, what proportion of x-values taken from a normal population will have z-scores in the given interval?

5. Let variable x have a normal distribution with mean $\mu = 16.2$ and standard deviation $\sigma = 3.6$. In each of the following, find the proportion of the scores of the population that satisfy the given inequality(ies) and interpret each answer in terms of area:

    a. $16.2 < x < 20$ **b.** $15 < x < 20$
    c. $18 < x < 19.5$ **d.** $x < 15.8$

6. Same as for exercise 5, but with $\mu = 105$, $\sigma = 7$, and the following:

    a. $99 < x < 105$ **b.** $100 < x < 120$
    c. $91 < x < 99.3$ **d.** $107.5 < x$

7. For each part of exercise 5, if one x-value is chosen at random, what is the probability that it satisfies the given inequality(ies)?

8. For each part of exercise 6, if one x-value is chosen at random, what is the probability that it satisfies the given inequality(ies)?

9. For the normal distribution of exercise 5, find the x-value corresponding to each of the following z-scores:

    a. $z = 0$ **b.** $z = 1.01$ **c.** $z = -2.23$

10. For the normal distribution of exercise 6, find the x-value corresponding to each of the following z-scores:

    a. $z = 0$ **b.** $z = 2.5$ **c.** $z = -.86$

*The next four exercises use the term **percentile**: If k is an integer, $1 \le k \le 100$, the kth percentile is the score such that at most k% of the scores are smaller and at most $(100 - k)$% of the scores are larger.*

11. Given the normal distribution of exercise 5, determine the score for the following percentiles:

    a. 50th   **b.** 90th

12. Given the normal distribution of exercise 6, determine the score for the following percentiles:

    a. 40th   **b.** 95th

13. Consider the histograms of exercise 5 in Exercises 6.3. Is it likely that the population of all student enrollments has a normal distribution? If so, estimate the mean μ, standard deviation σ, and 75th percentile score.

14. Consider the histograms of exercise 6 in Exercises 6.3. Is it likely that the population of all auto speeds has a normal distribution? If so, estimate the mean μ, standard deviation σ, and 75th percentile score.

15. Suppose it is estimated that in a certain population IQs are normally distributed with mean $\mu = 101$ and standard deviation $\sigma = 14$. Find the proportion of that population with IQs between 110 and 125.

16. It is estimated that the number of study-hours per week of a particular population of college students is normally distributed with mean $\mu = 28.5$ and standard deviation $\sigma = 6.2$. What proportion of the population studies more than 40 hours per week?

17. In a computer batch processing operation, ***turnaround time*** is that time between program submission and the return of the program output. If in a given batch processing situation the population of turnaround times is normally distributed with mean $\mu = 52$ minutes and standard deviation $\sigma = 13$ minutes, what is the probability that a randomly selected program will be returned within 30 minutes of submission?

18. Assume that the body weights of a certain population are normally distributed with mean $\mu = 147.5$ pounds and standard deviation $\sigma = 28.1$ pounds. Beyond what body weight would a person be among the heaviest 5% of the population?

6.5
Functions and Graphs

At a few places earlier in this text, we used the idea of a variable, particularly in the sections on algorithm development. A ***variable*** is any symbol or combination of symbols used to represent different values at different times. Variables that always assume numbers as values are called ***numeric***; nonnumeric variables are called ***string variables***.

Example 6.5.1

Variables

Variable	Representing	Possible values
N	Integer	$-9, 0, 2, 17, 1000$
AGE	Positive integer	5, 16, 73
SLRY	Positive number (for salary)	726.42, 305.10, 400.00
NAME	Letters (in a name)	JOHN, LILLIAN
ADDR	Letters and numbers (in an address)	72 SUNRISE BLVD. 6504 MOTT STREET
DATE	Numbers and other characters (in a date)	4-16-83, 15 JAN 40

As written, the first three of these are numeric variables, and the last three are string variables. However, if one wanted the values of SLRY to include the dollar symbol, as in $726.42, $305.10 or $400.00, then SLRY would also be a string variable. ∎

Definition

A *function* f is a correspondence between two sets A and B such that to each element of the first set A there corresponds one and only one element in second set B. Sets A and B are called the ***domain*** and ***range*** of f, respectively.

Notation: $f: A \rightarrow B$

For $x \in A$, $y = f(x)$ represents the element of B to which x corresponds.

In this context we sometimes call x and y the ***independent*** and ***dependent*** variables, respectively. The notation $f: A \rightarrow B$ will be used only when it is necessary to clarify the domain and range of f.

Example 6.5.2

Function: One Independent Variable

a. Suppose a person earns $3.50 for each hour worked. If n represents the total number of hours worked, then

$$G(n) = 3.50 * n$$

is the gross pay function. A few functional values are

$$G(10) = 35 \qquad G(15) = 52.50 \qquad G(40) = 140.00$$

b. A salesperson earns $200 base pay and $0.70 on each item sold. If n represents the number of items sold, then

$$GP(n) = 200 + .70 * n$$

is the gross pay function. Some functional values are

$$GP(50) = 235 \qquad GP(100) = 270 \qquad GP(165) = 315.50$$

c. Let D be the set of decimal digits $0, 1, \ldots, 9$, and T the set of all four-bit strings. Define the function

$$f: D \rightarrow T \quad \text{by} \quad f(x) = \text{the binary equivalent of } x$$

Since the domain D has only ten elements, we can easily list all the functional values:

$$f(0) = 0000 \quad f(1) = 0001$$
$$f(2) = 0010 \quad f(3) = 0011$$
$$f(4) = 0100 \quad f(5) = 0101$$
$$f(6) = 0110 \quad f(7) = 0111$$
$$f(8) = 1000 \quad f(9) = 1001$$

d. Let R be the set of all residents of a given town and S the set of all street addresses in that town. Define the function

$$f: R \rightarrow S \quad \text{by} \quad f(x) = \text{street address of } x$$

Typical functional values might look like

$$f(\text{J. T. Cosgrove}) = 27 \text{ Cathedral St.}$$
$$f(\text{M. Swanson}) \quad = 402 \text{ Second Ave.}$$

Here we assume that no two residents have exactly the same name. ∎

Observe that reversing the correspondence in (d) does not produce a function. Because it is likely that at least one individual address will have more than one resident, the "one and only one" requirement in the definition is not fulfilled.

Example 6.5.3

Function: Two Independent Variables

a. A real estate agent's total salary is the sum of the base pay and the varying commission. If BASE and COMM represent these respective quantities, then

$$GP(\text{BASE, COMM}) = \text{BASE} + \text{COMM}$$

is the gross pay function, where BASE and COMM are both independent variables. Two functional values are

$$GP(350, 611.20) = 961.20 \quad GP(475, 502) = 977$$

b. A certain company hires a number of employees at different hourly rates, and they work different numbers of hours per week. For a particular employee who works H hours at pay rate R,

$$GP(H, R) = H * R$$

is the gross pay function.

If we want to take into account time-and-a-half pay for hours worked over 35, then the gross pay function must be defined in two pieces:

$$GPO(H, R) = \begin{cases} H * R & \text{if } H \leq 35 \\ 35 * R + 1.5 * R * (H - 35) & \text{if } H > 35 \end{cases}$$

For instance,

$$GP(42, 3.75) \quad = 42 * 3.75 = 157.50$$
$$GPO(42, 3.75) = 35 * 3.75 + 1.5 * 3.75 * 7 = 170.62$$
$$GPO(30, 3.75) = 30 * 3.75 = 112.50 \quad \blacksquare$$

So called **built-in** (**library** or **standard**) functions are available in virtually every computer language. These are pre-defined algorithms that have been written as subprograms within the compiler of a particular language. While the collection of such library functions will vary from one compiler to the next (even different compilers for the same language!), nonetheless most contain commonly used numeric functions (such as rounding, truncation, square root) and string manipulation functions (such as length of a word). Often exponential, logarithmic, and trigonometric functions are also available. In addition, most languages provide the programmer with the capacity to create **user-defined functions**, which facilitate repeated calculations that use the same formula. Such a function must first be defined by the programmer as part of the program, and then it may be accessed later as often as needed.

Example 6.5.4 **Built-In Functions**

Here are a few functions similar to ones found in many computer compilers, where

R^+ is the set of nonnegative real numbers,
Z^+ is the set of nonnegative integers, and
W is the set of all character strings.

Function	Definition	Examples
a. ABS: $\mathbb{R} \to \mathbb{R}$	$\text{ABS}(x) = |x|$	$\text{ABS}(7) \quad = |7| \quad = 7$ $\text{ABS}(-2.3) = |-2.3| = 2.3$
b. SQR: $R^+ \to R^+$	$\text{SQR}(x) = \sqrt{x}$	$\text{SQR}(4) \quad = \sqrt{4} \quad = 2$ $\text{SQR}(.25) = \sqrt{.25} = .5$
c. MAX: $\mathbb{R} \times \mathbb{R} \to \mathbb{R}$	$\text{MAX}(x, y) = $ larger of x and y	$\text{MAX}(7, 2.9) \quad = 7$ $\text{MAX}(-12, 1.2) = 1.2$
d. LEN: $W \to Z^+$	$\text{LEN}(w) = $ number of characters (including spaces) in w	$\text{LEN}(\text{GRANITE}) \quad = 7$ $\text{LEN}(\text{WORD}) \quad = 4$ $\text{LEN}(\text{GOOD BYE}) = 8$
e. IND: $W \times W \to Z^+$	$\text{IND}(w, z) = $ starting position of first occurrence of string z within string w	$\text{IND}(\text{HELLO, E}) \quad = 2$ $\text{IND}(\text{HELLO, B}) \quad = 0$ $\text{IND}(\text{HELLO, LO}) = 4$
f. ROUND: $\mathbb{R} \to \mathbb{Z}$	$\text{ROUND}(x) = $ value of x rounded to nearest integer	$\text{ROUND}(7.902) = 8$ $\text{ROUND}(-6.4) = -6$ $\text{ROUND}(29) \quad = 29$

$\blacksquare$

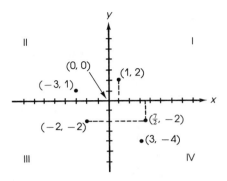

Figure 42

Functions of a single numeric variable can often be represented geometrically. Consider the Cartesian plane,[†] which establishes a correspondence between points and pairs of numbers and divides the plane into four quadrants, I, II, III, and IV (see Figure 42). The horizontal and vertical lines with coordinates are called the **x-axis** and **y-axis**, respectively.

Definition

> Let A and B be sets of numbers. The **graph of function** $f: A \rightarrow B$ is the set
>
> $$\{(x, f(x)) \mid x \in A\}$$

Example 6.5.5

Graphs

In Figures 43, 44, and 45, we plot several points for three of the functions encountered earlier in this section.

a. $G(n) = 3.50 * n$

n	$G(n)$
0	0
10	35
15	52.50
40	140

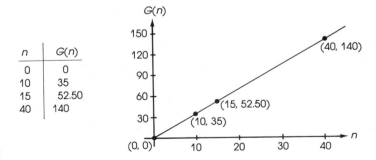

Figure 43

† Attributed to the renowned French philosopher René Descartes (1596–1650).

b. $GP(n) = 200 + .70 * n$

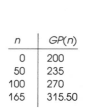

n	GP(n)
0	200
50	235
100	270
165	315.50

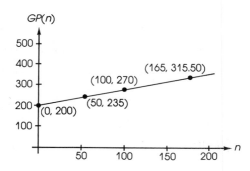

Figure 44

c. $ABS(x) = |x|$

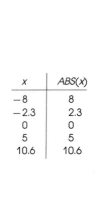

x	ABS(x)
−8	8
−2.3	2.3
0	0
5	5
10.6	10.6

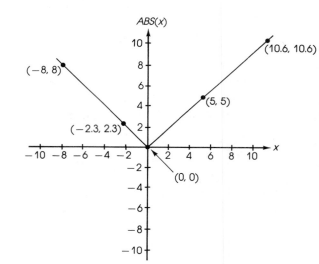

Figure 45

In each case, any additional points that are plotted will fall on the drawn line. ∎

Example 6.5.6

Graph: Postage Function

Suppose a certain parcel delivery service charges 20¢ per ounce or fraction thereof. Let w be the weight in ounces and $PST(w)$ the appropriate postage. Then $PST(w)$ is always some multiple of 20; some sample costs are plotted in Figure 46. Plotting additional points will lead to the steplike graph shown. Each small open circle indicates that that particular point is *not* part of the graph. ∎

Example 6.5.7

Maximum Value

A certain profit function $P(n)$ is given by the formula $P(n) = 890n - n^2$, where n is the number of units sold. The graph of $P(n)$ is a parabola opening

w	PST(w)
.5	20
1	20
1.1	40
1.9	40
4	80
4.001	100
5	100

PST(w)

(4.001, 100)
(5, 100)
(1.1, 40)
(1.9, 40)
(4, 80)
(.5, 20) (1, 20)

Figure 46

downward with n-intercepts

$$0 = 890n - n^2$$
$$0 = n(890 - n)$$

$$\boxed{n_1 = 0} \quad \text{and} \quad \boxed{n_2 = 890}$$

The vertex of this graph occurs midway between these intercepts, at

$$n = \frac{n_1 + n_2}{2} = \frac{0 + 890}{2} = 445$$

which yields

$$P(445) = 890(445) - 445^2 = 198025$$

as the maximum value of the profit function. Its graph is given in Figure 47.

■

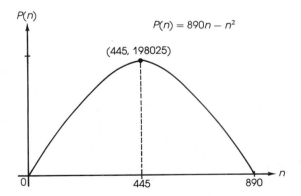

P(n)

$$P(n) = 890n - n^2$$

(445, 198025)

445   890   n

Figure 47

Obviously you cannot plot all of the infinitely many points on a curve individually. In applications, you are often interested in points in a certain vicinity of the plane.

Example 6.5.8   **Algorithm: Plotting Particular Points**

Devise an algorithm to plot all the first quadrant points (x, y) on the graph of

$$y = 3x - 10$$

where x is an integer and $y \leq 1000$.

The graph is a line rising from left to right (because the slope is positive) with y-intercept -10. To find the x-intercept:

$$3x - 10 = 0$$
$$3x = 10$$
$$x = \tfrac{10}{3} = 3\tfrac{1}{3}$$

Now, the y-values are restricted by 1000, and

$$3x - 10 = 1000$$
$$3x = 1010$$
$$x = \tfrac{1010}{3} = 336\tfrac{2}{3}$$

Figure 48 summarizes these results. This work tells us we must plot all points (x, y) where x is an integer and

$$4 \leq x \leq 336$$

The algorithm exhibited in Figure 49 will plot the desired 333 points.   ∎

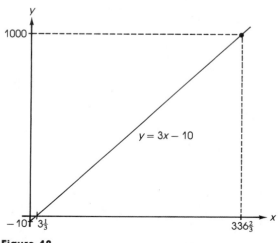

Figure 48

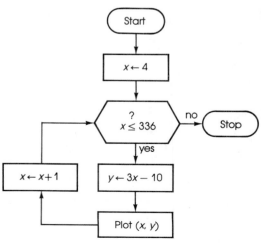

Figure 49

Example 6.5.9 **Algorithm: Plotting Particular Points**

Devise an algorithm to plot all third quadrant points (x, y) on the graph of

$$y = x^2 + 15x - 43000$$

where x is an integer.

The graph is a parabola opening upward. To find the x-intercepts, set $y = 0$:

$$0 = x^2 + 15x - 43000$$
$$0 = (x + 215)(x - 200)$$

so $x + 215 = 0$ $x - 200 = 0$

$$\boxed{x = -215} \text{ and } \boxed{x = 200}$$

For the y-intercept, set $x = 0$:

$$y = 0^2 + 15(0) - 43000$$

$$\boxed{y = -43000}$$

The graph in Figure 50 indicates that we must plot all points with integral x-coordinates ranging between -215 and 0. The algorithm to accomplish this is given in Figure 51. ∎

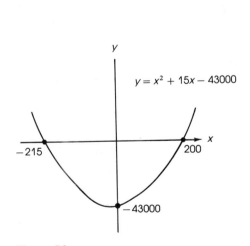

Figure 50

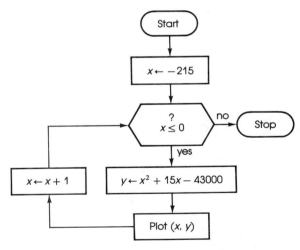

Figure 51

EXERCISES 6.5

1. For each of the following functions compute functional values when the independent variable equals -3, 0, $\frac{1}{4}$, and 2.

   a. $f(x) = 12x$ **b.** $g(s) = 5s + 2$
   c. $H(t) = 3 - 2t$ **d.** $F(x) = 7$

2. For each of the following functions, compute functional values when the independent variable equals $-\frac{13}{5}$, -1, 0, and 3.

   a. $F(s) = -5s$ **b.** $H(u) = -3u + 1$
   c. $V(w) = 4 + w$ **d.** $T(x) = -3$

3. In each case, specify the domain and range of the described function and state the defining equation.

a. A store gives a 10% discount off the marked price M. Define the function RP that gives the reduced price in terms of M.

b. As a Christmas bonus, an employer gives each employee $100 plus 5% of his or her weekly salary S. Determine the function B that gives the bonus in terms of S.

4. In each case, specify the domain and range of the described function and state the defining equation.

a. A state sales tax is 6% of the selling price P. Find the function TP that expresses the total price in terms of P.

b. An employee's weekly salary S is reduced by 26% for taxes and $50 for retirement. Express the function NET that gives the net salary in terms of S.

5. In each case, state the defining equation for the described function and compute two functional values of your choice.

a. A store gives D% discount off the marked price M. Define the function RP that gives the reduced price in terms of D and M.

b. As a Christmas bonus, an employer gives each employee a certain amount A plus 5% of his or her weekly salary S. Determine the function B that gives the bonus in terms of A and S.

c. A certain payroll is computed by adding $200 to an individual's commission C and deducting taxes T. Express the function NS that yields the net salary in terms of C and T.

d. Let D be the distance traveled by an object going at constant speed R for T hours. Express D as a function of variables R and T.

6. In each case, state the defining equation for the described function and compute two functional values of your choice.

a. A state sales tax is S% of the selling price P. Find the function TP that expresses the total price in terms of S and P.

b. An employee's weekly salary S is reduced by 26% for taxes and R for retirement. Express the function NET that gives the net salary in terms of S and R.

c. Let P be the perimeter of a rectangle of dimensions L and W. Express P as a function of L and W.

7. Using the built-in functions defined in Example 6.5.4, compute:

a. ABS(0), ABS(−10.9)

b. SQR(.09), SQR(100)

c. LEN(12BT), LEN(GEORGE)

d. IND(EXERCISE, E), IND(EXERCISE, ER)

e. MAX(−9.01, −9), MAX(−3, 2.01)

f. ROUND(89.6014), ROUND(−17)

8. Using the built-in functions defined in Example 6.5.4, compute:

a. ABS(−1), ABS(2)

b. SQR($\frac{1}{4}$), SQR(6.25)

c. LEN(LEN), LEN(GO TO)

d. IND(GOOD BYE, O), IND(GOOD BYE, GOODY)

e. MAX(16, 16), MAX(−10, −9.2)

f. ROUND(−6.49), ROUND(3000.801)

9. Plot each of the following points in the Cartesian plane:

$$O(0, 0) \quad A(4, 3) \quad B(−1, 5) \quad C(−3, −1) \quad D(5, −4)$$

10. Plot each of the following points in the Cartesian plane:

$$O(0, 0) \quad A(1, 6) \quad B(−7, 5) \quad C(−2, −\tfrac{1}{2}) \quad D(12, 5)$$

11. For each part of exercise 1, plot five points on the graph of the given function. Can you guess where other points will lie?

12. For each part of exercise 2, plot five points on the graph of the given function. Can you guess where other points will lie?

13. In each part of exercise 3, draw the graph of the function.

14. In each part of exercise 4, draw the graph of the function.

15. In each case, graph the given function.

a. $f(x) = \begin{cases} 1, & x < -1 \\ -5, & -1 \le x \le 3 \\ 2, & x > 3 \end{cases}$

b. $g(x) = \begin{cases} 2x - 5, & x < 3 \\ 4 - x, & x \ge 3 \end{cases}$

16. In each case, graph the given function.

a. $f(x) = \begin{cases} -2, & x \le 0 \\ 2, & x > 0 \end{cases}$ **b.** $g(x) = \begin{cases} 2x, & x < 0 \\ 1, & x = 0 \\ 2, & x > 0 \end{cases}$

17. Which curves in Figure 52 are graphs of functions with domain $[1, 2]$? (An explanation of the interval notation $[a, b]$ is given in Exercises 3.1.)

a.

b.

c.

Figure 52

18. Which curves in Figure 53 are graphs of functions with domain $[1, 2]$?

a.

b.

c.

Figure 53

19. Review the results of the two preceding exercises and then formulate a "vertical-line test" to determine if a given graph defines a function with domain $[a, b]$.

20. Assume the availability of a function $\mathbf{BLK}(x, y)$ that blackens a square of side length 1 unit centered at point (x, y) in the Cartesian plane. In each of the following, construct an algorithm that will produce the given figure(s):

   a. The letter X centered at the origin
   b. The figure $\square$ in the first quadrant
   c. Your initials, placed in the first quadrant

21. Let function ROOF be defined by

$$\text{ROOF}(x) = \text{smallest integer greater than or equal to } x, \text{ where } x \text{ is any number.}$$

a. Compute

   ROOF(5.2)   ROOF(6)   ROOF(-3.1)

b. Use ROOF to describe the postage function of Example 6.5.6.

22. Can you graph the functions of exercises 5 and 6 in the Cartesian plane? Why or why not?

23. Revise the algorithm in Example 6.5.8 so that it plots all points (x, y) where x is an integer and $-200 \leq y \leq 200$.

24. For each of the following conditions, draw the flow-chart for an algorithm to plot the desired points (x, y) on the graph of $y = 50 - 2x$:

   a. $x = 0, \frac{1}{2}, 1, \frac{3}{2}, 2, \frac{5}{2}, \ldots, 100$
   b. For some given positive number N, $-N < y < N$; x an integer.

25. Suppose a cost-per-item function $C(n)$ is determined to be

$$C(n) = 2n^2 - 40n + 225$$

where n is the number of items produced. Find the minimum value of C.

26. Construct an algorithm to plot all points (x, y) above the x-axis on the graph of

$$y = 30000 - 200x - x^2 \qquad x \text{ an integer}$$

27. Revise the algorithm in Example 6.5.9 so that all third-quadrant points (x, y) are plotted, where

$$x = \frac{k}{2} \qquad k \text{ an integer}$$

28. Suppose it is known that for $0 \leq x \leq 100$ the graph of function $f(x)$ rises, peaks, and then falls. Construct an algorithm that produces that value of x, accurate to the nearest hundredth, which gives the maximum value for $f(x)$ in the given interval.

6.6

Math Induction

Suppose we want to convince someone that the sum of the first k positive integers, or *natural numbers*,

$$1, 2, 3, 4, \ldots, k$$

is given by the formula

$$1 + 2 + 3 + 4 + \cdots + k = \frac{k(k+1)}{2}$$

One way might be to construct an algorithm of the type indicated in Figure 54 and execute it for several (large!) values of N. Whenever this is done for a given N and we receive a positive response to the last question (FLAG = 0?), then we know that the formula does hold for all $k = 1, 2, 3, \ldots, N$.

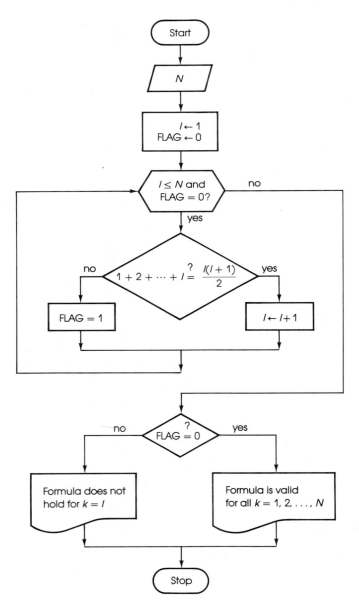

Figure 54

However, no matter how many times (finitely many) we execute this algorithm or how large (finitely large) a value we substitute for N and still receive a positive response, we cannot conclude on that basis alone that the formula is valid for all of the infinitely many values

$$k = 1, 2, 3, 4, \ldots$$

The *principle of math induction* (PMI) is a tool that allows us to deal with this type of situation. When this principle is properly applied, we can conclude that a given statement $P(k)$ concerning positive integer k is true for all of the infinitely many positive integers. Simply put, the PMI says the following. Suppose it can be shown that: (1) a statement $\mathscr{P}$ is valid for the positive integer 1; and (2) whenever $\mathscr{P}$ is valid for one positive integer, it is valid for the succeeding integer. Then statement $\mathscr{P}$ is valid for every positive integer.

Principle of Math Induction

Let $\mathscr{P}(k)$ be a statement involving a positive integer k.
If

1. $\mathscr{P}(1)$ is true, and
2. for any positive integer k,

$$\mathscr{P}(k) \text{ is true} \rightarrow \mathscr{P}(k + 1) \text{ is true,}$$

then $\mathscr{P}(k)$ is true for all $k = 1, 2, 3, \ldots$.

In mathematics, the PMI is accepted as valid. If upon a second (and third, and ...) reading you still have difficulty saying, "Yes, I accept that," do not be disheartened. In set theory, you can prove that the PMI is logically equivalent to the *well ordering principle*, which states that every nonempty set of positive integers contains a least element. Most people have less difficulty accepting this equivalent formulation of the PMI.

With the PMI at our disposal, we can now prove that the above formula holds for all positive integers k.

Example 6.6.1   **PMI: Equation**

Let $\mathscr{P}(k)$ be the equation

$$1 + 2 + \cdots + k = \frac{k(k + 1)}{2}$$

where k is a positive integer.

1. For $k = 1$,

$$1 = \frac{1(1 + 1)}{2}$$

is true; so $\mathscr{P}(1)$ is true.

2. Suppose $\mathscr{P}(k)$ is true, so that

$$1 + 2 + \cdots + k = \frac{k(k + 1)}{2}$$

is true. From this assumption, we hope to deduce that $\mathscr{P}(k + 1)$ is true; that is,

$$1 + 2 + \cdots + (k + 1) = \frac{(k + 1)(k + 2)}{2}$$

Now

$$1 + 2 + \cdots + (k + 1)$$

$= 1 + 2 + \cdots + k + (k + 1)$		Notation
$= (1 + 2 + \cdots + k) + (k + 1)$		Associativity of addition
$= \dfrac{k(k + 1)}{2} + (k + 1)$		$\mathscr{P}(k)$ true
$= \dfrac{k(k + 1)}{2} + \dfrac{2(k + 1)}{2}$		2 is common denominator
$= \dfrac{k(k + 1) + 2(k + 1)}{2}$		Addition of fractions
$= \dfrac{(k + 1)(k + 2)}{2}$		Factor $(k + 1)$ out of numerator

In summary, we have just verified that

$\mathscr{P}(k)$ is true $\rightarrow \mathscr{P}(k + 1)$ is true.

Since we have demonstrated that parts (1) and (2) of the PMI hold, $\mathscr{P}(k)$ must be true for all $k = 1, 2, \ldots$; that is,

$$1 + 2 + \cdots + k = \frac{k(k + 1)}{2}$$

for all $k = 1, 2, \ldots$. ∎

The next two examples show that the PMI can be used for statements other than equations.

Example 6.6.2   **PMI: Divisibility**

Let $\mathscr{P}(k)$ be the statement

$(k^2 + k)$ is an even integer.

1. For $k = 1$, $1^2 + 1 = 2$ is even; so $\mathscr{P}(1)$ is true.

2. Suppose $\mathscr{P}(k)$ is true, so that

$k^2 + k$ is an even integer.

We hope to be able to deduce that $\mathscr{P}(k + 1)$ is true; that is,

$$(k + 1)^2 + (k + 1) \text{ is an even integer.}$$

To this goal, we calculate as follows:

$$(k + 1)^2 + (k + 1) = k^2 + 2k + 1 + k + 1 \qquad \text{Algebra}$$
$$= (k^2 + k) + (2k + 2) \qquad \text{Commutativity and associativity of addition}$$

$$= \underbrace{(k^2 + k)}_{\substack{\text{Even since} \\ \mathscr{P}(k) \text{ is true}}} + \underbrace{2(k + 1)}_{\text{Even}} \qquad \text{Factor out 2}$$

$$\underbrace{}_{\text{Even}} \qquad \text{Because the sum of two even integers is also even}$$

In summary, this shows that

$$\mathscr{P}(k) \text{ is true} \to \mathscr{P}(k + 1) \text{ is true.}$$

Since parts (1) and (2) of the PMI hold, it follows that $\mathscr{P}(k)$ is true for all k; that is,

$$(k^2 + k) \text{ is even}$$

for all $k = 1, 2, \ldots$. ∎

Example **6.6.3**

PMI: Inequality

Let $\mathscr{P}(k)$ be the inequality

$$5^k \geq 5k$$

1. For $k = 1$, $5^1 \geq 5(1)$ is true; so $\mathscr{P}(1)$ is true,

2. Suppose $\mathscr{P}(k)$ is true, so that

$$5^k \geq 5k \text{ is true.}$$

We now intend to show that $\mathscr{P}(k + 1)$ is also true; that is,

$$5^{(k + 1)} \geq 5(k + 1)$$

Hence,

$$5^{(k + 1)} = (5^k)(5^1) \qquad \text{Exponent rule}$$
$$\geq (5k)(5) \qquad \mathscr{P}(k) \text{ true}$$
$$= 5k + 5k + 5k + 5k + 5k \qquad \text{Notation}$$
$$\geq 5k + 5 \qquad k \geq 1$$
$$= 5(k + 1) \qquad \text{Factor out 5}$$

The preceding work verifies that

$$\mathscr{P}(k) \text{ is true} \rightarrow \mathscr{P}(k + 1) \text{ is true.}$$

So, (1) and (2) of the PMI have been verified, and we conclude that $\mathscr{P}(k)$ is true for all k; that is,

$$5^k \geq 5k$$

for all $k = 1, 2, \dots$. ∎

We conclude this section with proofs of two formulas that were used in the first two sections of this chapter on math of finance.

Example 6.6.4   **PMI: Compound Amount Formula**

Suppose that principal P is deposited for N periods in a bank that pays rate of interest R per period. Using the PMI, we shall show that for any positive integer N, the deposit value S after N periods is given by the formula

$$S = P(1 + R)^N$$

Note in this formula that P and R are fixed, given numbers.

For positive integer N, let $\mathscr{P}(N)$ represent the formula

$$S = P(1 + R)^N$$

1. For $N = 1$,

$$S = P(1 + R)^1 = P + PR$$

does equal the deposit value after 1 period, so $\mathscr{P}(1)$ is true.

2. Suppose $\mathscr{P}(N)$ is true, so that $P(1 + R)^N$ equals the deposit value after N periods. We hope to deduce that $\mathscr{P}(N + 1)$ is true; that is, $P(1 + R)^{N+1}$ equals the deposit value after $(N + 1)$ periods. To this end, if S is the deposit value after $(N + 1)$ periods, we get

$$S = \underbrace{P(1 + R)^N}_{\substack{\text{Deposit value at} \\ \text{beginning of period}}} + \underbrace{[P(1 + R)^N]R}_{\substack{\text{Interest earned} \\ \text{during period}}}$$

$$\begin{aligned} &= P(1 + R)^N(1 + R) \quad &&\text{Factor out } P(1 + R)^N \\ &= P(1 + R)^{N+1} \quad &&\text{Exponent rule} \end{aligned}$$

This work verifies that

$$\mathscr{P}(N) \text{ is true} \rightarrow \mathscr{P}(N + 1) \text{ is true.}$$

Therefore, parts (1) and (2) of the PMI have been confirmed, and it follows that $\mathscr{P}(N)$ is true for all N; that is,

$$S = P(1 + R)^N$$

gives the deposit value S at the end of any period $N = 1, 2, \dots$. ∎

Example 6.6.5

PMI: Series Compound Amount Formula

Suppose fixed payment A is made at the end of each of N periods, where R is the rate of interest per period. We shall use the PMI to show that for any positive integer N, the total deposit value S_N at the time of the Nth payment is given by

$$S_N = A + A(1 + R) + A(1 + R)^2 + \cdots + A(1 + R)^{N-1}$$

Again, note that A and R are given, fixed values.

Now for a positive integer N, let $\mathscr{P}(N)$ designate the formula

$$S_N = A + A(1 + R) + \cdots + A(1 + R)^{N-1}$$

1. For $N = 1$, this is a little tricky! The given formula really should have been expressed as

$$S_N = \sum_{k=1}^{N} A(1 + R)^{k-1}$$

so when $N = 1$, we get

$$S_1 = \sum_{k=1}^{1} A(1 + R)^{k-1} = A(1 + R)^0 = A$$

which does equal the deposit value at the end of the first period, when the first payment A is made. Therefore, $\mathscr{P}(1)$ is true.

2. Suppose $\mathscr{P}(N)$ is true; it is our task to demonstrate that $\mathscr{P}(N + 1)$ is true. Therefore, we have

$$S_{N+1} = \underbrace{S_N}_{\substack{\text{Deposit value at} \\ \text{beginning of period}}} + \underbrace{S_N R}_{\substack{\text{Interest earned} \\ \text{during period}}} + \underbrace{A}_{\substack{\text{Payment made at} \\ \text{end of period}}}$$

$$= S_N(1 + R) + A \qquad\qquad \text{Factor out } S_N$$

$$= A + S_N(1 + R) \qquad\qquad \text{Commutativity of addition}$$

$$= A + [A + A(1 + R) + \cdots + A(1 + R)^{N-1}](1 + R) \qquad \mathscr{P}(k) \text{ is true}$$

$$= A + A(1 + R) + A(1 + R)^2 + \cdots + A(1 + R)^{N} \qquad \text{Algebra}$$

To summarize, we have shown

$$\mathscr{P}(N) \text{ is true} \to \mathscr{P}(N + 1) \text{ is true}$$

Hence, parts (1) and (2) of the PMI have been confirmed, and it follows that $\mathscr{P}(N)$ is true for all N; that is,

$$S_N = A + A(1 + R) + \cdots + A(1 + R)^{N-1}$$

gives the total deposit value at the end of any period $N = 1, 2, \ldots$. ∎

One final word. In the context of computer science, the technique of math induction can be used to verify certain algorithms involving the positive integers.

EXERCISES 6.6

In the first 18 exercises, use the PMI to prove that the given statement is valid for all positive integers $k = 1, 2, \ldots$.

1. $\displaystyle\sum_{n=1}^{k} n^2 = \frac{k(k+1)(2k+1)}{6}$

2. $\displaystyle\sum_{n=1}^{k} (2n-1) = k^2$

3. $\displaystyle\sum_{n=1}^{k} (4n-3) = k(2k-1)$

4. $\displaystyle\sum_{n=1}^{k} (4n-1) = k(2k+1)$

5. $\displaystyle\sum_{n=1}^{k} \frac{1}{n(n+1)} = \frac{k}{k+1}$

6. $\displaystyle\sum_{n=1}^{k} \frac{1}{(2n-1)(2n+1)} = \frac{k}{2k+1}$

7. $1 + 2 + 2^2 + \cdots + 2^{k-1} = 2^k - 1$

8. $1 + 3 + 3^2 + \cdots + 3^{k-1} = 3^k - 1$

9. $(k^3 + 2k)$ is a multiple of 3.

10. $(k^2 - k + 2)$ is a multiple of 2.

11. $(a - b)$ is a factor of $a^k - b^k$.

12. $(a + b)$ is a factor of $a^{2k-1} + b^{2k-1}$.

13. $3^k \geq 3k$

14. $4^k \geq 4k$

15. For $a > 1$, $a^k > 1$.

16. For $0 < a < 1$, $0 < a^k < 1$.

17. $(ab)^k = a^k b^k$

18. $\left(\dfrac{a}{b}\right)^k = \dfrac{a^k}{b^k}$ $b \neq 0$

19. Let I be a fixed positive integer. Revise the wording of the PMI so that the conclusion reads: "Then $\mathscr{P}(k)$ is true for all $k = I, I + 1, I + 2, \ldots$."

20. Use the reworded PMI of exercise 19 to prove each of the following:
   a. $2^k > 2k$ for all $k \geq 3$
   b. $2^k > 2(k + 1)$ for all $k \geq 4$

CHAPTER 7

Introduction to Advanced Topics

Those of you who wish to continue the study of computer-related math may follow this course with one in what is called *discrete math*, *discrete structures*, or *applied abstract algebra*. This chapter is intended to be a mere appetizer for such an entrée.

In that type of course, you are apt to delve into several topics already discussed in this text, as well as subjects like semigroup, group, and quotient structures; morphisms; graphs and trees; posets and lattices; finite state machines and computability; coding; and formal languages and grammars. Such topics require skill in using complex symbolism, and at times they require an advanced level of abstract thinking.

This chapter presents formal definitions and introductory examples in four of these areas—namely, graphs and trees, semigroups, finite state machines, and formal languages and grammars. Each section concludes with a brief statement of additional subjects that could be pursued and, in some cases, the statement (without explanation!) of an important related result. To do any more with reasonable detail would require another volume.

7.1

Graphs and Trees

Twice in this text it was natural to use diagrams of a *directed graph* (Example 3.4.9 in the context of a binary relation) and a *tree graph* (Examples 3.5.9 and 3.5.10, which deal with counting problems). In numerous other computer applications, these kinds of diagrams are useful. In this section we shall present additional examples. However, first we shall formally define these terms in a fashion that is typical of a more advanced discrete math course.

The word *graph* is used here in a different sense from the graph of a function discussed in Section 6.5.

Definition

A *graph* is a pair

$$\langle N, A \rangle$$

consisting of a set N whose elements are called *nodes* (or *vertices*) and a set A of *unordered* pairs of nodes called *arcs* (or *edges*).

A graph with a finite number of nodes and arcs is called *finite*, and every finite graph has an obvious geometric representation.

Example 7.1.1

Finite Graph

The graph $\langle N, A \rangle$ where

$$N = \{B, C, D, E\} \quad \text{and} \quad A = \{\{B, C\}, \{C, B\}, \{C, D\}, \{D, D\}\}$$

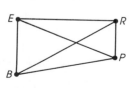

Figure 1

is drawn in Figure 1, where the nodes are identified with points and arcs with curves or line segments. Note that the arcs are *unordered* pairs, and so we could have written equivalently

$$A = \{\{B, C\}, \{B, C\}, \{D, C\}, \{D, D\}\} \qquad \blacksquare$$

With this type of geometric illustration, it is quite natural to use certain other terms. Two nodes that determine an arc are called the **endpoints of the arc**. (Above, B and C are endpoints of two arcs, and C and D are endpoints of one arc.) An arc with the same node as its two endpoints is called a **loop** (arc $\{D, D\}$ is the only loop above). Two distinct nodes are called **adjacent nodes** if they are the endpoints of some arc of the graph. (In the preceding example, B and C are adjacent nodes, as are C and D.) A node that is adjacent to no other node is called **isolated** (node E is the only isolated node above).

A **path** from node B_0 to node B_k is a finite string of arcs of the form

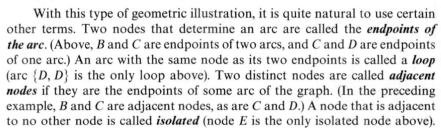

$$\{B_0, B_1\}, \{B_1, B_2\}, \{B_2, B_3\}, \ldots, \{B_{k-1}, B_k\}$$

The **length of a path** is the number of arcs in that path. (In Figure 1, $\{D, C\}$, $\{C, B\}$ is a path of length 2.) A graph is **connected** if for each pair of distinct nodes there is a path that joins these nodes. (The graph above is not connected since no path joins node E with any other node.) Finally, a **cycle** is a path of the form

$$\{B, B_1\}, \{B_1, B_2\}, \{B_2, B_3\}, \ldots, \{B_{k-1}, B\}$$

where $B_i \neq B_j$ for $i \neq j$. For instance, in the preceding example $\{B, C\}$, $\{C, B\}$ is a cycle, and the loop $\{D, D\}$ is a (trivial) cycle. A graph with at least one cycle is called **cyclic**.

Example 7.1.2

Finite Graph: Map

Consider the graph $\langle N, A \rangle$ where $N = \{B, E, P, R\}$, representing Belmont, CA; Eugene, OR; Philadelphia, PA; and Round Lake, NY; respectively, and arcs $A = \{\{B, E\}, \{B, P\}, \{B, R\}, \{E, P\}, \{E, R\}, \{P, R\}\}$. A graph that takes into account the relative positions of these locations is shown in Figure 2. Note that any two nodes are adjacent, and there are no isolated nodes. The graph is connected and cyclic ($\{E, P\}, \{P, B\}, \{B, E\}$ is one of several cycles), yet is loop-free. $\blacksquare$

Figure 2

Now we examine graphs in which the ordering of the pairs of nodes is important.

Definition

> A *directed graph* (or *digraph*) is a graph $\langle N, A \rangle$ whose arcs are *ordered* pairs of nodes.

For a digraph $\langle N, A \rangle$, the set of arcs A is a relation (in the sense of Section 3.4) on the set of nodes N. Consistent with prior notation, we'll write

(B, C)

for the ordered pair of nodes B and C, and we shall call B the *initial node* and C the *terminal node*; geometrically, we connect B and C with a directed segment:

$B \bullet \!\!\longrightarrow\!\!\bullet C$

Example 7.1.3   **Directed Graph**

The digraph with nodes $N = \{1, 2, 3, 4, 5\}$ and arcs $A = \{(1, 2), (3, 2), (2, 5), (4, 1), (1, 1), (4, 3), (4, 4)\}$ is drawn in Figure 3. It has two loops, so it is cyclic; it is not connected, since no path joins 1 and 3.   ∎

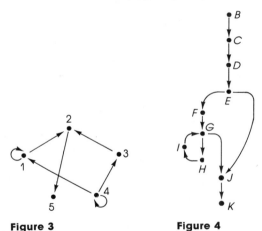

Figure 3   **Figure 4**

Example 7.1.4   **Digraph: Flowchart**

So far in this text we have seen many algorithms described by means of flowcharts. If we collapse the geometric shapes in any flowchart into points and leave the connecting arrows intact, the result is a digraph. For instance, if we do this with Figure 12 associated with Example 3.6.6, we get the digraph given in Figure 4. This graph has ten nodes, identified by the ten letters B through K, and 11 arcs. It is connected, loop-free, and cyclic. This example shows that it is possible to discuss, analyze, and classify certain properties of flowcharts by using digraphs.   ∎

There is another type of graph, namely a *tree*, that has many important uses within computer science. We shall conclude this section by defining *tree* and giving several illustrations.

Definition

A *tree* is a graph that is connected but has no cycles.

Remember that no cycles mean no loops, but not vice versa.

Example 7.1.5

Tree: Data Structure

Suppose a company wishes to create a computer data file to store the following 11 items (or *fields*) of information for each employee:

Name (NAME)   Social Security tax (FICA)
Address (ADDR)   Federal tax (FED)
Position (POSN)   State tax (ST)
Gross salary (GPAY)   Insurance payment (INS)
Number of dependents (#DEP)   Other deductions (OTH)
Social Security number (SS#)

One possible way of organizing this information into a tree structure is shown in Figure 5. Observe that in this figure, the 11 fields of information appear as *leaves* (that is, bottom nodes) on the tree, and the nodes

EMPL (employee)
ID INFO (identifying information)
SAL INFO (salary information)
DEDUCT (deductions)

were introduced to help organize the given fields. This is an example of a *rooted tree* with the node EMPL being the root. (Unlike real trees, mathematical trees commonly have a root at the top and leaves at the bottom.) This tree or data structure may help in identifying the variables in a computer program that processes this information—such as a program that outputs weekly paychecks for all employees. ∎

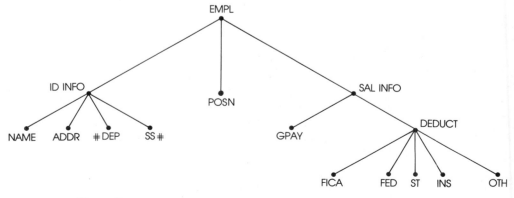

Figure 5

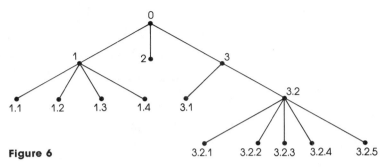

Figure 6

In the preceding example, the order in which we placed the leaves on the ID INFO node or the DEDUCT node was of no importance. However, in some applications arcs emanating from a particular node must be ordered, and the usual method is to order them *left to right*. Utilizing the same basic tree structure of Figure 5, we illustrate this left-to-right ordering in Figure 6 by numbering the nodes. Such a tree is called an ***ordered tree***.

Example 7.1.6

Ordered Tree: Arithmetic Expressions

Recall the order of executing arithmetic operations listed in Section 1.5; Namely, first remove all parentheses and then perform exponentiations, multiplications and divisions, and additions and subtractions, in that order. By using ordered trees in which the arcs emanating from a given node are read left to right, Figure 7 shows how several different arithmetic expressions can be represented. Notice that the leaves are the given variables or numbers and

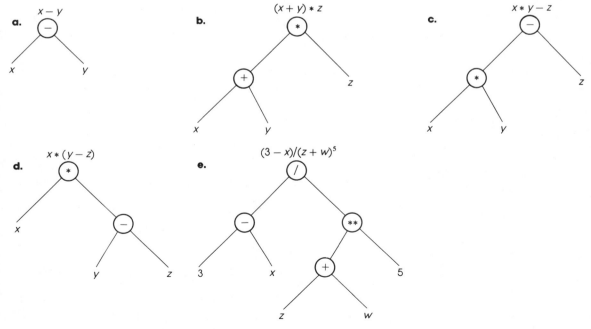

Figure 7

the other nodes are the binary operators. These trees are called **binary trees**, since there are at most two arcs (called the **left branch** and **right branch**) emanating from any given node. ▮

Example 7.1.7   **Ordered Tree: Binary Sort**

The flowchart in Figure 8 gives an algorithm for placing one word WORD in its correct alphabetical position in a list of ordered words, using a binary tree with a root node at the top. In that flowchart, the symbol $>$ means "succeed in alphabetical ordering."

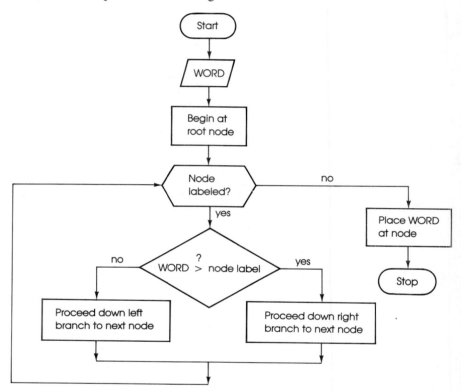

Figure 8

For instance, beginning with the unordered list of names

BARBARA JIM ANNE KERRY CLARE MAURA CHATHAM

and applying the algorithm to the first word BARBARA gives the root node

BARBARA

Applying it to the next word JIM yields

to the following name ANNE gives

and to the next word KERRY yields

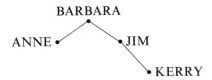

Continuing this procedure with the last three names gives the binary tree shown in Figure 9. Observe that reading the nodes left to right, independently of vertical position, gives the sorted list

ANNE BARBARA CHATHAM CLARE JIM KERRY MAURA

∎

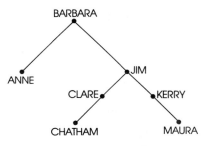

Figure 9

There are many other types of graphs (such as complete, simple, nonplanar, and regular) that you would pursue in a more comprehensive presentation. In addition, it is sometimes helpful to be able to transform the information held by a graph into numeric form suitable for computer processing (see exercises 13 and 14).

EXERCISES **7.1**

For the graphs in the first two exercises, list all nodes, arcs, loops, and cycles, and determine if they are connected.

1.

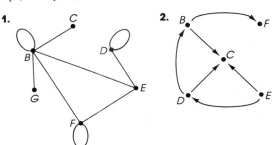

2.

In the next two exercises, draw the associated graph. Then list all loops and cycles, and determine if it is connected.

3. $N = \{B, C, D, E, F\}$
$A = \{(B, C), (B, E), (E, D), (D, B)\}$

4. $N = \{B, C, D, E\}$
$A = \{\{E, B\}, \{B, C\}, \{C, D\}, \{B, D\}, \{B, B\}, \{C, E\}\}$

5. Use the method of Example 7.1.4 to draw the digraph associated with the flowchart of Figure 10 of Chapter 2.

6. Same as exercise 5, but using Figure 16 of Chapter 2.

7. Suppose a college wishes to create a computer data file to store the following fields of information on each student:

Name
Address
Birthdate
Social Security number
Sex
Major
Minor
Credits accumulated
Number of courses passed
Number of courses failed

Construct a tree diagram that organizes these data.

8. A professional basketball club wants to create a data file to store the following information on each player:

Name	Assists/career
Birthdate	Rebounds/career
College	Blocks/career
Height	Points/last season
Weight	Assists/last season
Hand speed	Rebounds/last season
Foot speed	Blocks/last season
Points/career	

Draw a tree diagram that will assist in organizing these data.

9. Construct an ordered binary tree for each of the following algebraic expressions:

a. y/x **b.** $y - x/z$
c. $(y - x)/z$ **d.** $x^2 + 3y/z$
e. $(x + y)/3 - 4w^3$

10. Construct an ordered binary tree for each of the following algebraic expressions:

a. $y - 3z$ **b.** $x/y - 5w$
c. $x/(y - 5)w$ **d.** $x/((y - 5)w)$
e. $w^3 - (7 - x)/z + (y - 1)^5$

In the next two exercises, use the algorithm given in Example 7.1.7 to place the given list of words, read left to right, in alphabetical order. Draw the associated binary tree.

11. Four score and seven years ago

12. Joseph, Lillian, Lorraine, John, James, Sugar

*Let $\langle N, A \rangle$ be a finite graph with nodes $N = \{B_1, B_2, \ldots, B_n\}$. The **binary adjacency matrix** of this graph is defined to be the $n \times n$ matrix (a_{ij}) given by*

$$a_{ij} = \begin{cases} 1 & \text{if } \{B_i B_j\} \in A \\ 0 & \text{if } \{B_i B_j\} \notin A \end{cases}$$

13. Construct the binary adjacency matrix for the graph of exercise 1.

14. Construct the binary adjacency matrix for the graph of exercise 4.

15. Use the trees drawn in Figure 7 to write Polish notation (prefix form) for each of the five associated arithmetic expressions. (*Note*: Polish notation was illustrated in Section 1.5.)

16. The following recursive definition[†] of a tree can be shown to be equivalent to the definition given in the text:

---

† Taken from J.-P. Tremblay and R. B. Bunt, *An Introduction to Computer Science: An Algorithmic Approach* (New York: McGraw-Hill, 1979), p. 556.

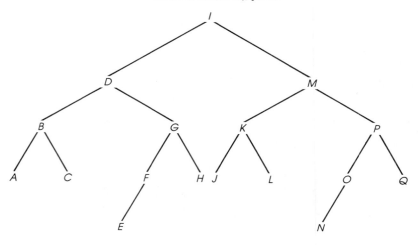

Figure 10

A *tree* is a finite set of one or more nodes such that:

1. there is a particular node called the root, and

2. the remaining nodes, if any, can be placed in disjoint subsets, each of which is itself a tree.

Verify that each of the trees in the last three examples of this section satisfies this alternate definition.

17. *Search of an Ordered Binary Tree.* Consider the ordered binary tree drawn in Figure 10, which contains the first 17 letters of the alphabet.

a. How many nodes must be examined, starting at the root, to locate the letter *E*? *J*? *G*? *M*? *I*?

b. Can you generalize the results of part (a) for searching an ordered binary tree with *m* levels of nodes?

7.2

Semigroups

Although you may not have encountered the term *semigroup* before, we have all worked with mathematical systems that have the properties of a semigroup. From our first days of working with numbers, we learn that the sum of two counting numbers

$$m + n$$

is also a counting number. In addition, if we want to compute the sum

$$m + n + p$$

it makes no difference whether we work from right to left or left to right; that is,

$$m + (n + p) = (m + n) + p$$

These are the two distinguishing properties of a semigroup. That is, a semigroup is a set S and a binary operation $\circ$ that have the following properties:

1. *Closure*: For any $a, b \in S$,

$$a \circ b \in S$$

2. *Associativity*: For any $a, b, c \in S$,

$$a \circ (b \circ c) = (a \circ b) \circ c$$

If we use the Cartesian product notation introduced in Section 3.4 and the functional notation of Section 6.5, the definition can be succinctly stated as follows:

Definition

A *semigroup* is a pair

$$\langle S, \circ \rangle$$

consisting of a set S and an associative binary operation

$$\circ : S \times S \to S$$

The number of elements in a semigroup is called the *order* of the semigroup, and a semigroup of finite order is called a *finite semigroup*. Given semigroup $\langle S, \circ \rangle$, if there exists an element $e \in S$ with the property that

$$e \circ a = a = a \circ e$$

for each $a \in S$, then e is called an *identity for semigroup* $\langle S, \circ \rangle$ and $\langle S, \circ \rangle$ is called a *monoid*.

Example 7.2.1 **Semigroups**

Recall the familiar sets $\mathbb{Z}$, $\mathbb{Q}$, and $\mathbb{R}$ of integers, rational numbers, and real numbers, respectively.

a. For the binary operation $+$ of usual addition, each of the pairs

$$\langle \mathbb{Z}, + \rangle \quad \langle \mathbb{Q}, + \rangle \quad \langle \mathbb{R}, + \rangle$$

is a semigroup. In fact, each of these is a monoid with identity $e = 0$.
b. With the binary operation $*$ of usual multiplication, each of the pairs

$$\langle \mathbb{Z}, * \rangle \quad \langle \mathbb{Q}, * \rangle \quad \langle \mathbb{R}, * \rangle$$

is a semigroup. Also, each of these is a monoid with identity $e = 1$. ∎

Example 7.2.2 **Semigroup**

If E is the set of all even integers

$$\{\ldots, -6, -4, -2, 0, 2, 4, 6, \ldots\}$$

and $+$ and $*$ are the usual addition and multiplication, respectively, then $\langle E, + \rangle$ is a monoid with identity $e = 0$ and $\langle E, * \rangle$ is a semigroup but *not* a monoid, since E does not contain a number that is an identity with respect to $*$. ∎

Example 7.2.3 **Nonsemigroup**

If $-$ is usual subtraction, then the pair $\langle \mathbb{Z}, - \rangle$ is closed since

$$a - b \in \mathbb{Z} \quad \text{for any } a, b \in \mathbb{Z}$$

However, this pair is not associative since, for example,

$$1 - (2 - 3) \neq (1 - 2) - 3$$

Hence, $\langle \mathbb{Z}, - \rangle$ is *not* a semigroup. ∎

Example 7.2.4 **Free Semigroup of a Set**

Let S be a set and S^* the set of all finite-length strings of elements from S. We say that elements $a, b \in S^*$ are *equal*, written $a = b$, if the strings a and b have the same length and have the same element of S in corresponding positions. For instance, if $S = \{x, y, z\}$ then

$$xzxy = xzxy \qquad xzzy \neq xzxy \qquad xzy \neq zy$$

Now define the binary operation, called *concatenation* or *adjoining*, on $S*$ as follows: For $a, b \in S*$

$$a \mathbin{\varheartsuit} b = ab$$

Then, for example,

$$x \mathbin{\varheartsuit} y = xy$$
$$xy \mathbin{\varheartsuit} zy = xyzy$$
$$y \mathbin{\varheartsuit} yz = yyz$$

The pair

$$\langle S*, \mathbin{\varheartsuit} \rangle$$

can be shown to possess the two properties of a semigroup (the reader may want to verify these!), and it is called the *free semigroup generated by set S*. Also, if you augment the set $S*$ with the "null character" ϵ having the property that for any $a \in B*$

$$a \mathbin{\varheartsuit} \epsilon = \epsilon \mathbin{\varheartsuit} a = a$$

then this free semigroup becomes a monoid with identity ϵ. Henceforth in this text, when we refer to $S*$, we shall assume that ϵ has been adjoined.

In particular, if we apply this example to the set

$$B = \{0, 1\}$$

then the free semigroup

$$\langle B*, \mathbin{\varheartsuit} \rangle$$

is the *semigroup of all binary strings*.   ∎

In the preceding examples, we have stated that certain pairs are semigroups without verifying that the properties of closure and associativity actually hold. In the next example, we explicitly verify these properties.

Example **7.2.5**   Semigroup

Let S be any set, and define binary operation $\circ$ as follows: For any $a, b \in S$

$$a \circ b = a$$

That is, the "product" of any two elements equals the element on the left.

   1. Closure: For any $a, b \in S$

$$a \circ b = a \in S$$

so $a \circ b \in S$

   2. Associativity: For any $a, b, c \in S$

$$a \circ (b \circ c) = a \circ b = a$$
$$(a \circ b) \circ c = a \circ c = a$$

so

$$a \circ (b \circ c) = (a \circ b) \circ c$$

Therefore, $\langle S, \circ \rangle$ is a semigroup. ∎

Example 7.2.6

Semigroup of Integers mod 4

Consider the semigroup $\langle \mathbb{Z}, + \rangle$ presented in Example 7.2.1. Before proceeding further, you should review Example 3.4.13 and its associated diagram, which we shall redraw here as Figure 11. There we discussed the equivalence relation R defined on $\mathbb{Z}$ by

$a \, R \, b$ means $a \equiv b(\text{mod } 4)$

and the associated partition of $\mathbb{Z}$ into four equivalence classes, as shown in Figure 11. This equivalence relation can be shown (see exercise 24) to have a special property: If a, b, c, $d \in \mathbb{Z}$ and

$a \, R \, b$ and $c \, R \, d$

then

$$(a + c) \, R \, (b + d)$$

This property means that if you choose *any two* of the four equivalence classes shown in Figure 11, the sum of any element in one class with any element in the other class will always lie in the same class. For instance, if we choose the classes

$$[1] = \{\ldots, -7, -3, 1, 5, 9, \ldots\}$$
$$[2] = \{\ldots, -6, -2, 2, 6, 10, \ldots\}$$

then all of the sums

$$1 + 2 \quad 1 + 6 \quad 1 + 10 \quad 1 - 2 \quad 5 + 2 \quad 5 + 6 \quad 5 - 2 \quad \text{and so on}$$

will lie in the class

$$[3] = \{\ldots, -5, -1, 3, 7, 11, \ldots\}$$

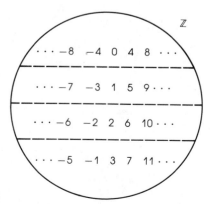

Figure 11

An equivalence relation possessing this special property is called a *congruence*, and the classes are called *congruence classes*. The significance of this is that we can now consider the four classes [0], [1], [2], and [3] as elements themselves and define a new addition $+_4$ of these classes in a natural way:

$$[a] +_4 [b] = [a + b]$$

Identifying the four classes by the symbol

$$Z_4 = \{[0], [1], [2], [3]\}$$

then the pair

$$\langle Z_4, +_4 \rangle$$

is a semigroup of order 4, called the *semigroup of integers mod 4*. The operation $+_4$ is called *addition mod 4*.

For a finite semigroup with few elements, like $\langle Z_4, +_4 \rangle$, the semigroup is often described by a *semigroup table* listing all possible products or sums. The semigroup table for $\langle Z_4, +_4 \rangle$ is shown in Table 1, where the elements of Z_4 are listed in the left column as well as the top row. To illustrate the convention used in completing the table, the result of

$$[1] +_4 [2]$$

is placed in the row identified by [1] and the column identified by [2].   ■

Table 1

$+_4$	[0]	[1]	[2]	[3]
[0]	[0]	[1]	[2]	[3]
[1]	[1]	[2]	[3]	[0]
[2]	[2]	[3]	[0]	[1]
[3]	[3]	[0]	[1]	[2]

Continued study in this area leads to other important algebraic structures such as groups, rings, fields, vector spaces, quotient spaces, and functions (called homomorphisms) relating to these. Typical theorems that we could state and prove in this context are as follows:

1. There is a one-to-one correspondence between congruences on a semigroup and homomorphic images of that semigroup.

2. There is a one-to-one correspondence between homomorphic images of a semigroup and quotient semigroups formed from that semigroup.

As we shall see in the next two sections, theorems of this type regarding semigroups can be used to express important properties of finite state machines and grammars. Therefore, in computer science the notion of semigroup provides a vehicle for organizing, expressing, and perhaps illuminating certain important ideas.

EXERCISES 7.2

In exercises 1 through 16, determine whether the given pair is a semigroup, monoid, or neither. The symbols $+$, $-$, $*$, *and* $/$ *represent usual addition, subtraction, multiplication, and division, respectively, and*

$$A^+ = \{k \in A \,|\, k > 0\} \qquad A_0^+ = \{k \in A \,|\, k \geq 0\}$$

1. $\langle \mathbb{Z}^+, + \rangle$ **2.** $\langle \mathbb{Z}^+, * \rangle$

3. $\langle \mathbb{Z}_0^+, * \rangle$ **4.** $\langle \mathbb{Z}_0^+, + \rangle$

5. $\langle \mathbb{Z}^+, / \rangle$ **6.** $\langle \mathbb{Z}^+, - \rangle$

7. $\langle \mathbb{Q}^+, + \rangle$ **8.** $\langle \mathbb{Q}^+, * \rangle$

9. $\langle \mathbb{Q}^+, - \rangle$ **10.** $\langle \mathbb{Q}^+, / \rangle$

11. $\langle \mathbb{R}^+, + \rangle$ **12.** $\langle \mathbb{R}^+, * \rangle$

13. $\langle \mathbb{R}_0^+, / \rangle$ **14.** $\langle \mathbb{R}^+, / \rangle$

15. $\langle \mathbb{R}_0^+, + \rangle$ **16.** $\langle \mathbb{R}^+, - \rangle$

17. In reference to the free semigroup on a set (see Example 7.2.4), complete the following table:

$\circ$	x	xx	$yxxy$
y			
yx			
yxy			

18. In reference to the semigroup of binary strings (see Example 7.2.4), complete the following table:

$\circ$	1	10	1101
0			
10010			

19. Let V be the set of all vectors of length 3 (in the sense of Chapter 5), and let $+$ and $\cdot$ be the addition and inner product of vectors, respectively.

a. Is $\langle V, + \rangle$ a semigroup?
b. Is $\langle V, \cdot \rangle$ a semigroup?

20. Let M be the set of all 3×3 real matrices, and let $+$ and $\cdot$ be usual matrix addition and multiplication, respectively.

a. Is $\langle M, + \rangle$ a semigroup?
b. Is $\langle M, \cdot \rangle$ a semigroup?

21. Using Example 7.2.6 as a model, construct the semigroup table for $\langle Z_5, +_5 \rangle$.

22. Using Example 7.2.6 as a model, construct the semigroup table for $\langle Z_6, +_6 \rangle$.

23. A semigroup $\langle S, \circ \rangle$ is called *commutative* if

$$a \circ b = b \circ a \qquad \text{for all } a, b \in S$$

Of those pairs in exercises 19 through 22 that are semigroups, which are commutative?

24. Verify that the equivalence relation R discussed in Example 7.2.6 is a congruence. That is, verify the following: If $a, b, c, d \in \mathbb{Z}$ and

$$a R b \quad c R d$$

then

$$(a + c) R (b + d)$$

(*Hint:* Use the fact that $a R b$ means $a \equiv b \pmod 4$, which in turn means that $a - b = 4k$ for some $k \in \mathbb{Z}$.]

25. For each positive integer n, let

$$Z^n = \{k \in Z^+ \,|\, k \text{ has exactly } n \text{ digits}\}$$

where $Z^+ = \{k \in \mathbb{Z} \,|\, k > 0\}$. So, in effect, Z^n is the set of all positive integers with exactly n digits. Does the partition $\{Z^1, Z^2, \ldots\}$ form a congruence on semigroup $\langle Z^+, * \rangle$, where $*$ is usual integer multiplication?

7.3

Finite State Machines

While the phrase *finite state machine* may be new to most of you, nevertheless everyone has had experience in operating many such machines. To illustrate, consider a conventional two-slice "pop-up" toaster with a dial that allows the user to select one of three settings—light (L), medium (M), or dark (D). At any instant the toaster is in one of four possible conditions, or ***internal states***:

$$S = \{OFF, L, M, D\}$$

The user has the choice of one of two *inputs*

$$I = \{1, 2\}$$

where 1 and 2 represent one and two slices of bread, respectively. Having chosen an input from I and a setting from S (that is, having chosen an element of $I \times S$), there are eight possible outputs:

(1, OFF) → one slice untoasted ($1U$)
(2, OFF) → two slices untoasted ($2U$)
(1, L)   → one slice light ($1L$)
(2, L)   → two slices light ($2L$)
(1, M)   → one slice medium ($1M$)
(2, M)   → two slices medium ($2M$)
(1, D)   → one slice dark ($1D$)
(2, D)   → two slices dark ($2D$)

Here we'll call

$$O = \{1U, 2U, 1L, 2L, 1M, 2M, 1D, 2D\}$$

the set of *outputs*, and the function

$$f_o: I \times S \to O$$

defined by the arrows above the *current-output function*. Also, for safety purposes, it's important that the toaster goes OFF after completing its task; so we define the *next-state function*

$$f_S: I \times S \to S$$
$$f_S(i, s) = \text{OFF} \qquad \text{for any } (i, s) \in I \times S$$

Therefore, we have described this particular toaster in terms of five parameters

$$\langle S, I, O, f_S, f_o \rangle$$

This is an example of what is called a *finite state machine*.

Moreover, this toaster machine can be represented geometrically by the digraph given in Figure 12, called the *state graph of the machine*. In this graph, the nodes represent the states S, the connecting arrows define the next-state function f_S, and the symbols associated with the arrows of the form

(input)/(output)

define the input set I and the output set O, as well as the current-output function. For instance, the portion of the graph in Figure 13 tells us that

$$\{1, 2\} \subseteq I \qquad \{1L, 2L\} \subseteq O$$
$$f_S(1, L) = \text{OFF} \qquad f_S(2, L) = \text{OFF}$$

and

$$f_o(1, L) = 1L \qquad f_o(2, L) = 2L$$

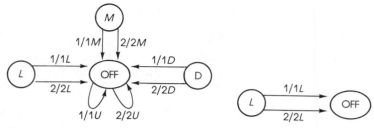

Figure 12   **Figure 13**

The functions f_S and f_O can be summarized in tabular form as shown in Table 2. These are called the *state table* and *output table* for the machine, respectively, and they provide convenient ways of defining the functions f_S and f_O if the sets I and S have only a few elements.

Table 2

f_S	OFF	L	M	D
1	OFF	OFF	OFF	OFF
2	OFF	OFF	OFF	OFF

State Table

f_O	OFF	L	M	D
1	$1U$	$1L$	$1M$	$1D$
2	$2U$	$2L$	$2M$	$2D$

Output Table

Definition

A *finite state machine* (FSM) is a quintuple

$$\langle S, I, O, f_S, f_O \rangle$$

where

   S is a finite set (whose elements are called *internal states*),
   I is a finite set (whose elements are called *inputs*),
   O is a finite set (whose elements are called *outputs*),
   $f_S: I \times S \to S$ is a function (called the *next-state function*), and
   $f_O: I \times S \to O$ is a function (called the *current-output function*).

Note that the concept of time or sequence is implicit in this definition, as indicated by the progression

$$\left.\begin{array}{l}\text{Initial state} \\ \text{Initial input}\end{array}\right\} \to \left\{\begin{array}{l}\text{Output} \\ \text{Next state} \\ \text{Next input}\end{array}\right\} \to \left\{\begin{array}{l}\text{Output} \\ \text{Next state} \\ \text{Next input}\end{array}\right\} \to \text{and so on}$$

Sometimes a particular machine should always be started in a certain state, called the *initial state*, and designated s_0.

Example 7.3.1   **Parity Check Machine**

Consider the FSM $\langle S, I, O, f_S, f_O \rangle$ defined by

$$S = O = \{\text{even, odd}\}$$
$$I = \{1, 0\}$$
$$f_S = f_O \quad \text{are defined as follows:}$$

$f_S = f_O$	Even	Odd
0	even	odd
1	odd	even

The state graph of this machine is given in Figure 14.

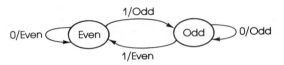

Figure 14

Continuing with this example, suppose that this particular FSM with initial state

$$s_0 = \text{even}$$

was placed in a black box; then it is fed, one symbol at a time, a finite string of input symbols, called an ***input tape***. The diagram in Figure 15 depicts such a black box and summarizes both internal and external conditions at a given instant.

Machine

Input tape ⟶	Input symbol	Current state	Current output	Output tape ⟶
		Next state		

Figure 15

We can use this diagram to indicate how the machine acts on each symbol of any input tape. Let's take tape 101101 as an example (see Figure 16). Here the last (leftmost) output symbol is even, indicating that there is an even number of 1 bits on the input tape; that is, this input tape has even parity. It takes only a few such sample tapes to see that the final output is always the parity of the input tape. This conceptual device, called a ***parity check machine***, accomplishes the same task as the parity check circuit of Example 4.5.1. (The purpose of a parity check was discussed in Section 3.8.)   ∎

If we replaced even by 0 and odd by 1 in the preceding example, the resulting machine could be described as a ***mod 2 counter***, that is, a machine whose final output is the number, modulo 2, of 1 bits on the input tape. The next example is a generalization of this idea.

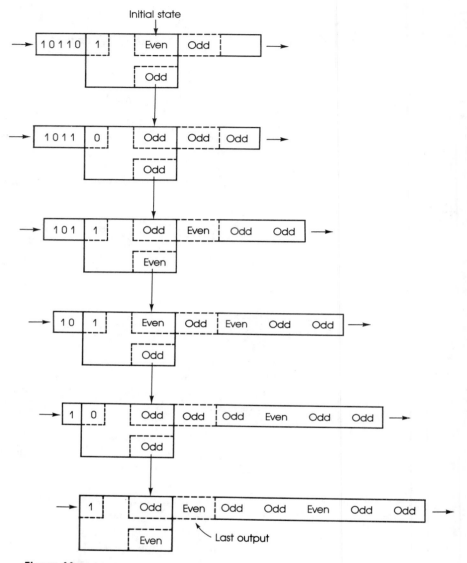

Figure 16

Example 7.3.2 **Mod 3 Counter Machine**

Sometimes it is easier to derive the state graph of a particular process and use it to describe the five parameters of the associated FSM. In Figure 17 we have drawn a state graph in which

$$S = O = \{0, 1, 2\}$$

represents the integers mod 3,

$$I = \{0, 1\}$$

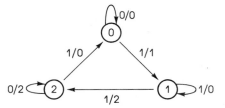

Figure 17

and the connecting arrows and their labels mimic addition mod 3. From this graph we see that $f_S = f_O$ is given by the following table:

$f_S = f_O$	0	1	2
0	0	1	2
1	1	2	0

Just as in the preceding example, let's consider this machine with initial state $s_0 = 0$ to have been placed in a black box, ready to process bit by bit the symbols on an input tape. Figure 18 shows the step-by-step results for the sample input tape

1011011

The last (leftmost) output symbol 2 means that there are 2(modulo 3) 1 bits on this input tape. We call this machine

$$\langle S, I, O, f_S, f_O \rangle$$

a *mod 3 counter*. ■

Example 7.3.3   **Binary Adder Machine**

In Figure 19 we have drawn a state graph to express what happens when we add two binary digits and a carry digit. The symbols $C0$ and $C1$ represent the states of carry 0 and carry 1, respectively. For clarity in that figure, we have used the form

$$\frac{\text{(input)}}{\text{(output)}} \quad \text{rather than} \quad \text{(input)/(output)}$$

In terms of a machine, we have

$$S = \{C0, C1\}$$
$$I = \{0+0, 0+1, 1+0, 1+1\}$$
$$O = \{0, 1\}$$

f_S	$C0$	$C1$		f_O	$C0$	$C1$
$0+0$	$C0$	$C0$		$0+0$	0	1
$0+1$	$C0$	$C1$		$0+1$	1	0
$1+0$	$C0$	$C1$		$1+0$	1	0
$1+1$	$C1$	$C1$		$1+1$	0	1

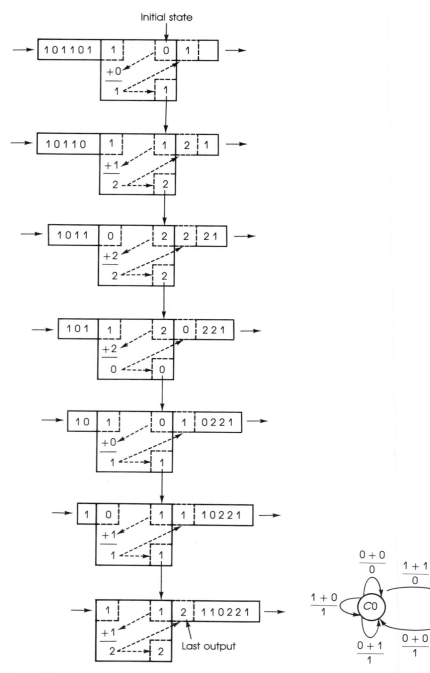

Figure 18

Figure 19

If we initialize this machine in state $s_0 = C0$ and feed the input tapes

   1110   and   100

we obtain the results shown in Figure 20. The final output tape, when read from left to right, gives the sum

   $1110 + 100$

of the binary numbers on the original input tapes. So this FSM, called a **binary adder**, accomplishes the same task as the full adder circuit constructed in Example 4.5.3. ∎

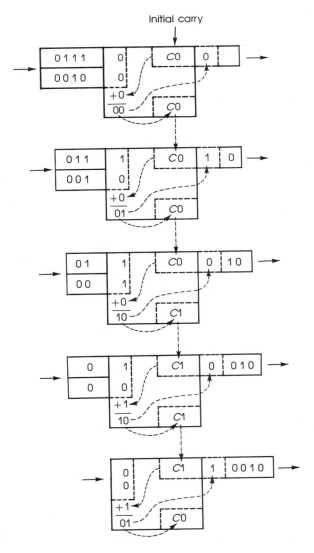

Figure 20

Example 7.3.4   **Machine of a Semigroup**

Let $\langle T, \circ \rangle$ be any finite semigroup. If we let

$$T = S = I = O$$

and define f_S and f_O by

$$f_S(a, b) = f_O(a, b) = a \circ b \qquad \text{for all } a, b \in T$$

then the quintuple

$$\langle T, T, T, f_S, f_O \rangle$$

is called the **machine of the semigroup** $\langle T, \circ \rangle$. If $a \neq a \circ b$, then one portion of the state graph would appear as

Conversely, given an FSM, there is a way of defining the **semigroup of the machine** (see exercise 20).   ∎

Should you continue studying FSMs, a likely next step would be to define a state/output FSM, where the output is determined *only* by its internal state. An important result here is:

Any FSM can be "rebuilt" into a state/output FSM.

Another important topic is that of sets which are recognizable by an FSM. If I is the set of input symbols of an FSM, then I^* represents the set of all (finite) strings of symbols in I (recall the notation of Example 7.2.4). That is, I^* consists of all input tapes, and $\langle I^*, \circ \rangle$ is the free semigroup generated by I. Loosely speaking, a subset of I^* is **recognizable by an FSM** if it consists of all input tapes that produce a particular output symbol. By using the notion of congruence mentioned in Section 7.2, the following can be proved:

A subset $L \subseteq I^*$ is recognizable by an FSM *if and only if* L is a union of classes of a finite congruence on semigroup $\langle I^*, \circ \rangle$.

This illustrates how the vocabulary of semigroups can be used to express properties related to FSMs.

EXERCISES 7.3

1. Taking Figure 16 as a model, process the input tape 1101110 using the parity check machine with $s_0 = $ even.

2. Taking Figure 16 as a model, process the input tape 1110100 using the parity check machine with $s_0 = $ even.

3. Taking Figure 18 as a model, process the input tape 101010 using the mod 3 counter machine with $s_0 = 0$.

4. Taking Figure 18 as a model, process the input tape 11101 using the mod 3 counter machine with $s_0 = 0$.

5. Taking Figure 20 as a model, compute the sum

$$101 + 100$$

using the binary adder machine with $s_0 = C0$.

6. Taking Figure 20 as a model, compute the sum

$$1011 + 101$$

using the binary adder machine with $s_0 = C0$.

7. a. Draw the state graph for the FSM $\langle S, I, O, f_S, f_O \rangle$ given the following information:

$$S = \{A, B, C, D, E\} \quad I = \{0, 1\} \quad O = \{x, y, z\}$$

f_S	A	B	C	D	E
0	D	B	C	C	D
1	E	C	E	B	C

f_O	A	B	C	D	E
0	x	x	x	z	z
1	y	y	y	y	x

b. Using the black box model of the text, determine the output tape if this machine is fed the input tape 1011, where $s_0 = A$.

8. a. Draw the state graph for the FSM $\langle S, I, O, f_S, f_O \rangle$ given the following information:

$$S = \{A, B, C, D\} \quad I = \{1\} \quad O = \{Y, N\}$$

f_S	A	B	C	D
1	B	C	A	A

f_O	A	B	C	D
1	Y	N	N	Y

b. Using the black box model of the text, determine the output tape if this machine is fed the input tape 11111, where $s_0 = D$.

9. The state graph of FSM $\langle S, I, O, f_S, f_O \rangle$ is given in Figure 21. Explicitly describe the sets S, I, and O and the functions f_S and f_O.

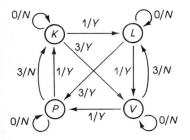

Figure 21

10. The state graph of FSM $\langle S, I, O, f_S, f_O \rangle$ is given in Figure 22. Explicitly describe the sets S, I, and O and the functions f_S and f_O.

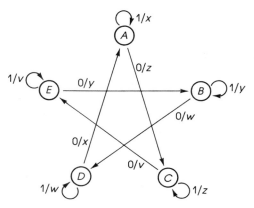

Figure 22

In the next six exercises, construct the described FSM, including its state graph.

11. Machine that determines the parity (even or odd) of the number of 0 bits in a binary string

12. Machine that counts modulo 4 the number of 1 bits in a binary string

13. Machine that will process a binary string and output YES if the last two bits processed were 11 and NO otherwise

14. Machine that will process a binary string and output YES if the last three bits processed were 111 and NO otherwise.

15. Machine of the semigroup $\langle Z_4, +_4 \rangle$ of integers mod 4

16. Machine of the semigroup $\langle Z_5, +_5 \rangle$ of integers mod 5

17. Draw a state graph for an FSM that reads in letters from the alphabet one by one and outputs 1 when the word CAT (in the order C, A, and T) has just been read and 0 otherwise.

18. Draw a state graph for an FSM that reads in precisely four decimal digits and outputs ACCEPT when 3109 is read and REJECT otherwise.

19. Choose a household appliance other than a toaster and describe it as an FSM.

20. Let $M = \langle S, I, O, f_S, f_O \rangle$ be an FSM and let $t \in I^*$. Define function $g_t: S \rightarrow S$ by $g_t(s) = s'$, where s' is the final state of M after t is applied symbol by symbol to M, which has initial state s. If

$$T = \{g_t: S \to S \mid t \in I^*\}$$

and binary operation $\circ$ is composition of functions, show that $\langle T, \circ \rangle$ is a semigroup. (This is called the *semigroup of the machine M*.)

7.4

Languages and Grammars

Each of us became familiar with the terms *language* and *grammar* in our early schooling. At that time we learned how to join letters of the alphabet into certain combinations called *words* and to combine certain words according to grammatical rules into *sentences*.

For instance, with just three letters,

$$a \quad s \quad t$$

some of the possible letter combinations are

$$a \quad s \quad t \quad as \quad ast \quad taa \quad tas \quad sat \quad aast \quad sats \quad tttt$$

There are infinitely many possibilities here, since repeated letters are allowed, and all are considered words according to the general definition that follows. However, only a few, like *as* and *sat*, are considered to be words in the English language.

Definition

> **1.** An **alphabet** (or **vocabulary**) A is a finite, nonempty set of symbols called **letters**.
>
> **2.** A **word over alphabet** A is a finite-length string of letters from A.
>
> **3.** Any collection of words over alphabet A is called a **language over alphabet** A.

Remember that the notation A^* is used to represent all finite-length strings of elements in set A, so a language over alphabet A is just a subset of A^*. The symbol ϵ denotes the **empty word**—that is, the word with no letters. If $a_1 a_2 \cdots a_k$ is a word with letters $a_i \in A$, then we say that this word **contains** the letters $a_1, a_2, \ldots, a_k$.

Example 7.4.1

Language

a. Consider the alphabet $A = \{d, g, o, 5, \theta\}$. Then

$$d \quad dg \quad o\theta \quad 555 \quad go\theta dd5 \quad \epsilon$$

are some of the infinitely many words over A, and

$$L_1 = \{do, go, god, good, dog\}$$
$$L_2 = \{d, g, 5g, \theta 5o, dg\}$$
$$L_3 = \{5, 55, 555, 5555, 55555, \ldots\}$$
$$L_4 = \{w \in A^* \mid \text{word } w \text{ has length} \geq 4\}$$

are four examples of languages over A.

b. For $A = \{0, 1\}$,

$L_1 = A^*$ is the language over A of all (finite-length) binary strings

$L_2 = \{000, 100, 010, 001, 110, 101, 011, 111\}$ is the language over A of all binary strings of length 3 ∎

Example 7.4.2

Semigroup of an Alphabet

Let A be an alphabet. Then A^* is the set of all words over A, and the semigroup $\langle A^*, \bigcirc \rangle$ generated by A is called the ***semigroup of alphabet A***. In particular, when $A = \{0, 1\}$, then the semigroup of A is the semigroup of all binary strings (see Example 7.2.4). ∎

Using this notion of language, we can construct rules of grammar that will enable us to form sentences. The tree graph in Figure 23 is one way of exhibiting the structure of the sentence "The huge black bear slowly lifted its head." Here the root of the tree (or ***start symbol***) is the word *sentence*, the leaves of the tree (or ***terminals***) are the words of the given sentence, and the intermediate node labels describe the steps to get from one to the other. This figure motivates the definition below.

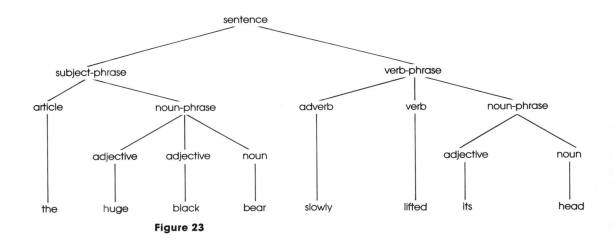

Figure 23

Definition

A ***grammar*** is a 4-tuple

$$\langle V, V_T, s, P \rangle$$

where

V is a vocabulary,
V_T is a nonempty subset of V (whose elements are called ***terminals***),
$s \in V - V_T$ (called the ***start symbol***), and
$P = \{(\alpha, \beta) \mid \alpha, \beta \in V^*$ where α contains at least one nonterminal symbol$\}$ is a finite set (whose elements are called ***productions***).

A production (α, β) is ordinarily denoted by the symbol $\alpha \to \beta$; these can be associated with the lines of the tree in Figure 23.

Example 7.4.3

Grammar

If

$$V = \{a, b, c, s\}$$
$$V_T = \{a, b\}$$
$$P = \{s \xrightarrow{1} as,\ bc \xrightarrow{2} b,\ s \xrightarrow{3} a\}$$

then $\langle V, V_T, s, P \rangle$ is a grammar. (In this example and the next, we number the productions for the sake of identification.)

Now let's use the productions to transform words over V with nonterminals into words over V with terminals, if possible:

$$abc = a(bc) \xrightarrow{2} \boxed{ab}$$

$$bca = (bc)a \xrightarrow{2} \boxed{ba}$$

ab cannot be transformed (all terminals)

cb cannot be transformed (no applicable production)

$$as \xrightarrow{3} \boxed{aa}$$

$$as \xrightarrow{1} a(as) = (aa)s \xrightarrow{3} (aa)a = \boxed{aaa}$$

$$as \xrightarrow{1} a(as) = (aa)s \xrightarrow{1} (aa)as = (aaa)s \xrightarrow{3} (aaa)a = \boxed{aaaa}$$

Observe that by reapplying the production $s \to as$, as can be transformed into a string of the symbol a of any length ≥ 2. This work shows us that we can think of productions as *rewriting rules*. ∎

In the next example, the vocabulary consists of words rather than letters.

Example 7.4.4

Grammar

Let

$$V = \{\text{three, two, dogs, barked, loudly, sentence, noun-phrase,}$$
$$\text{verb-phrase, adjective, noun, adverb, verb}\}$$

$$V_T = \{\text{three, two, dogs, barked, loudly}\}$$

$$s = \text{sentence}$$

$$P = \{\text{sentence} \xrightarrow{1} \text{noun-phrase   verb-phrase}$$

$$\text{noun-phrase} \xrightarrow{2} \text{adjective   noun}$$

$$\text{verb-phrase} \xrightarrow{\;3\;} \text{verb adverb}$$

$$\text{adjective} \xrightarrow{\;4\;} \text{two}$$

$$\text{adjective} \xrightarrow{\;5\;} \text{three}$$

$$\text{noun} \xrightarrow{\;6\;} \text{dogs}$$

$$\text{verb} \xrightarrow{\;7\;} \text{barked}$$

$$\text{adverb} \xrightarrow{\;8\;} \text{loudly}\}$$

Then $\langle V, V_T, s, P \rangle$ is a grammar that can be used to derive the sentence

"Two dogs barked loudly"

as follows:

$$\text{sentence} \xrightarrow{\;1\;} \text{noun-phrase verb-phrase}$$

$$\xrightarrow{\;2\;} \text{adjective noun verb-phrase}$$

$$\xrightarrow{\;3\;} \text{adjective noun verb adverb}$$

$$\xrightarrow{\;4\;} \text{two noun verb adverb}$$

$$\xrightarrow{\;6\;} \text{two dogs verb adverb}$$

$$\xrightarrow{\;7\;} \text{two dogs barked adverb}$$

$$\xrightarrow{\;8\;} \text{two dogs barked loudly}$$

Of course, the sentence

"Three dogs barked loudly"

could also be derived by applying production 5 in place of production 4. ∎

If we were to pursue the study of grammars further, the next step would be to define the notion of the *language generated by a grammar*, or the so-called *formal languages*[†] (see exercises 11 through 14). Also, by considering productions with different properties, we can define grammars that are called Type 0 (the grammar defined above), Type 1 (or *context-sensitive*), Type 2 (or *context-free*), and Type 3 (or *regular*). These same terms are also applied to the languages generated by the particular grammars. Then it is possible to prove results that connect semigroups, FSMs, and languages. For instance, the following can be shown:

Language L over alphabet A is regular *if and only if*

$L \subseteq A^*$ is recognizable by an FSM *if and only if*

L is a union of classes of a finite congruence on semigroup $\langle A^*, \frown \rangle$.

[†] The programming language ALGOL satisfies the definition of a formal language.

EXERCISES **7.4**

1. Given the alphabet $A = \{e, d, p, s\}$, construct a language over A with the following:

   a. Exactly 4 words
   b. Exactly 12 words, each of length 5
   c. Infinitely many words

2. Given the alphabet $A = \{e, p, t\}$, construct a language over A with the following:

   a. Exactly 6 words
   b. More than 10 words, each of length ≥ 4
   c. Infinitely many words

3. Describe the semigroup of the alphabet $\{n, o\}$.

4. Describe the semigroup of the alphabet $\{1\}$.

5. Using Figure 23 as a model, draw a tree that exhibits the grammatical structure of the sentence "Everyone attended the final game."

6. Using Figure 23 as a model, draw a tree that exhibits the grammatical structure of the sentence "Sixteen girls and fifteen boys graduated this past June."

7. Given the grammar $\langle V, V_T, s, P \rangle$ where

$$V = \{x, y, s\}$$
$$V_T = \{x, y\}$$
$$P = \{s \to xsy, xs \to x\}$$

rewrite each of the following words so as to contain all terminals, if possible:

   a. sy **b.** xs **c.** $xsyy$

8. Given the grammar $\langle V, V_T, s, P \rangle$ where

$$V = \{a, b, c, s\}$$
$$V_T = \{b, c\}$$
$$P = \{s \to sc, s \to b, ab \to c\}$$

rewrite each of the following words so as to contain all terminals, if possible:

   a. abc **b.** cab **c.** bca **d.** asb

9. Construct a grammar from which the sentence given in exercise 5 can be derived. (*Hint:* See Example 7.4.4.)

10. Construct a grammar from which the sentence given in exercise 6 can be derived. (*Hint:* See Example 7.4.4.)

Definitions: Let $G = \langle V, V_T, s, P \rangle$ be a grammar.

   1. If w and α are words over V and $w = \alpha \subset \beta$ for some word β, we say that w **contains an instance of** α.

   2. If w_1 and w_2 are words over V and there exists a production $\alpha \to \beta$ such that

      a. w_1 contains an instance of α and
      b. w_2 is obtained from w_1 by replacing each instance of α with β,

then we say that w_1 **directly generates** w_2, written $w_1 \Rightarrow w_2$.

   3. If $w_1, w_2, \ldots, w_n$ are words over V and

$$w_1 \Rightarrow w_2, w_2 \Rightarrow w_3, \ldots, w_{n-1} \Rightarrow w_n$$

then we say that w_1 **generates** w_n, written $w_1 \Rrightarrow w_n$.

   4. $L(G) = \{w \in V_T^* \mid s \Rrightarrow w\}$

is called the **language generated by grammar** G.

In each of the last four exercises, describe the language generated by the given grammar.

11. $V = \{0, 1, s\}$ $V_T = \{0, 1\}$ $P = \{s \to ss, s \to 101\}$

12. $V = \{0, 1, s\}$ $V_T = \{0, 1\}$ $P = \{s \to 101s, s \to 101\}$

13. $V = \{x, y, s\}$ $V_T = \{x, y\}$
$$P = \{s \to xy, s \to xs, s \to sy\}$$

14. $V = \{$the, girl, boy, loves, dog, cat, sentence, noun-phrase, verb-phrase, article, noun, verb$\}$

$V_T = \{$the, girl, boy, loves, dog, cat$\}$

$s = $ sentence

$P = \{$sentence $\to$ noun-phrase verb-phrase,
    noun-phrase $\to$ article noun,
    verb-phrase $\to$ verb noun-phrase,
    article $\to$ the
    noun $\to$ girl
    noun $\to$ boy
    noun $\to$ cat
    noun $\to$ dog
    verb $\to$ loves$\}$

Tables

1. Compound Amount Factor $\qquad (1 + R)^N$

2. Series Compound Amount Factor $\qquad \dfrac{(1 + R)^N - 1}{R}$

3. Series Present Value Factor $\qquad \dfrac{(1 + R)^N - 1}{R(1 + R)^N}$

4. Standard Normal Distribution

Table 1 Compound Amount Factor $(1 + R)^N$

N	.01	.02	.03	.04	.05	.06	.07	.08
1	1.01	1.02	1.03	1.04	1.05	1.06	1.07	1.08
2	1.0201	1.0404	1.0609	1.0816	1.1025	1.1236	1.1449	1.1664
3	1.0303	1.06121	1.09273	1.12486	1.15762	1.19102	1.22504	1.25971
4	1.0406	1.08243	1.12551	1.16986	1.21551	1.26248	1.3108	1.36049
5	1.05101	1.10408	1.15927	1.21665	1.27628	1.33823	1.40255	1.46933
6	1.06152	1.12616	1.19405	1.26532	1.3401	1.41852	1.50073	1.58687
7	1.07214	1.14869	1.22987	1.31593	1.4071	1.50363	1.60578	1.71382
8	1.08286	1.17166	1.26677	1.36857	1.47746	1.59385	1.71819	1.85093
9	1.09369	1.19509	1.30477	1.42331	1.55133	1.68948	1.83846	1.999
10	1.10462	1.21899	1.34392	1.48024	1.62889	1.79085	1.96715	2.15892
11	1.11567	1.24337	1.38423	1.53945	1.71034	1.8983	2.10485	2.33164
12	1.12683	1.26824	1.42576	1.60103	1.79586	2.0122	2.25219	2.51817
13	1.13809	1.29361	1.46853	1.66507	1.88565	2.13293	2.40985	2.71962
14	1.14947	1.31948	1.51259	1.73168	1.97993	2.2609	2.57853	2.93719
15	1.16097	1.34587	1.55797	1.80094	2.07893	2.39656	2.75903	3.17217
16	1.17258	1.37279	1.60471	1.87298	2.18287	2.54035	2.95216	3.42594
17	1.1843	1.40024	1.65285	1.9479	2.29202	2.69277	3.15882	3.70002
18	1.19615	1.42825	1.70243	2.02582	2.40662	2.85434	3.37993	3.99602
19	1.20811	1.45681	1.75351	2.10685	2.52695	3.0256	3.61653	4.3157
20	1.22019	1.48595	1.80611	2.19112	2.6533	3.20714	3.86968	4.66096
21	1.23239	1.51567	1.86029	2.27877	2.78596	3.39956	4.14056	5.03383
22	1.24472	1.54598	1.9161	2.36992	2.92526	3.60354	4.4304	5.43654
23	1.25716	1.5769	1.97359	2.46472	3.07152	3.81975	4.74053	5.87146
24	1.26973	1.60844	2.03279	2.5633	3.2251	4.04893	5.07237	6.34118
25	1.28243	1.64061	2.09378	2.66584	3.38635	4.29187	5.42743	6.84848
26	1.29526	1.67342	2.15659	2.77247	3.55567	4.54938	5.80735	7.39635
27	1.30821	1.70689	2.22129	2.88337	3.73346	4.82235	6.21387	7.98806
28	1.32129	1.74102	2.28793	2.9987	3.92013	5.11169	6.64884	8.62711
29	1.3345	1.77584	2.35657	3.11865	4.11614	5.41839	7.11426	9.31727
30	1.34785	1.81136	2.42726	3.2434	4.32194	5.74349	7.61226	10.0627
31	1.36133	1.84759	2.50008	3.37313	4.53804	6.0881	8.14511	10.8677
32	1.37494	1.88454	2.57508	3.50806	4.76494	6.45339	8.71527	11.7371
33	1.38869	1.92223	2.65234	3.64838	5.00319	6.84059	9.32534	12.676
34	1.40258	1.96068	2.73191	3.79432	5.25335	7.25103	9.97811	13.6901
35	1.4166	1.99989	2.81386	3.94609	5.51602	7.68609	10.6766	14.7853
36	1.43077	2.03989	2.89828	4.10393	5.79182	8.14725	11.4239	15.9682
37	1.44508	2.08069	2.98523	4.26809	6.08141	8.63609	12.2236	17.2456
38	1.45953	2.1223	3.07478	4.43881	6.38548	9.15425	13.0793	18.6253
39	1.47412	2.16474	3.16703	4.61637	6.70475	9.70351	13.9948	20.1153
40	1.48886	2.20804	3.26204	4.80102	7.03999	10.2857	14.9745	21.7245
41	1.50375	2.2522	3.3599	4.99306	7.39199	10.9029	16.0227	23.4625
42	1.51879	2.29724	3.4607	5.19278	7.76159	11.557	17.1443	25.3395
43	1.53398	2.34319	3.56452	5.4005	8.14967	12.2505	18.3444	27.3666
44	1.54932	2.39005	3.67145	5.61652	8.55715	12.9855	19.6285	29.556
45	1.56481	2.43785	3.7816	5.84118	8.98501	13.7646	21.0025	31.9204
46	1.58046	2.48661	3.89504	6.07482	9.43426	14.5905	22.4726	34.4741
47	1.59626	2.53634	4.0119	6.31782	9.90597	15.4659	24.0457	37.232
48	1.61223	2.58707	4.13225	6.57053	10.4013	16.3939	25.7289	40.2106
49	1.62835	2.63881	4.25622	6.83335	10.9213	17.3775	27.5299	43.4274
50	1.64463	2.69159	4.38391	7.10668	11.4674	18.4202	29.457	46.9016
51	1.66108	2.74542	4.51542	7.39095	12.0408	19.5254	31.519	50.6537
52	1.67769	2.80033	4.65089	7.68659	12.6428	20.6969	33.7253	54.706
53	1.69447	2.85633	4.79041	7.99405	13.2749	21.9387	36.0861	59.0825
54	1.71141	2.91346	4.93412	8.31381	13.9387	23.255	38.6122	63.8091
55	1.72852	2.97173	5.08215	8.64637	14.6356	24.6503	41.315	68.9139
56	1.74581	3.03117	5.23461	8.99222	15.3674	26.1293	44.2071	74.427
57	1.76327	3.09179	5.39165	9.35191	16.1358	27.6971	47.3015	80.3811
58	1.7809	3.15362	5.5534	9.72599	16.9426	29.3589	50.6127	86.8116
59	1.79871	3.2167	5.72	10.115	17.7897	31.1205	54.1555	93.7565
60	1.8167	3.28103	5.8916	10.5196	18.6792	32.9877	57.9464	101.257

Table 2 Series Compound Amount Factor $\dfrac{(1 + R)^N - 1}{R}$

				R				
N	.01	.02	.03	.04	.05	.06	.07	.08
1	1	1	1	1	1	1	1	1
2	2.01	2.02	2.03	2.04	2.05	2.06	2.07	2.08
3	3.0301	3.0604	3.0909	3.1216	3.1525	3.1836	3.2149	3.2464
4	4.0604	4.12161	4.18363	4.24646	4.31012	4.37462	4.43994	4.50611
5	5.10101	5.20404	5.30914	5.41632	5.52563	5.63709	5.75074	5.8666
6	6.15202	6.30812	6.46841	6.63298	6.80191	6.97532	7.15329	7.33593
7	7.21354	7.43428	7.66246	7.89829	8.14201	8.39384	8.65402	8.9228
8	8.28567	8.58297	8.89234	9.21423	9.54911	9.89747	10.2598	10.6366
9	9.36853	9.75463	10.1591	10.5828	11.0266	11.4913	11.978	12.4876
10	10.4622	10.9497	11.4639	12.0061	12.5779	13.1808	13.8164	14.4866
11	11.5668	12.1687	12.8078	13.4864	14.2068	14.9716	15.7836	16.6455
12	12.6825	13 4121	14.192	15.0258	15.9171	16.8699	17.8885	18.9771
13	13.8093	14.6803	15.6178	16.6268	17.713	18.8821	20.1406	21.4953
14	14.9474	15.9739	17.0863	18.2919	19.5986	21.0151	22.5505	24.2149
15	16.0969	17.2934	18.5989	20.0236	21.5786	23.276	25.129	27.1521
16	17.2579	18.6393	20.1569	21.8245	23.6575	25.6725	27.8881	30.3243
17	18.4304	20.0121	21.7616	23.6975	25.8404	28.2129	30.8402	33.7502
18	19.6147	21.4123	23.4144	25.6454	28.1324	30.9057	33.999	37.4502
19	20.8109	22.8406	25.1169	27.6712	30.539	33.76	37.379	41.4463
20	22.019	24.2974	26.8704	29.7781	33.066	36.7856	40.9955	45.762
21	23.2392	25.7833	28.6765	31.9692	35.7193	39.9927	44.8652	50.4229
22	24.4716	27.299	30.5368	34.248	38.5052	43.3923	49.0057	55.4568
23	25.7163	28.845	32.4529	36.6179	41.4305	46.9958	53.4361	60.8933
24	26.9735	30.4219	34.4265	39.0826	44.502	50.8156	58.1767	66.7648
25	28.2432	32.0303	36.4593	41.6459	47.7271	54.8645	63.249	73.1059
26	29.5256	33.6709	38.553	44.3117	51.1135	59.1564	68.6765	79.9544
27	30.8209	35.3443	40.7096	47.0842	54.6691	63.7058	74.4838	87.3508
28	32.1291	37.0512	42.9309	49.9676	58.4026	68.5281	80.6977	95.3388
29	33.4504	38.7922	45.2189	52.9663	62.3227	73.6398	87.3465	103.966
30	34.7849	40.5681	47.5754	56.0849	66.4388	79.0582	94.4608	113.283
31	36.1327	42.3794	50.0027	59.3283	70.7608	84.8017	102.073	123.346
32	37.4941	44.227	52.5028	62.7015	75.2988	90.8898	110.218	134.214
33	38.869	46.1116	55.0778	66.2095	80.0638	97.3432	118.933	145.951
34	40.2577	48.0338	57.7302	69.8579	85.067	104.184	128.259	158.627
35	41.6603	49.9945	60.4621	73.6522	90.3203	111.435	138.237	172.317
36	43.0769	51.9944	63.2759	77.5983	95.8363	119.121	148.913	187.102
37	44.5076	54.0343	66.1742	81.7022	101.628	127.268	160.337	203.07
38	45.9527	56.1149	69.1594	85.9703	107.71	135.904	172.561	220.316
39	47.4123	58.2372	72.2342	90.4091	114.095	145.058	185.64	238.941
40	48.8864	60.402	75.4013	95.0255	120.8	154.762	199.635	259.057
41	50.3752	62.61	78.6633	99.8265	127.84	165.048	214.61	280.781
42	51.879	64.8622	82.0232	104.82	135.232	175.951	230.632	304.244
43	53.3978	67.1595	85.4839	110.012	142.993	187.508	247.776	329.583
44	54.9318	69.5027	89.0484	115.413	151.143	199.758	266.121	356.95
45	56.4811	71.8927	92.7199	121.029	159.7	212.744	285.749	386.506
46	58.0459	74.3306	96.5015	126.871	168.685	226.508	306.752	418.426
47	59.6263	76.8172	100.397	132.945	178.119	241.099	329.224	452.9
48	61.2226	79.3535	104.408	139.263	188.025	256.565	353.27	490.132
49	62.8348	81.9406	108.541	145.834	198.427	272.958	378.999	530.343
50	64.4632	84.5794	112.797	152.667	209.348	290.336	406.529	573.77
51	66.1078	87.271	117.181	159.774	220.815	308.756	435.986	620.672
52	67.7689	90.0164	121.696	167.165	232.856	328.281	467.505	671.326
53	69.4466	92.8167	126.347	174.851	245.499	348.978	501.23	726.032
54	71.141	95.6731	131.137	182.845	258.774	370.917	537.316	785.114
55	72.8525	98.5865	136.072	191.159	272.713	394.172	575.929	848.923
56	74.581	101.558	141.154	199.806	287.348	418.822	617.244	917.837
57	76.3268	104.589	146.388	208.798	302.716	444.952	661.451	992.264
58	78.0901	107.681	151.78	218.15	318.851	472.649	708.752	1072.65
59	79.871	110.835	157.333	227.876	335.794	502.008	759.365	1159.46
60	81.6697	114.052	163.053	237.991	353.584	533.128	813.52	1253.21

Table 3 Series Present Value Factor $\dfrac{(1 + R)^N - 1}{R(1 + R)^N}$

				R				
N	.01	.02	.03	.04	.05	.06	.07	.08
1	.990099	.980392	.970874	.961538	.952381	.943396	.934579	.925926
2	1.9704	1.94156	1.91347	1.88609	1.85941	1.83339	1.80802	1.78326
3	2.94099	2.88388	2.82861	2.77509	2.72325	2.67301	2.62432	2.5771
4	3.90197	3.80773	3.7171	3.6299	3.54595	3.46511	3.38721	3.31213
5	4.85343	4.71346	4.57971	4.45182	4.32948	4.21236	4.1002	3.99271
6	5.79548	5.60143	5.41719	5.24214	5.07569	4.91732	4.76654	4.62288
7	6.72819	6.47199	6.23028	6.00205	5.78637	5.58238	5.38929	5.20637
8	7.65168	7.32548	7.01969	6.73274	6.46321	6.20979	5.9713	5.74664
9	8.56602	8.16224	7.78611	7.43533	7.10782	6.80169	6.51523	6.24689
10	9.4713	8.98259	8.5302	8.1109	7.72173	7.36009	7.02358	6.71008
11	10.3676	9.78685	9.25262	8.76048	8.30641	7.88687	7.49867	7.13896
12	11.2551	10.5753	9.954	9.38507	8.86325	8.38384	7.94269	7.53608
13	12.1337	11.3484	10.635	9.98565	9.39357	8.85268	8.35765	7.90378
14	13.0037	12.1062	11.2961	10.5631	9.89864	9.29498	8.74547	8.24424
15	13.8651	12.8493	11.9379	11.1184	10.3797	9.71225	9.10791	8.55948
16	14.7179	13.5777	12.5611	11.6523	10.8378	10.1059	9.44665	8.85137
17	15.5623	14.2919	13.1661	12.1657	11.2741	10.4773	9.76322	9.12164
18	16.3983	14.992	13.7535	12.6593	11.6896	10.8276	10.0591	9.37189
19	17.226	15.6785	14.3238	13.1339	12.0853	11.1581	10.3356	9.6036
20	18.0456	16.3514	14.8775	13.5903	12.4622	11.4699	10.594	9.81815
21	18.857	17.0112	15.415	14.0292	12.8212	11.7641	10.8355	10.0168
22	19.6604	17.658	15.9369	14.4511	13.163	12.0416	11.0612	10.2007
23	20.4558	18.2922	16.4436	14.8568	13.4886	12.3034	11.2722	10.3711
24	21.2434	18.9139	16.9355	15.247	13.7986	12.5504	11.4693	10.5288
25	22.0232	19.5235	17.4131	15.6221	14.0939	12.7834	11.6536	10.6748
26	22.7952	20.121	17.8768	15.9828	14.3752	13.0032	11.8258	10.81
27	23.5596	20.7069	18.327	16.3296	14.643	13.2105	11.9867	10.9352
28	24.3164	21.2813	18.7641	16.6631	14.8981	13.4062	12.1371	11.0511
29	25.0658	21.8444	19.1885	16.9837	15.1411	13.5907	12.2777	11.1584
30	25.8077	22.3965	19.6004	17.292	15.3725	13.7648	12.409	11.2578
31	26.5423	22.9377	20.0004	17.5885	15.5928	13.9291	12.5318	11.3498
32	27.2696	23.4683	20.3888	17.8736	15.8027	14.084	12.6466	11.435
33	27.9897	23.9886	20.7658	18.1476	16.0025	14.2302	12.7538	11.5139
34	28.7027	24.4986	21.1318	18.4112	16.1929	14.3681	12.854	11.5869
35	29.4086	24.9986	21.4872	18.6646	16.3742	14.4982	12.9477	11.6546
36	30.1075	25.4888	21.8323	18.9083	16.5469	14.621	13.0352	11.7172
37	30.7995	25.9695	22.1672	19.1426	16.7113	14.7368	13.117	11.7752
38	31.4847	26.4406	22.4925	19.3679	16.8679	14.846	13.1935	11.8289
39	32.163	26.9026	22.8082	19.5845	17.017	14.9491	13.2649	11.8786
40	32.8347	27.3555	23.1148	19.7928	17.1591	15.0463	13.3317	11.9246
41	33.4997	27.7995	23.4124	19.9931	17.2944	15.138	13.3941	11.9672
42	34.1581	28.2348	23.7014	20.1856	17.4232	15.2245	13.4524	12.0067
43	34.81	28.6616	23.9819	20.3708	17.5459	15.3062	13.507	12.0432
44	35.4555	29.08	24.2543	20.5488	17.6628	15.3832	13.5579	12.0771
45	36.0945	29.4902	24.5187	20.72	17.7741	15.4558	13.6055	12.1084
46	36.7272	29.8923	24.7754	20.8847	17.8801	15.5244	13.65	12.1374
47	37.3537	30.2866	25.0247	21.0429	17.981	15.589	13.6916	12.1643
48	37.974	30.6731	25.2667	21.1951	18.0772	15.65	13.7305	12.1891
49	38.5881	31.0521	25.5017	21.3415	18.1687	15.7076	13.7668	12.2122
50	39.1961	31.4236	25.7298	21.4822	18.2559	15.7619	13.8007	12.2335
51	39.7981	31.7878	25.9512	21.6175	18.339	15.8131	13.8325	12.2532
52	40.3942	32.1449	26.1662	21.7476	18.4181	15.8614	13.8621	12.2715
53	40.9844	32.495	26.375	21.8727	18.4934	15.907	13.8898	12.2884
54	41.5687	32.8383	26.5777	21.993	18.5651	15.95	13.9157	12.3041
55	42.1472	33.1748	26.7744	22.1086	18.6335	15.9905	13.9399	12.3186
56	42.72	33.5047	26.9655	22.2198	18.6985	16.0288	13.9626	12.3321
57	43.2871	33.8281	27.1509	22.3267	18.7605	16.0649	13.9837	12.3445
58	43.8486	34.1452	27.331	22.4296	18.8195	16.099	14.0035	12.356
59	44.4046	34.4561	27.5058	22.5284	18.8758	16.1311	14.0219	12.3667
60	44.955	34.7609	27.6756	22.6235	18.9293	16.1614	14.0392	12.3766

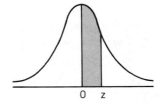

Table 4 Standard Normal Distribution

z	.00	.01	.02	.03	.04	.05	.06	.07	.08	.09
.0	.0000	.0040	.0080	.0120	.0160	.0199	.0239	.0279	.0319	.0359
.1	.0398	.0438	.0478	.0517	.0557	.0596	.0636	.0675	.0714	.0753
.2	.0793	.0832	.0871	.0910	.0948	.0987	.1026	.1064	.1103	.1141
.3	.1179	.1217	.1255	.1293	.1331	.1368	.1406	.1443	.1480	.1517
.4	.1554	.1591	.1628	.1664	.1700	.1736	.1772	.1808	.1844	.1879
.5	.1915	.1950	.1985	.2019	.2054	.2088	.2123	.2157	.2190	.2224
.6	.2257	.2291	.2324	.2357	.2389	.2422	.2454	.2486	.2518	.2549
.7	.2580	.2612	.2642	.2673	.2704	.2734	.2764	.2794	.2823	.2852
.8	.2881	.2910	.2939	.2967	.2995	.3023	.3051	.3078	.3106	.3133
.9	.3159	.3186	.3212	.3238	.3264	.3289	.3315	.3340	.3365	.3389
1.0	.3413	.3438	.3461	.3485	.3508	.3531	.3554	.3577	.3599	.3621
1.1	.3643	.3665	.3686	.3708	.3729	.3749	.3770	.3790	.3810	.3830
1.2	.3849	.3869	.3888	.3907	.3925	.3944	.3962	.3980	.3997	.4015
1.3	.4032	.4049	.4066	.4082	.4099	.4115	.4131	.4147	.4162	.4177
1.4	.4192	.4207	.4222	.4236	.4251	.4265	.4279	.4292	.4306	.4319
1.5	.4332	.4345	.4357	.4370	.4382	.4394	.4406	.4418	.4429	.4441
1.6	.4452	.4463	.4474	.4484	.4495	.4505	.4515	.4525	.4535	.4545
1.7	.4554	.4564	.4573	.4582	.4591	.4599	.4608	.4616	.4625	.4633
1.8	.4641	.4649	.4656	.4664	.4671	.4678	.4686	.4693	.4699	.4706
1.9	.4713	.4719	.4726	.4732	.4738	.4744	.4750	.4756	.4761	.4767
2.0	.4772	.4778	.4783	.4788	.4793	.4798	.4803	.4808	.4812	.4817
2.1	.4821	.4826	.4830	.4834	.4838	.4842	.4846	.4850	.4854	.4857
2.2	.4861	.4864	.4868	.4871	.4875	.4878	.4881	.4884	.4887	.4890
2.3	.4893	.4896	.4898	.4901	.4904	.4906	.4909	.4911	.4913	.4916
2.4	.4918	.4920	.4922	.4925	.4927	.4929	.4931	.4932	.4934	.4936
2.5	.4938	.4940	.4941	.4943	.4945	.4946	.4948	.4949	.4951	.4952
2.6	.4953	.4955	.4956	.4957	.4959	.4960	.4961	.4962	.4963	.4964
2.7	.4965	.4966	.4967	.4968	.4969	.4970	.4971	.4972	.4973	.4974
2.8	.4974	.4975	.4976	.4977	.4977	.4978	.4979	.4979	.4980	.4981
2.9	.4881	.4982	.4982	.4983	.4984	.4984	.4985	.4985	.4986	.4986
3.0	.49865	.4987	.4987	.4988	.4988	.4989	.4989	.4989	.4990	.4990

From Frank S. Budnick, *Applied Mathematics for Business, Economics, and the Social Sciences* (Second Edition), (McGraw-Hill Book Co., 1983), p. 373. Reprinted with permission.

Answers to Odd-Numbered Exercises

CHAPTER 1

Section 1.1

1.

COUNT	INT	DEP
1	60	2060
2	61.80	2121.80

3. a.

NAME	HOURS	SAL	COUNT
ALBRIGHT	35	175	1
DEMITRAS	40	180	2

b.

ALBRIGHT	175
DEMITRAS	180

5. KIM is a child   **7.** ALLIE is oldest

9.

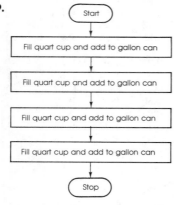

11. B = black paint
W = white paint

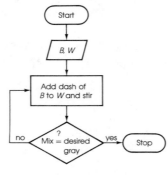

13. PP = purchase price
C = change from \$1
Q = # quarters in change
D = # dimes in change
N = # nickels in change
P = # pennies in change

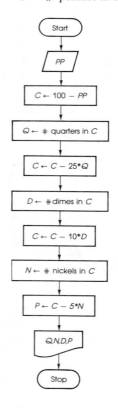

15.

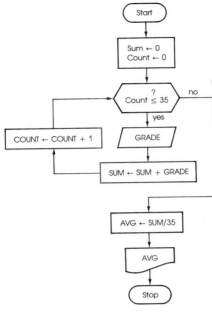

17.

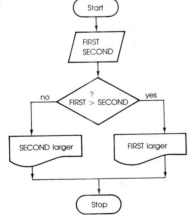

19.

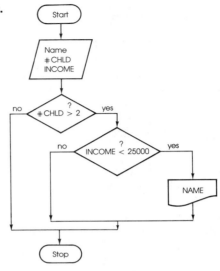

21.

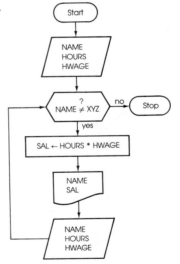

Section 1.2

1. ALBRIGHT   175
    DEMITRAS   191.25

3. The work mimics the mechanics of

$$
\begin{array}{r}
{\scriptstyle 1\ \ 1}\\
6504\\
+\,1687\\
\hline
8191
\end{array}
$$

5.

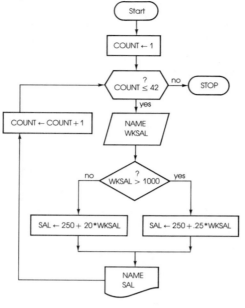

7. Same as Figure 15 of Example 1.2.3, but adding the "read N" symbol to top of flowchart and replacing 500 by N.

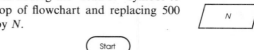

9.

11.

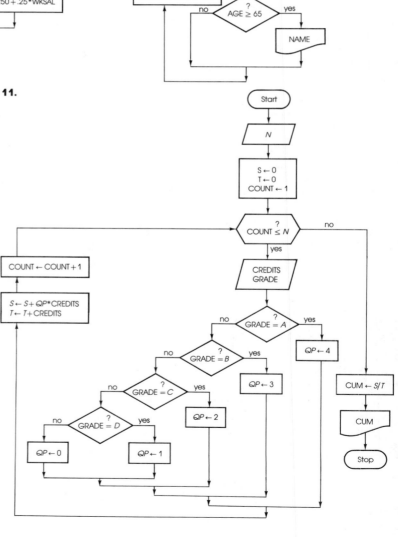

13.

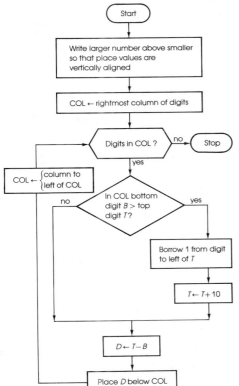

15.

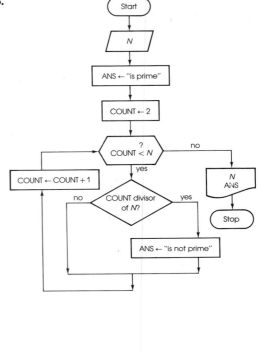

Section 1.3

1. a. Hundreds   **b.** Hundredths   **c.** Millions
d. Thousandths   **e.** Units   **f.** Tenths

3. a. $7*10^3 + 4*10^2 + 6*10^1 + 2*10^0$
b. $6*10^2 + 8*10^1 + 2*10^0 + 9*10^{-2}$
c. $9*10^7 + 6*10^6$
d. $1*10^{-4} + 3*10^{-5}$
e. $7*10^1 + 6*10^0 + 6*10^{-1}$
f. $-[1*10^3 + 4*10^2 + 3*10^0 + 1*10^{-1}]$

5. a. $.1792E+3$ **b.** $.45E+6$ **c.** $.1865E+1$
d. $.902E0$ **e.** $.571E-1$ **f.** $.1001E-3$
g. $-.102E+3$ **h.** $-.1E+1$

7. a. 24.1   **b.** 90000   **c.** 5.026   **d.** .5
e. .0062   **f.** .00000040189   **g.** -720
h. $-.00005$

9. a. $1.792E+2$ **b.** $4.5E+5$ **c.** $1.865E0$
d. $9.02E-1$ **e.** $5.71E-2$ **f.** $1.001E-4$
g. $-1.02E+2$ **h.** $-1.0E0$

11.

Fixed	Floating	Scientific notation
5	$.5E+1$	$5.0E0$
1742.69	$.174269E+4$	$1.74269E+3$
63,029,001	$.63029001E+8$	$6.3029001E+7$
.42	$.42E0$	$4.2E-1$
.00601	$.601E-2$	$6.01E-3$
6000	$.6E+4$	$6.0E+3$
.00006	$.6E-4$	$6.0E-5$
411,200	$.4112E+6$	$4.112E+5$
.00102	$.102E-2$	$1.02E-3$
$-69,987,000$	$-.69987E+8$	$-6.9987E+7$

13. a. Sum is $.521721E+12$; difference is $.520279E+12$; product is $.37564E+21$; quotient is $.7226E+3$.
b. Sum is $.1999E-6$; difference is $.25E-8$; product is $.9988E-14$; quotient is $.1025E+1$.

15. a. Sum is $5.21721E+11$; difference is $5.20279E+11$; product is $3.7564E+20$; quotient is $7.226E+2$.

b. Sum is 1.999E − 7; difference is 2.5E − 9; product is 9.988E − 15; quotient is 1.025E0.

17. Product is 2409.74; quotient is 1,195,070,400,000,000.

19. Because the registers in digital computers can hold (that is, represent) only finitely many digits of a given number

Section 1.4

1. a. 273.551, 273.5   **b.** −3.274, −3.2
c. 62.032, 62.0   **d.** .000, .0

3. a. .695043E + 1   **b.** .420055E − 4
c. .112233E0   **d.** .500000E − 11

5. a. 273.551, 273.6   **b.** −3.275, −3.3
c. 62.033, 62.0   **d.** .000, .0

7. a. .695043E + 1   **b.** .420056E − 4
c. .112233E0   **d.** .500001E − 11

9. a. 66.34   **b.** 66.96   **c.** 66.33   **d.** 65.32

11. For DIV = 748 and DOR = 7, REM = 6.

13. For NUM = 4.60983, ANS = 4.610 when $n = 3$ and ANS = 4.6098 when $n = 4$.

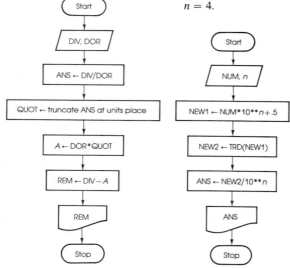

15. a. ABS = .064; REL = .906%
b. ABS = −.036; REL = −.51%

17. a. ABS = .0005E + 4; REL = .118%
b. ABS = −.0005E + 4; REL = −.118%

19. a. ABS = $\dfrac{.0004}{6}$; REL = .04%

b. ABS = $\dfrac{.000004}{6}$; REL = .0004%

21. a. ABS = .0004; REL = .04%
b. ABS = .000004; REL = .0004%

23. ABS = −.0000008E + 7; REL = −400%

25. a. ABS = −.62388; REL = −.94%
b. ABS = 1.01612; REL = 1.53%

27. Two times

29. Underscored digits are significant;
425.16, 63.025, .07020, 1.0, 1, .83129E − 12
.543E + 3, .71902E0, .1E + 15, .100E − 3

31. a. 430, 63, .070, not possible, not possible, .83E − 12, .54E + 3, .72E0, not possible, not possible
b. 400, 60, .07, 1, 1, .8E − 12, .5E + 3, .7E0, .1E + 15, .1E − 3

Section 1.5

1. a. 27   **b.** 48   **c.** 27   **d.** 34   **e.** $\frac{6}{5}$
f. $\frac{3}{10}$

3. a. −1.76875   **b.** 25.6

5. Underscore indicates unnecessary parenthesis:
a. All needed   **b.** 5 ∗ (6 ∗∗ 2)   **c.** All needed
d. 4 − (5)/(6 + (3 ∗∗ 2))

7. a. False   **b.** True   **c.** True   **d.** False

9. a. / − 7,3,4   **b.** ∗∗a − bcd   **c.** / − − xyz + wv

11. a. 6∗(4−2)   **b.** (a+b)/c
c. (u+v)∗[w/x − y + y∗z]

C H A P T E R   2

Section 2.1

1. a. $7*10 + 9*1$, $4*10000 + 3*1$,
$9*10 + 1*1 + 0*.1 + 8*.01 + 5*.001$,
$1*1000 + 0*100 + 1*10 + 1*1 + 1*.1$

b. $1*2 + 1*1$,
$1*32 + 1*16 + 0*8 + 0*4 + 1*2 + 0*1$,
$1*.125$, $1*4 + 0*2 + 0*1 + 1*.5$

3.

Odometer	Binary	Decimal
001110	1110	14
001111	1111	15
010000	10000	16
010001	10001	17
010010	10010	18
010011	10011	19
010100	10100	20

5. 63 miles

7. a. 5   **b.** 26   **c.** 181
d. 990   **e.** 6.5   **f.** 2.171875

9. a. 1000110   **b.** 10110010001
c. .101   **d.** .0100101

11. a. 10010011.1   **b.** 1.001
c. 1110101110.10101   **d.** 100101111.$\overline{0011}$
e. 10.1$\overline{0110}$ **f.** 110111.111

13.

POS	1	2	3	4	5	6	7
BIT_{POS}	0	0	0	1	1	0	1
DECINT	0	0	0	8	24	24	88

The decimal equivalent of BININT is 88.

15. For
$DECINT = 57$,
the output left
to right is 111001.

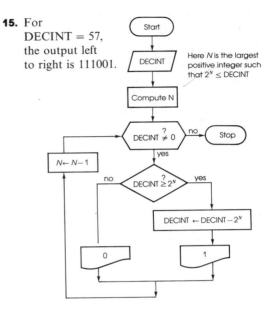

Here N is the largest
positive integer such
that $2^N \le DECINT$

17. a. 5, 26, 106

b.

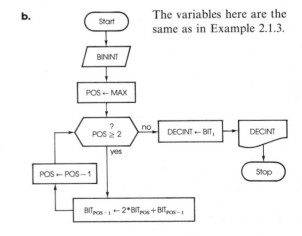

The variables here are the
same as in Example 2.1.3.

Section 2.2

1. 111   **3.** 10011   **5.** 10100   **7.** 100000

9. 1.011   **11.** 10110   **13.** 110011   **15.** 11.10

17. .101   **19.** 101   **21.** 110   **23.** 1010

25. .1001   **27.** 11.1   **29.** 1010   **31.** 110010101

33. 1001.011   **35.** 1111.1

37. Quotient $= 1001$, remainder $= 10$

39. Quotient $= 1111$, remainder $= 1$

41. 10.1   **43.** .11   **45.** 1010

47. a. 00   **b.** 010   **c.** 00101
d. 011111   **e.** 0.010   **f.** 010.110

51. a. 1100   **b.** .001011   **c.** .111   **d.** .01001

53. a. $.1E+2$ **b.** $.101E+3$
c. $.101E+1$ **d.** $.11E-2$

55. a. Sum: $.100011E + 2$; difference: $.11001E + 1$;
Product: $.1001011E0$; Quotient: $.11E+3$
b. Sum: $.10100E0$; Difference: $.1010E-1$;
Product: $.1001011E-3$; Quotient: $.11E+2$

57.

POS	a_{POS}	b_{POS}	c_{POS}	c_{POS+1}	SUM	S_{POS}
1	0	0	0	0	0	0
2	1	1	0	1	10	0
3	1	0	1	1	10	0
4	0	0	1	0	1	1
5	1	1	0	1	10	0
6	1	0	1	1	10	0
7	0	0	1	0	1	1

The sum $a + b$ is the rightmost column read bottom to top.

59.

Section 2.3

1.

Octal	Decimal	Octal	Decimal
42	34	55	45
43	35	56	46
44	36	57	47
45	37	60	48
46	38	61	49
47	39	62	50
50	40	63	51
51	41	64	52
52	42	65	53
53	43	66	54
54	44	67	55

3. 2, 4, 6, 10, 12, 14, 16, 20, 22, 24, 26, 30, 32, 34, 36, 40, 42, 44, 46, 50

5. a. $5*8 + 0*1$ **b.** $1*64 + 1*8 + 1*1$
c $2*.125 + 4*.015625$
d. $4*512 + 0*64 + 0*8 + 7*1 + 1*.125$
e. $2*8^1 + 6*8^0 + 7*8^{-1} + 7*8^{-2} + 0*8^{-3} + 6*8^{-4}$

7. 262,143

9. a. 13   **b.** 454   **c.** 1311
d. .015625   **e.** 63.875   **f.** 585.14258

11. a. 61   **b.** 363   **c.** 13562   **d.** 300715

13. a. 111001   **b.** 100011010   **c.** 1000111111100
d. .010101   **e.** 1010.110
f. 101000111.000000110011

15. a. 32   **b.** 265   **c.** 1736
d. .3   **e.** 16.4   **f.** 2.13

17. a. 122.4   **b.** 1177.7
c. 22272.1   **d.** $157.0\overline{6314}$

19.

	Binary		
Binary	101.1	1000001.001	1001001101100
Octal	5.4	101.1	11154
Decimal	5.5	65.125	4716

23. 505   **25.** 5100   **27.** 16.32

29. 14135.011   **31.** 341   **33.** 4707

35. .070   **37.** 547.261

39.

*	0	1	2	3	4	5	6	7
0	0	0	0	0	0	0	0	0
1	0	1	2	3	4	5	6	7
2	0	2	4	6	10	12	14	16
3	0	3	6	11	14	17	22	25
4	0	4	10	14	20	24	30	34
5	0	5	12	17	24	31	36	43
6	0	6	14	22	30	36	44	52
7	0	7	16	25	34	43	52	61

41. a. Quotient 47, remainder 4
   b. Quotient 24, remainder 10
   c. Quotient 76, remainder 255

43. a. $156 = 19*8 + 4$
   $= (2*8 + 3)*8 + 4$
   $= ((0*8 + 2)*8 + 3)*8 + 4 = 234_{oct}$

   $9085 = 1135*8 + 5$
   $= (141*8 + 7)*8 + 5$
   $= ((17*8 + 5)*8 + 7)*8 + 5$
   $= (((2*8 + 1)*8 + 5)*8 + 7)*8 + 5$
   $= ((((0*8 + 2)*8 + 1)*8 + 5)*8 + 7)*8$
   $+ 5$
   $= 21575_{oct}$

b.

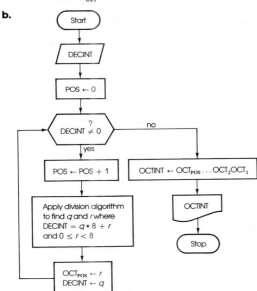

45. a. For BININT $= 10011001$, then MAX $= 8$; and

POS	POS + 3	BIT_{POS}	BIT_{POS+1}
1	4	1	0
4	7	1	1

BIT_{POS+2}	SUB	OCT_{SUB}
0	1	1
0	2	3

b. From bottom half, $OCT_3 = 2$
c. 231

47.

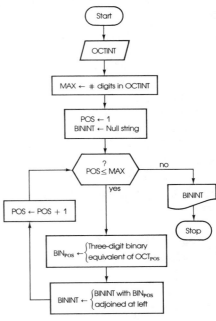

49. a. Always   **b.** Always

Section 2.4

1.

Hex	Decimal	Hex	Decimal
22	34	2D	45
23	35	2E	46
24	36	2F	47
25	37	30	48
26	38	31	49
27	39	32	50
28	40	33	51
29	41	34	52
2A	42	35	53
2B	43	36	54
2C	44	37	55

3. 2, 4, 6, 8, A, C, E, 10, 12, 14, 16, 18, 1A, 1C, 1E, 20, 22, 24, 26, 28

5. a. $5*16 + 0*1$ **b.** $4*256 + D*16 + 1*1$
   c. $2*16^{-1} + B*16^{-2}$ **d.** $D*1 + 0*16^{-1}$
   $+ 7*16^{-2}$
   e. $F*16 + 6*1 + 9*16^{-1} + 0*16^{-2} + E*16^{-3}$

7. 16777215

9. a. 23   **b.** 2272   **c.** 43981
   d. .00390625   **e.** 15.9375   **f.** 273.04711914

11. a. 20   **b.** 63   **c.** FD   **d.** 3E8

*	0	1	2	3	4	5	6	7	8	9	A	B	C	D	E	F
0	0	0	0	0	0	0	0	0	0	0	0	0	0	0	0	0
1	0	1	2	3	4	5	6	7	8	9	A	B	C	D	E	F
2	0	2	4	6	8	A	C	E	10	12	14	16	18	1A	1C	1E
3	0	3	6	9	C	F	12	15	18	1B	1E	21	24	27	2A	2D
4	0	4	8	C	10	14	18	1C	20	24	28	2C	30	34	38	3C
5	0	5	A	F	14	19	1E	23	28	2D	32	37	3C	41	46	4B
6	0	6	C	12	18	1E	24	2A	30	36	3C	42	48	4E	54	5A
7	0	7	E	15	1C	23	2A	31	38	3F	46	4D	54	5B	62	69
8	0	8	10	18	20	28	30	38	40	48	50	58	60	68	70	78
9	0	9	12	1B	24	2D	36	3F	48	51	5A	63	6C	75	7E	87
A	0	A	14	1E	28	32	3C	46	50	5A	64	6E	78	82	8C	96
B	0	B	16	21	2C	37	42	4D	58	63	6E	79	84	8F	9A	A5
C	0	C	18	24	30	3C	48	54	60	6C	78	84	90	9C	A8	B4
D	0	D	1A	27	34	41	4E	5B	68	75	82	8F	9C	A9	B6	C3
E	0	E	1C	2A	38	46	54	62	70	7E	8C	9A	A8	B6	C4	D2
F	0	F	1E	2D	3C	4B	5A	69	78	87	96	A5	B4	C3	D2	E1

13. a. 10011011　　**b.** 10000110010
　　c. 1110000010100101　　**d.** .1011
　　e. 1101.00011001　　**f.** 100000001111.00001100

15. a. 1A　　**b.** B5　　**c.** 3DE
　　d. 7　　**e.** 5.A　　**f.** 2.2C

17. a. 52.8　　**b.** 27F.E
　　c. 24BA.2　　**d.** 6F.$1\overline{9}$

19.

Binary	11.01	11101110011	101101.0001
Decimal	3.25	1907	45.0625
Hex	3.4	773	2D.1

23. 97A8　　**25.** 1033E

27. 21.59　　**29.** 86

31. CFEFF　　**33.** 9CC.EF

35. See the table at the top of the page.

37. a. Quotient 97, remainder 4
　　b. Quotient 65, remainder 3
　　c. Quotient 197, remainder 2B

39. a. $156 = 9*16 + C$
　　　　$= (0*16 + 9)*16 + C = 9C_{hex}$
　　$9085 = 567*16 + D$
　　　　$= (35*16 + 7)*16 + D$
　　　　$= ((2*16 + 3)*16 + 7)*16 + D$
　　　　$= (((0*16 + 2)*16 + 3)*16 + 7)*16 + D$
　　　　$= 237D_{hex}$

Section 2.5

1. a. 110001　　**b.** 31　　**c.** 0100 1001　　**d.** 49

3. a. 10001010111　　**b.** 457
　　c. 0001 0001 0001 0001　　**d.** 1111

5. a. 58　　**b.** 71　　**c.** 256

b. Same as flowchart in exercise 43(b) of Exercises 2.3, except replace 8 by 16, OCT_{POS} by HEX_{POS}, and OCTINT by HEXINT.

41. a. For $BININT = 11111000101$, then $MAX = 11$; and

POS	POS + 4	BIT_{POS}	BIT_{POS+1}	BIT_{POS+2}
1	5	1	0	1
5	9	0	0	1

BIT_{POS+3}	SUM	SUB	HEX_{SUB}
0	5	1	5
1	12	2	C

b. From bottom half, $HEX_3 = 7$　　**c.** 7C5

43. Same as flowchart in exercise 47 of Exercises 2.3 except replace OCTINT by HEXINT, OCT_{POS} by HEX_{POS}, and the word *three* by *four*.

45. a. Always　　**b.** Always

7. a. 33　　**b.** 18

9. a. 110001 100101 100101 110101
　　b. 61　45　45　65
　　c. 11000001 11001110 11001110 11000101
　　d. C1　CE　CE　C5

+	1	2	3	4	5	6	7	8	9	10	11	12
1	2	3	4	5	6	7	8	9	10	11	12	1
2	3	4	5	6	7	8	9	10	11	12	1	2
3	4	5	6	7	8	9	10	11	12	1	2	3
4	5	6	7	8	9	10	11	12	1	2	3	4
5	6	7	8	9	10	11	12	1	2	3	4	5
6	7	8	9	10	11	12	1	2	3	4	5	6
7	8	9	10	11	12	1	2	3	4	5	6	7
8	9	10	11	12	1	2	3	4	5	6	7	8
9	10	11	12	1	2	3	4	5	6	7	8	9
10	11	12	1	2	3	4	5	6	7	8	9	10
11	12	1	2	3	4	5	6	7	8	9	10	11
12	1	2	3	4	5	6	7	8	9	10	11	12

e. 11000001 11010101 11010101 11000101
f. C1   D5   D5   C5

11. a. 000001 000010 111011 000110 010000 000101
b. 01   02   73   06   20   05
c. 10110001 10110010 10101110 10110110 10101011 10110101
d. B1   B2   AE   B6   AB   B5

e. 11110001 11110010 01001011 11110110 01001110 11110101
f. F1   F2   4B   F6   4E   F5

13. 11110100 11111000, 01001000

15. 11110001 11110101 11110110, 000101010110

17. 38   **19.** 93

Section 2.6

1. a. Yes   **b.** Yes   **c.** No   **d.** Yes   **e.** No
f. Yes

3. a. False   **b.** True   **c.** False   **d.** True
e. False   **f.** False

5. 0, 1, 2, 3, 4

+	0	1	2	3	4
0	0	1	2	3	4
1	1	2	3	4	0
2	2	3	4	0	1
3	3	4	0	1	2
4	4	0	1	2	3

7. 1, 2, 3, 4, 5, 6, 7, 8, 9, 10, 11, 12 = 0
See the table at the top of the page.

9. a. False   **b.** True   **c.** True   **d.** True
e. False   **f.** False

11. 100010   **13.** -100010

19. a.

*	0	1	2	3	4
0	0	0	0	0	0
1	0	1	2	3	4
2	0	2	4	1	3
3	0	3	1	4	2
4	0	4	3	2	1

b. $x = 1$ **c.** Yes; $x = 2, 3$

Section 2.7

1. a. 6,144   **b.** 32,768   **c.** 49,152   **d.** 262,144

3.

Bits	Bytes	KB	MB
5000	625	.625	.000625
40,000	5000	5	.005
4,000,000	500,000	500	.5
400,000,000	50,000,000	50,000	50

5. Approximately 4

7. Additions: 4,000,000; subtractions: 1,000,000; multiplications: 333,333; divisions: 142,857

9.

Seconds	Microseconds	Nanoseconds
.002	2000	2,000,000
.0000006	.6	600
.000000042	.042	42

11. 98.8 in./sec   **13.** 20 disks   **15.** 9091 sec

C H A P T E R   3

Section 3.1

1. a. n, a, m, e   **b.** 2, 3, . . . , 13
   c. 2, 3, . . . , 13   **d.** a, l, g, o, r, i, t, h, m
   e. 3, −3   **f.** 10, 11, 12, . . .
   g. $\dfrac{2}{1}, \dfrac{2}{-1}, \dfrac{2}{2}, \dfrac{2}{-2}, \dfrac{2}{3}, \dfrac{2}{-3}, \ldots$

3. a. {1, 2, 3, 4, 5, 6, 7}   **b.** {0, 1, 2, 3, 4, 5, 6, 7}
   c. {a, e, i, o, u}   **d.** {January, June, November}
   e. ϕ **f.** {1, 3, 5, . . .}
   g. Not possible to list

5. a. $\{k \mid k \in \mathbb{Z} \text{ and } 0 < k < 8\}$
   b. $\{k \mid k \in \mathbb{Z} \text{ and } 0 \le k < 8\}$
   c. $\{l \mid l \text{ is a vowel in the alphabet}\}$
   d. $\{m \mid m \text{ is a month whose name contains the letter } n\}$
   e. $\{w \mid w \text{ was a woman and } w \text{ was president of the United States}\}$
   f. $\{k \mid k \in \mathbb{Z}, k > 0, \text{ and } k \text{ odd}\}$
   g. $\{x \mid x \in \mathbb{R} \text{ and } 0 < x < 1\}$

7. a. {Kennedy, Johnson, Nixon, Ford, Carter, Reagan}
   b. $\{k \mid k \in \mathbb{Z} \text{ and } k \text{ odd}\}$

Section 3.2

1. a. {−5, 0, 1, 2, 3, 4, 5, 10}   **b.** {0, 1, 2, 3, 4, 5}
   c. {−5, 0, 1, 5, 10}   **d.** {0, 5}   **e.** {1}
   f. ϕ **g.** {1, 2, 3, 4}   **h.** {−5, 10}
   i. {0, 2, 3, 4, 5}   **j.** ϕ
   k. {−5, 0, 5, 10}   **l.** {1}
   m. {−5, −4, −3, −2, −1, 6, 7, 8, 9, 10}
   n. {−4, −3, −2, −1, 1, 2, 3, 4, 6, 7, 8, 9}
   o. {−5, −4, −3, −2, −1, 0, 2, 3, 4, 5, 6, 7, 8, 9, 10}

3. a. Same as Figure 3 in Chapter 3, except reverse yes and no words on question "$x \in B$?".
   b. Same as Figure 4 in Chapter 3, except move "Print" statement leftward to first downward arrow.

c. $\{10^n \mid n \in \mathbb{Z}\}$
d. {0, 1}

9. a. True   **b.** False   **c.** True   **d.** False
   e. False   **f.** True   **g.** True

11. U = all employees

13. $R_{>4}$ = {Bashful, Doc, Happy, Hondo, Snoopy}
   $R_{>5}$ = {Hondo, Snoopy}
   $R_{>6} = \phi$
   $H_{>45}$ = {Dopey, Happy}
   $H_{<50} = H_{>30}$ = set of all ten names

15. a. $\{x \mid x \in \mathbb{R} \text{ and } -1 < x < 2\}$
   b. $\{x \mid x \in \mathbb{R} \text{ and } \frac{1}{2} \le x < 5\}$
   c. $\{x \mid x \in \mathbb{R} \text{ and } 0 < x \le \frac{12}{5}\}$
   d. $\{x \mid x \in \mathbb{R} \text{ and } -5 \le x \le 6\}$
   e. $\{x \mid x \in \mathbb{R} \text{ and } 2 \le x \le 1\} = \phi$

17. a. $[0, 9]$ **b.** $(-\sqrt{2}, 15]$ **c.** $(-2, -1)$

c.

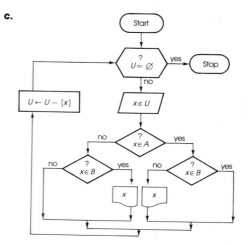

5. a.

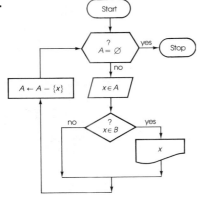

b.

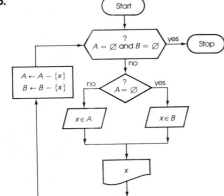

7. Because if U has infinitely many elements, then the loop in each flowchart is traversed *ad infinitum*, and it may be that *none* of the elements of A or B are ever output (for example, $U = [0, 10]$, $A = [1, 3]$, $B = [2, 4]$).

9. a. Same as flowchart shown in exercise 5(a).
   b. Same as flowchart shown in exercise 5(b).

11. a.

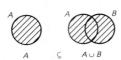

Section 3.3

3. a. A **b.** $A \cap B$ **c.** B

9. Same flowchart as shown in Figure 10 of Example 3.3.8 except move the Print statement to the first downward arrow to the left of its position in Figure 10. The associated Boolean property is $S \cap \bar{H} = \bar{H} \cap S$.

b.

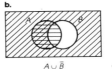   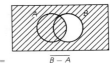

$A \cup \bar{B}$ $=$ $\overline{B - A}$

c.

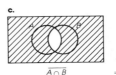   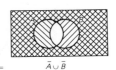

$\overline{A \cap B}$ $=$ $\bar{A} \cup \bar{B}$

d.

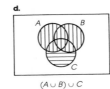   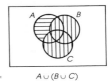

$(A \cup B) \cup C$ $=$ $A \cup (B \cup C)$

15. a. All names
   b. {Bashful, Bonzo, Doc, Grumpy, Hondo, Sneezy, Snoopy}
   c. {Doc, Snoopy}
   d. {Bonzo, Doc, Hondo, Snoopy}

17. a. $R_{<5} \cap H_{>40}$ **b.** $R_{>5} \cup R_{<4}$
   c. $H_{>40} \cap H_{<50}$ **d.** $\bar{H}_{>40}$

19. a. $[0, 1]$ **b.** $(0, 1]$ **c.** $[0, 1)$ **d.** $(0, 1)$
   e. $[-5, 4]$ **f.** $(-5, 4]$ **g.** $[-5, 4)$
   h. $(-5, 4)$ **i.** $(1, 4)$ **j.** $[-5, 0]$

21. a. $\{0, 1\}$ **b.** ϕ
   c. $\left\{ \dfrac{m}{n} \,\middle|\, m, n \in \mathbb{Z}, m \geq 0, n > 0, m \leq n \right\}$
   d. $\left\{ \dfrac{m}{n} \,\middle|\, m, n \in \mathbb{Z}, m > 0, n > 0, m < n \right\}$
   e. $[0, 1]$ **f.** $(0, 1)$

11. The related Boolean property is

$$(S \cap H) \cup (S \cap I) = S \cap (H \cup I)$$

where I is the set of all persons with income below \$10000. See the flowchart at the top of the next page.

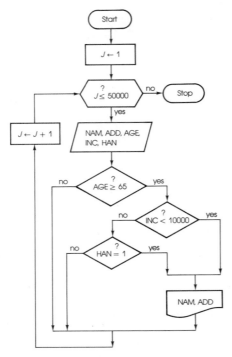

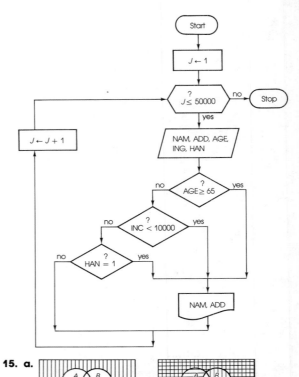

13. The related Boolean property is

$$S \cup H \cup I = S \cup I \cup H$$

where I is the set of all persons with income below $10000. See the flowchart above on the right.

15. a.

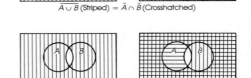

$\bar{A} \cup \bar{B}$ (Striped) $= \bar{A} \cap \bar{B}$ (Crosshatched)

$\overline{A \cap B}$ (Striped) $= \bar{A} \cup \bar{B}$ (Striped and crosshatched)

Section 3.4

1. a. Yes   **b.** No   **c.** No   **d.** No
   e. Yes   **f.** No

3. a. No   **b.** No   **c.** Yes

5. a.

b. $U - (A \cup B \cup C)$, $A - (B \cup C)$, $B - (A \cup C)$,
$C - (A \cup B)$, $(A \cap B) - C$, $(A \cap C) - B$,
$(B \cap C) - A$, $A \cap B \cap C$

7. a. $A \times B = \{(5, a), (5, b), (5, c), (5, d)\}$

$B \times B = \{(a, a), (a, b), (a, c), (a, d), (b, a),$
$\qquad (b, b), (b, c), (b, d), (c, a), (c, b),$
$\qquad (c, c), (c, d), (d, a), (d, b),$
$\qquad (d, c), (d, d)\}$

$B \times Z_2 = \{(a, 0), (a, 1), (b, 0), (b, 1), (c, 0),$
$\qquad (c, 1), (d, 0), (d, 1)\}$

$A \times A \times A = \{(5, 5, 5)\}$

b. $\{(5, k) \mid k \in \mathbb{Z}\}$

9. a.   **b.**   **c.**

11. a. No **b.** No **c.** No **d.** Yes **e.** Yes

13. a. **b.** **c.** **d.** **e.**

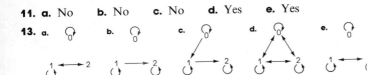

Section 3.5

1. 253 **3.** 133 **5.** 289,800 **7.** 3650

9. V76 and V80 take on values 0 or 1, which mean "did not vote" and "did vote," respectively.

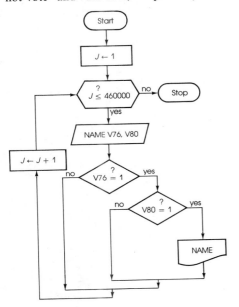

11. B (bus), T (train), and P (plane) take on values 0 or 1, which mean "no" and "yes," respectively.

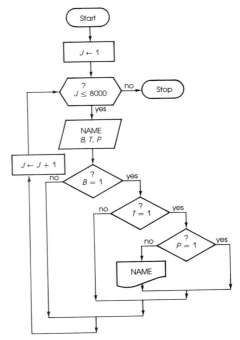

13. a. 21 **b.** 21 **c.** 49 **d.** 27
A has 128 subsets.

15. 26000

17. a. 512 **b.** 64 **c.** 12005

19. 1280 seconds

21. Take that portion of Figure 21 up to and including the column of three-bit strings.

23. The 20 possible outcomes from the tree graph are as follows:

AAA, BBB

AABA, ABAA, BAAA, BBAB, BABB, ABBB

AABBA, ABABA, ABBAA, BAABA, BABAA,
BBAAA, BBAAB, BABAB, BAABB, ABBAB,
ABABB, AABBB

25.
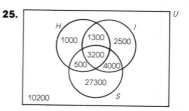

Section 3.6

1. a. 24   **b.** 5040   **c.** 20
   d. 840   **e.** 5   **f.** 70

3. a. {*AB, AC, AD, BA, BC, BD, CA, CB, CD, DA, DB, DC*}

   b. {*AB, AC, AD, AE, BC, BD, BE, CD, CE, DE*}

5. a. 3360   **b.** 720   **c.** 4080

7. 5040   **9.** 120   **11.** 792   **13.** 56

15.

Variable				Value				
F	8	8	8	8	8	8	8	8
C	1	2	3	4	5	6	7	8
FACT	1	1	2	6	24	120	720	5040

17.
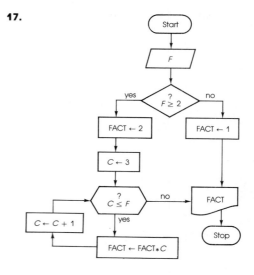

Section 3.7

1. a. {*H, T*}, $\frac{1}{2}$ **b.** {*H, T*}, $\frac{1}{2}$ **c.** {*H, T*}, 0
   d. {1, 2, 3, 4, 5, 6}, $\frac{1}{6}$ **e.** {1, 2, 3, 4, 5, 6}, 1
   f. {1, 2, 3, 4, 5, 6} × {1, 2, 3, 4, 5, 6}, $\frac{1}{12}$
   g. {1, 2, 3, 4, 5, 6}, $\frac{2}{3}$

In parts (h)–(l) the sample space is a deck of cards.

   h. $\frac{1}{13}$ **i.** $\frac{12}{13}$ **j.** $\frac{3}{13}$ **k.** $\frac{3}{4}$ **l.** $\frac{1}{26}$

3. a. 7/32   **b.** 35/128   **c.** 9/64

5.

	L	M	H	Total
D	.112	.288	.057	.457
I	.020	.073	.043	.136
R	.066	.125	.216	.407
Total	.198	.486	.316	1.000

a. .407   **b.** .316   **c.** .288   **d.** .020

7. a. .198   **b.** .486   **c.** .684
   d. 0, mutually exclusive

9. a. .407   **b.** .316   **c.** .507
   d. .216, not mutually exclusive

13. 1 to 2   **15.** 93 to 35

17. a. $.50   **b.** Yes, no

Section 3.8

1. a. $SS = \{HHHH,\ HHHT,\ HHTH,\ HTHH,$
$THHH,\ HHTT,\ HTTH,\ TTHH,\ THTH,$
$THHT,\ HTHT,\ TTTH,\ TTHT,\ THTT,$
$HTTT,\ TTTT\}$

$\frac{1}{16}$ $\frac{1}{16}$
$\frac{1}{8}$ $\frac{3}{8}$

b. $\frac{1}{2}$ $\frac{1}{2}$
$\frac{1}{2}$ $\frac{1}{2}$
$\frac{1}{4}$ $\frac{1}{8}$
$\frac{1}{16}$ $\frac{1}{16}$

Yes, they are independent.

3. a. $SS =$ set of all five-bit strings

$\frac{1}{32}$ $\frac{1}{2}$
$\frac{1}{16}$ $\frac{5}{16}$

b. $\frac{1}{2}$ $\frac{1}{2}$
$\frac{1}{2}$ $\frac{1}{2}$
$\frac{1}{2}$ $\frac{1}{4}$
$\frac{1}{8}$ $\frac{1}{16}$
$\frac{1}{32}$ $\frac{1}{32}$

Yes, they are independent.

5. a. $\{HHH,\ HHT,\ HTH,\ THH,$
$TTH,\ THT,\ HTT,\ TTT\}$
b. $\frac{1}{2},\ \frac{1}{4},\ \frac{1}{8}$
c. Yes

7. a. $\{MMM,\ MMF,\ MFM,\ FMM,$
$FFM,\ FMF,\ MFF,\ FFF\}$
b. $\frac{1}{2},\ \frac{1}{4},\ \frac{1}{4}$
c. No

9. a. $\{000,\ 001,\ 010,\ 100,\ 110,\ 101,\ 011,\ 111\}$
b. $\frac{1}{2},\ \frac{3}{4},\ \frac{3}{8}$
c. Yes

11.

x	Pr(x)
0	.03125
1	.15625
2	.3125
3	.3125
4	.15625
5	.03125

13.

x	Pr(x)
0	.01024
1	.0768
2	.2304
3	.3456
4	.2592
5	.07776

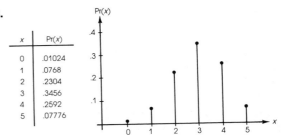

15.

x	Pr(x)
0	.00390625
1	.03125
2	.109375
3	.21875
4	.2734375
5	.21875
6	.109375
7	.03125
8	.00390625

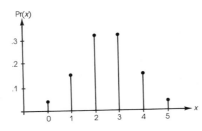

17.

x	Pr(x)
0	.16777216
1	.33554432
2	.29360128
3	.14680064
4	.0458752
5	.00917504
6	.00114688
7	.00008192
8	.00000256

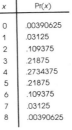

19. a. 0   **b.** 1   **c.** 0   **d.** 0

21.

x	P(x)
0	.13421773
1	.30198989
2	.30198989
3	.17616077
4	.06606029
5	.01651507
6	.00275251
7	.00029491
8	.00001843
9	.00000051

23. a. .83222784   **b.** .37082112   **25.** .33616

C H A P T E R   4

Section 4.1

1. a. Proposition   **b.** Proposition
c. Not proposition   **d.** Proposition
e. Proposition   **f.** Proposition
g. Not proposition   **h.** Not proposition

3. a. George Orwell wrote *1984* and William Shakespeare was a poet and dramatist. (True)
b. George Orwell wrote *1984* and Carl Sandburg is still alive. (False)
c. William Shakespeare was a poet and dramatist, and Carl Sandburg is still alive. (False)
d. George Orwell wrote *1984* or William Shakespeare was a poet and dramatist. (True)
e. George Orwell wrote *1984* or Carl Sandburg is still alive. (True)
f. William Shakespeare was a poet and dramatist, or Carl Sandburg is still alive. (True)

5. a. $p \wedge q$; truth table same as in Table 1
b. $p \vee q$; truth table same as in Table 1
c. $(p \wedge q) \vee (r \wedge s)$; truth table below

p	q	r	s	$p \wedge q$	$r \wedge s$	$(p \wedge q) \vee (r \wedge s)$
T	T	T	T	T	T	T
T	T	T	F	T	F	T
T	T	F	T	T	F	T
T	T	F	F	T	F	T
T	F	T	T	F	T	T
T	F	T	F	F	F	F
T	F	F	T	F	F	F
T	F	F	F	F	F	F
F	T	T	T	F	T	T
F	T	T	F	F	F	F
F	T	F	T	F	F	F
F	T	F	F	F	F	F
F	F	T	T	F	T	T
F	F	T	F	F	F	F
F	F	F	T	F	F	F
F	F	F	F	F	F	F

d. $(p \wedge q) \vee (r \wedge q)$; truth table below

p	q	r	$p \wedge q$	$r \wedge q$	$(p \wedge q) \vee (r \wedge q)$
T	T	T	T	T	T
T	T	F	T	F	T
T	F	T	F	F	F
T	F	F	F	F	F
F	T	T	F	T	T
F	T	F	F	F	F
F	F	T	F	F	F
F	F	F	F	F	F

e. $(p \wedge q) \vee (r \wedge s)$; truth table same as for part (c)

7. a.

p	$p \wedge p$
T	T
F	F

b.

p	q	$q \wedge p$	$p \vee (q \wedge p)$
T	T	T	T
T	F	F	T
F	T	F	F
F	F	F	F

c.

p	q	r	$p \wedge q$	$(p \wedge q) \wedge r$
T	T	T	T	T
T	T	F	T	F
T	F	T	F	F
T	F	F	F	F
F	T	T	F	F
F	T	F	F	F
F	F	T	F	F
F	F	F	F	F

d.

p	q	r	$q \vee r$	$p \vee (q \vee r)$
T	T	T	T	T
T	T	F	T	T
T	F	T	T	T
T	F	F	F	T
F	T	T	T	T
F	T	F	T	T
F	F	T	T	T
F	F	F	F	F

e.

p	q	r	$p \wedge r$	$q \vee r$	$(p \wedge r) \vee (q \vee r)$
T	T	T	T	T	T
T	T	F	F	T	T
T	F	T	T	T	T
T	F	F	F	F	F
F	T	T	F	T	T
F	T	F	F	T	T
F	F	T	F	T	T
F	F	F	F	F	F

f.

p	q	r	s	q∧r	p∧(q∧r)	[p∧(q∧r)]∨s
T	T	T	T	T	T	T
T	T	T	F	T	T	T
T	T	F	T	F	F	T
T	T	F	F	F	F	F
T	F	T	T	F	F	T
T	F	T	F	F	F	F
T	F	F	T	F	F	T
T	F	F	F	F	F	F
F	T	T	T	T	F	T
F	T	T	F	T	F	F
F	T	F	T	F	F	T
F	T	F	F	F	F	F
F	F	T	T	F	F	T
F	F	T	F	F	F	F
F	F	F	T	F	F	T
F	F	F	F	F	F	F

g.

p	q	r	s	p∧q	r∨s	(p∧q)∧(r∨s)
T	T	T	T	T	T	T
T	T	T	F	T	T	T
T	T	F	T	T	T	T
T	T	F	F	T	F	F
T	F	T	T	F	T	F
T	F	T	F	F	T	F
T	F	F	T	F	T	F
T	F	F	F	F	F	F
F	T	T	T	F	T	F
F	T	T	F	F	T	F
F	T	F	T	F	T	F
F	T	F	F	F	F	F
F	F	T	T	F	T	F
F	F	T	F	F	T	F
F	F	F	T	F	T	F
F	F	F	F	F	F	F

9. a. Off, on, off, on
b. Off, on, off, on
c. Off, on, on, off

11. a. b. c. d. e. f. g.

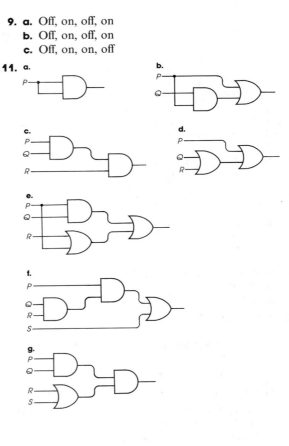

The output tables for these circuits are given at the bottom of the page and at the top of the next page.

a.

P	P·P
1	1
0	0

b.

P	Q	Q·P	P+(Q·P)
1	1	1	1
1	0	0	1
0	1	0	0
0	0	0	0

						c.	**d.**	**e.**
P	Q	R	P·Q	Q+R	P·R	(P·Q)·R	P+(Q+R)	(P·R)+(Q+R)
1	1	1	1	1	1	1	1	1
1	1	0	1	1	0	0	1	1
1	0	1	0	1	1	0	1	1
1	0	0	0	0	0	0	1	0
0	1	1	0	1	0	0	1	1
0	1	0	0	1	0	0	1	1
0	0	1	0	1	0	0	1	1
0	0	0	0	0	0	0	0	0

P	Q	R	S	$Q \cdot R$	$P \cdot (Q \cdot R)$	$P \cdot Q$	$R + S$	**f.** $[P \cdot (Q \cdot R)] + S$	**g.** $(P \cdot Q) \cdot (R + S)$
1	1	1	1	1	1	1	1	1	1
1	1	1	0	1	1	1	1	1	1
1	1	0	1	0	0	1	1	1	1
1	1	0	0	0	0	1	0	0	0
1	0	1	1	0	0	0	1	1	0
1	0	1	0	0	0	0	1	0	0
1	0	0	1	0	0	0	1	1	0
1	0	0	0	0	0	0	0	0	0
0	1	1	1	1	0	0	1	1	0
0	1	1	0	1	0	0	1	0	0
0	1	0	1	0	0	0	1	1	0
0	1	0	0	0	0	0	0	0	0
0	0	1	1	0	0	0	1	1	0
0	0	1	0	0	0	0	1	0	0
0	0	0	1	0	0	0	1	1	0
0	0	0	0	0	0	0	0	0	0

13. a. $P + P$ **b.** $(P \cdot Q) + P$ **c.** $(P + Q) \cdot R$
   d. $P \cdot [(Q \cdot R) + S]$
   e. $[(P \cdot Q) + (Q + R)] \cdot [(Q + R) + (R \cdot S)]$

15. $8, 16, 32, 64, 2^k$

Section 4.2

1. a. T **b.** F **c.** F **d.** T **e.** F

3. a. On   **b.** On   **c.** Off   **d.** On   **e.** Off

5. a. False; Thomas Eakins was not a twentieth-century American painter.
   b. False; Thomas Eakins was a twentieth-century American painter, but Georges Seurat was not a nineteenth-century French impressionist artist.
   c. True; Thomas Eakins was a twentieth-century American painter, or Georges Seurat was not a nineteenth-century French impressionist artist.
   d. False; Thomas Eakins was not a twentieth-century American painter, or Georges Seurat was not a nineteenth-century French impressionist artist.

7. a.

p	q	$\sim q$	$p \wedge \sim q$
T	T	F	F
T	F	T	T
F	T	F	F
F	F	T	F

b.

p	q	$\sim p$	$\sim q$	$\sim p \vee \sim q$
T	T	F	F	F
T	F	F	T	T
F	T	T	F	T
F	F	T	T	T

c.

p	q	r	$p \wedge q$	$\sim r$	$(p \wedge q) \vee \sim r$
T	T	T	T	F	T
T	T	F	T	T	T
T	F	T	F	F	F
T	F	F	F	T	T
F	T	T	F	F	F
F	T	F	F	T	T
F	F	T	F	F	F
F	F	F	F	T	T

7. d,e. See the tables at the top of the next page.

9. a.

P	Q	$\bar{Q}$	$P \cdot \bar{Q}$
1	1	0	0
1	0	1	1
0	1	0	0
0	0	1	0

b.

P	Q	$\bar{P}$	$\bar{Q}$	$\bar{P} + \bar{Q}$
1	1	0	0	0
1	0	0	1	1
0	1	1	0	1
0	0	1	1	1

7. d.

p	q	r	$p \wedge q$	$\sim(p \wedge q)$	$r \wedge q$	$\sim(r \wedge q)$	$\sim(p \wedge q) \vee \sim(r \wedge q)$
T	T	T	T	F	T	F	F
T	T	F	T	F	F	T	T
T	F	T	F	T	F	T	T
T	F	F	F	T	F	T	T
F	T	T	F	T	T	F	T
F	T	F	F	T	F	T	T
F	F	T	F	T	F	T	T
F	F	F	F	T	F	T	T

e.

p	q	r	s	$\sim q$	$p \wedge \sim q$	$\sim s$	$r \wedge \sim s$	$(p \wedge \sim q) \vee (r \wedge \sim s)$
T	T	T	T	F	F	F	F	F
T	T	T	F	F	F	T	T	T
T	T	F	T	F	F	F	F	F
T	T	F	F	F	F	T	F	F
T	F	T	T	T	T	F	F	T
T	F	T	F	T	T	T	T	T
T	F	F	T	T	T	F	F	T
T	F	F	F	T	T	T	F	T
F	T	T	T	F	F	F	F	F
F	T	T	F	F	F	T	T	T
F	T	F	T	F	F	F	F	F
F	T	F	F	F	F	T	F	F
F	F	T	T	T	F	F	F	F
F	F	T	F	T	F	T	T	T
F	F	F	T	T	F	F	F	F
F	F	F	F	T	F	T	F	F

9. c.

P	Q	R	$P \cdot Q$	$\bar{R}$	$(P \cdot Q) + \bar{R}$
1	1	1	1	0	1
1	1	0	1	1	1
1	0	1	0	0	0
1	0	0	0	1	1
0	1	1	0	0	0
0	1	0	0	1	1
0	0	1	0	0	0
0	0	0	0	1	1

d.

P	Q	R	$\bar{P}$	$\bar{Q}$	$\bar{P} \cdot \bar{Q}$	$\bar{Q} \cdot R$	$\bar{P} \cdot \bar{Q} + \bar{Q} \cdot R$
1	1	1	0	0	0	0	0
1	1	0	0	0	0	0	0
1	0	1	0	1	0	1	1
1	0	0	0	1	0	0	0
0	1	1	1	0	0	0	0
0	1	0	1	0	0	0	0
0	0	1	1	1	1	1	1
0	0	0	1	1	1	0	1

e.

P	Q	R	S	$\bar{Q}$	$\bar{R}$	$P + \bar{Q}$	$\bar{R} + S$	$(P + \bar{Q}) \cdot (\bar{R} + S)$
1	1	1	1	0	0	1	1	1
1	1	1	0	0	0	1	0	0
1	1	0	1	0	1	1	1	1
1	1	0	0	0	1	1	1	1
1	0	1	1	1	0	1	1	1
1	0	1	0	1	0	1	0	0
1	0	0	1	1	1	1	1	1
1	0	0	0	1	1	1	1	1
0	1	1	1	0	0	0	1	0
0	1	1	0	0	0	0	0	0
0	1	0	1	0	1	0	1	0
0	1	0	0	0	1	0	1	0
0	0	1	1	1	0	1	1	1
0	0	1	0	1	0	1	0	0
0	0	0	1	1	1	1	1	1
0	0	0	0	1	1	1	1	1

11. a.

p	q	$\sim p$	$\sim p \wedge q$
T	T	F	F
T	F	F	F
F	T	T	T
F	F	T	F

b.

p	q	$\sim p$	$\sim p \vee q$	$\sim(\sim p \vee q)$
T	T	F	T	F
T	F	F	F	T
F	T	T	T	F
F	F	T	T	F

c.

p	q	$\sim p$	$q \vee \sim p$	$p \wedge (q \vee \sim p)$
T	T	F	T	T
T	F	F	F	F
F	T	T	T	F
F	F	T	T	F

d.

p	q	$p \wedge q$	$\sim(p \wedge q)$	$(p \wedge q) \vee \sim(p \wedge q)$
T	T	T	F	T
T	F	F	T	T
F	T	F	T	T
F	F	F	T	T

e.

p	q	r	$p \wedge q$	$\sim(p \wedge q)$	$\sim(p \wedge q) \wedge r$
T	T	T	T	F	F
T	T	F	T	F	F
T	F	T	F	T	T
T	F	F	F	T	F
F	T	T	F	T	T
F	T	F	F	T	F
F	F	T	F	T	T
F	F	F	F	T	F

f.

p	q	r	$\sim q$	$p \wedge \sim q$	$r \wedge \sim q$	$(p \wedge \sim q) \vee (r \wedge \sim q)$
T	T	T	F	F	F	F
T	T	F	F	F	F	F
T	F	T	T	T	T	T
T	F	F	T	T	F	T
F	T	T	F	F	F	F
F	T	F	F	F	F	F
F	F	T	T	F	T	T
F	F	F	T	F	F	F

g.

p	q	r	s	$\sim q$	$\sim s$	$p \wedge \sim q$	$r \wedge \sim s$	$(p \wedge \sim q) \vee (r \wedge \sim s)$
T	T	T	T	F	F	F	F	F
T	T	T	F	F	T	F	T	T
T	T	F	T	F	F	F	F	F
T	T	F	F	F	T	F	F	F
T	F	T	T	T	F	T	F	T
T	F	T	F	T	T	T	T	T
T	F	F	T	T	F	T	F	T
T	F	F	F	T	T	T	F	T
F	T	T	T	F	F	F	F	F
F	T	T	F	F	T	F	T	T
F	T	F	T	F	F	F	F	F
F	T	F	F	F	T	F	F	F
F	F	T	T	T	F	F	F	F
F	F	T	F	T	T	F	T	T
F	F	F	T	T	F	F	F	F
F	F	F	F	T	T	F	F	F

13. a.

P	Q	$\bar{Q}$	$P + \bar{Q}$
1	1	0	1
1	0	1	1
0	1	0	0
0	0	1	1

b.

P	Q	$\bar{Q}$	$P \cdot \bar{Q}$	$\overline{P \cdot \bar{Q}}$
1	1	0	0	1
1	0	1	1	0
0	1	0	0	1
0	0	1	0	1

c.

P	Q	$\bar{P}$	$Q \cdot \bar{P}$	$P + Q \cdot \bar{P}$
1	1	0	0	1
1	0	0	0	1
0	1	1	1	1
0	0	1	0	0

d.

P	Q	$P + Q$	$\overline{P + Q}$	$(P + Q) \cdot (\overline{P + Q})$
1	1	1	0	0
1	0	1	0	0
0	1	1	0	0
0	0	0	1	0

e.

P	Q	R	$Q \cdot R$	$\overline{Q \cdot R}$	$P \cdot (\overline{Q \cdot R})$
1	1	1	1	0	0
1	1	0	0	1	1
1	0	1	0	1	1
1	0	0	0	1	1
0	1	1	1	0	0
0	1	0	0	1	0
0	0	1	0	1	0
0	0	0	0	1	0

f.

P	Q	R	$\bar{P}$	$\bar{P} + Q$	$\bar{P} + R$	$(\bar{P} + Q) \cdot (\bar{P} + R)$
1	1	1	0	1	1	1
1	1	0	0	1	0	0
1	0	1	0	0	1	0
1	0	0	0	0	0	0
0	1	1	1	1	1	1
0	1	0	1	1	1	1
0	0	1	1	1	1	1
0	0	0	1	1	1	1

g.

P	Q	R	S	$\bar{P}$	$\bar{S}$	$\bar{P} + Q$	$\bar{S} + R$	$(\bar{P} + Q) \cdot (\bar{S} + R)$
1	1	1	1	0	0	1	1	1
1	1	1	0	0	1	1	1	1
1	1	0	1	0	0	1	0	0
1	1	0	0	0	1	1	1	1
1	0	1	1	0	0	0	1	0
1	0	1	0	0	1	0	1	0
1	0	0	1	0	0	0	0	0
1	0	0	0	0	1	0	1	0
0	1	1	1	1	0	1	1	1
0	1	1	0	1	1	1	1	1
0	1	0	1	1	0	1	0	0
0	1	0	0	1	1	1	1	1
0	0	1	1	1	0	1	1	1
0	0	1	0	1	1	1	1	1
0	0	0	1	1	0	1	0	0
0	0	0	0	1	1	1	1	1

19. a.

P	Q	$P + Q$	$\overline{P + Q}$	$(P + Q) \cdot (\overline{P + Q})$
1	1	1	0	0
1	0	1	0	0
0	1	1	0	0
0	0	0	1	0

a.

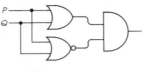

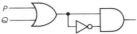

b.

P	Q	$P \cdot Q$	$\overline{P \cdot Q}$	$(P \cdot Q) \oplus (\overline{P \cdot Q})$
1	1	1	0	1
1	0	0	1	1
0	1	0	1	1
0	0	0	1	1

b.

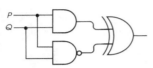

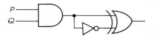

c.

P	Q	R	$Q \cdot R$	$\overline{Q \cdot R}$	$P + \overline{Q \cdot R}$
1	1	1	1	0	1
1	1	0	0	1	1
1	0	1	0	1	1
1	0	0	0	1	1
0	1	1	1	0	0
0	1	0	0	1	1
0	0	1	0	1	1
0	0	0	0	1	1

c.

d.

P	Q	$P \oplus Q$	$\overline{P \oplus Q}$	$(\overline{P \oplus Q}) \cdot P$
1	1	0	1	1
1	0	1	0	0
0	1	1	0	0
0	0	0	1	0

d.

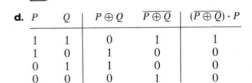

21.

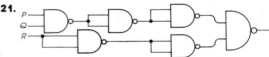

25.

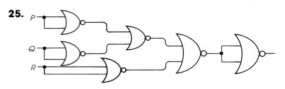

Section 4.3

9. $\overline{P + \overline{Q}} \cdot P \cdot Q$

Section 4.4

1. P **3.** PQ **5.** Q **7.** 1   **9.** PQ

11. $P\overline{Q}$ **13.** $PQR + \overline{Q}\overline{R}$ **15.** $P(Q + R)$

17. $\overline{P}Q + PR$ **19.** $P\overline{R} + \overline{Q}S$ **21.** $\overline{P} + \overline{Q}$

Section 4.5

1. a. Output 0 means there is an even number of 1 bits in the given string.

 b. Output 1 means there is an odd number of 1 bits in the given string.

 c. Output 0 means there is an even number of 1 bits in the given string.

 d. Output 1 means there is an odd number of 1 bits in the given string.

3. a. Output 0 means A string $\not\leq B$ string.
   b. Output 1 means A string $\leq B$ string.
   c. Output 0 means A string $\not\leq B$ string.
   d. Output 1 means A string $\leq B$ string.

5. $S_5S_4S_3S_2S_1$ equals:

  a. 01111   **b.** 01101   **c.** 10000   **d.** 10010

In each case, the output is the sum of the given A string and B string.

7. Two possible circuits are $\bar{A} + AB$ and $\bar{A} + B$.

A	B	Output
1	1	1
1	0	0
0	1	1
0	0	1

9. Circuit is $(\bar{A} + \bar{B})C$.

A	B	C	Output
1	1	1	0
1	1	0	0
1	0	1	1
1	0	0	0
0	1	1	1
0	1	0	0
0	0	1	1
0	0	0	0

11. a. $P\bar{Q} + \bar{P}\bar{Q}$, which simplifies to $\bar{Q}$.
   b. $PQ\bar{R} + P\bar{Q}\bar{R} + \bar{P}QR + \bar{P}\bar{Q}R$, which simplifies to $P \oplus R$.

13. Circuit for C is AB. Circuit for S is $A \oplus B$. Half-adder circuit is

A	B	C	S
1	1	1	0
1	0	0	1
0	1	0	1
0	0	0	0

15. The table with headings A, B, C, COUT, and S for this circuit is identical with the table for the full adder circuit given in Example 4.5.3.

17. The circuit is

$$PQR + P\bar{Q}\bar{R} + \bar{P}Q\bar{R} + \bar{P}\bar{Q}R$$

which simplifies to $P \oplus Q \oplus R$.

P	Q	R	Light
1	1	1	1 (on)
1	1	0	0 (off)
1	0	1	0
1	0	0	1
0	1	1	0
0	1	0	1
0	0	1	1
0	0	0	0

19. a.

P	Q	R	S	Light
1	1	1	1	0
1	1	1	0	0
1	1	0	1	0
1	1	0	0	1
1	0	1	1	0
1	0	1	0	1
1	0	0	1	1
1	0	0	0	0
0	1	1	1	0
0	1	1	0	1
0	1	0	1	1
0	1	0	0	0
0	0	1	1	1
0	0	1	0	0
0	0	0	1	0
0	0	0	0	0

$$(P \oplus S)(Q \oplus R) + \bar{P}QR\bar{S} + P\bar{Q}\bar{R}S$$

b.

P	Q	R	S	Light
1	1	1	1	1
1	1	1	0	1
1	1	0	1	1
1	1	0	0	0
1	0	1	1	1
1	0	1	0	0
1	0	0	1	0
1	0	0	0	0
0	1	1	1	1
0	1	1	0	0
0	1	0	1	0
0	1	0	0	0
0	0	1	1	0
0	0	1	0	0
0	0	0	1	0
0	0	0	0	0

$$PQ(R + S) + (P + Q)RS$$

Section 4.6

1. a.

p	q	$\sim p$	$\sim p \vee q$	$(\sim p \vee q) \to p$
T	T	F	T	T
T	F	F	F	T
F	T	T	T	F
F	F	T	T	F

b.

p	q	$p \to q$	$p \wedge q$	$(p \to q) \to (p \wedge q)$
T	T	T	T	T
T	F	F	F	T
F	T	T	F	F
F	F	T	F	F

c.

p	q	r	$p \wedge q$	$\sim r$	$(p \wedge q) \to \sim r$
T	T	T	T	F	F
T	T	F	T	T	T
T	F	T	F	F	T
T	F	F	F	T	T
F	T	T	F	F	T
F	T	F	F	T	T
F	F	T	F	F	T
F	F	F	F	T	T

d.

p	q	r	$p \vee q$	$r \to p$	$(p \vee q) \to (r \to p)$
T	T	T	T	T	T
T	T	F	T	T	T
T	F	T	T	T	T
T	F	F	T	T	T
F	T	T	T	F	F
F	T	F	T	T	T
F	F	T	F	F	T
F	F	F	F	T	T

e. See the table at the bottom of the page.

3. a.

p	q	$\sim q$	$p \to \sim q$
T	T	F	F
T	F	T	T
F	T	F	T
F	F	T	T

p = I leave.
q = I return.

b.

p	q	r	$p \vee q$	$(p \vee q) \to r$
T	T	T	T	T
T	T	F	T	F
T	F	T	T	T
T	F	F	T	F
F	T	T	T	T
F	T	F	T	F
F	F	T	F	T
F	F	F	F	T

p = You work hard.
q = You hit the lottery.
r = You will succeed.

c.

p	q	r	$q \wedge r$	$p \to (q \wedge r)$
T	T	T	T	T
T	T	F	F	F
T	F	T	F	F
T	F	F	F	F
F	T	T	T	T
F	T	F	F	T
F	F	T	F	T
F	F	F	F	T

p = increased productivity
q = increased wages
r = improved sales

d.

p	q	r	$\sim q$	$\sim q \to r$	$p \to (\sim q \to r)$
T	T	T	F	T	T
T	T	F	F	T	T
T	F	T	T	T	T
T	F	F	T	F	F
F	T	T	F	T	T
F	T	F	F	T	T
F	F	T	T	T	T
F	F	F	T	F	T

p = Remain in that job.
q = You complain.
r = You'll be accepted.

p	r	s	$\sim r$	$\sim s$	$p \wedge \sim r$	$r \vee \sim s$	$(p \wedge \sim r) \leftrightarrow (r \vee \sim s)$
T	T	T	F	F	F	T	F
T	T	F	F	T	F	T	F
T	F	T	T	F	T	F	F
T	F	F	T	T	T	T	T
F	T	T	F	F	F	T	F
F	T	F	F	T	F	T	F
F	F	T	T	F	F	F	T
F	F	F	T	T	F	T	F

e. $p \rightarrow q$ (truth table in text); $p =$ Person is an athlete. $q =$ Person is healthy.

f.

p	q	r	s	$p \wedge q$	$r \vee s$	$(p \wedge q) \rightarrow (r \vee s)$
T	T	T	T	T	T	T
T	T	T	F	T	T	T
T	T	F	T	T	T	T
T	T	F	F	T	F	F
T	F	T	T	F	T	T
T	F	T	F	F	T	T
T	F	F	T	F	T	T
T	F	F	F	F	F	T
F	T	T	T	F	T	T
F	T	T	F	F	T	T
F	T	F	T	F	T	T
F	T	F	F	F	F	T
F	F	T	T	F	T	T
F	F	T	F	F	T	T
F	F	F	T	F	T	T
F	F	F	F	F	F	T

$p =$ Today is Saturday. $r =$ I'll go to the beach.
$q =$ Weather is mild. $s =$ I'll wax the car.

g.

p	q	$\sim q$	$p \leftrightarrow \sim q$
T	T	F	F
T	F	T	T
F	T	F	T
F	F	T	F

$p =$ We will win.
$q =$ We fumble the ball.

5. a. Either I will not return or I don't leave.
b. Either you will succeed, or you don't work hard and don't hit the lottery.

c. Either there will be increased wages and improved sales, or productivity does not increase.

7. The contrapositive, converse, and inverse are, respectively:
a. If John doesn't win the election, then he didn't win the primary.
If John wins the election, then he won the primary.
If John didn't win the primary, then he won't win the election.
b. If you don't diet, then you won't lose weight.
If you diet, then you will lose weight.
If you don't lose weight, then you don't diet.
c. Nonretention implies nonpromotion.
Retention implies promotion.
Nonpromotion implies nonretention.

9. a. If $x > y$, then exchange x and y.
b. If $a > 1$ then $b \leftarrow 2$ and $c \leftarrow a$;
else $b \leftarrow a$ and $c \leftarrow 2$.
c. If $b \neq 0$ and $m > 5$, then $m \leftarrow (m - 1)$;
else $m \leftarrow (m + 1)$.
d. If $m < 0$, then $A \leftarrow 3$;
else

if $m > 10$,   then $A \leftarrow 2$;
else $A \leftarrow 1$.

13. a. No   **b.** No

15. The *only* instance in which $p \rightarrow q$ is false is when p is true and q is false.

Section 4.7

1. Valid   **3.** Invalid   **5.** Invalid   **7.** Valid

9. Premises: $\begin{cases} p \rightarrow q \\ p \vee q \end{cases}$ **11.** Premises: $\begin{cases} p \rightarrow q \\ r \rightarrow s \\ p \vee r \end{cases}$

Conclusion: p
Invalid

Conclusion: $q \vee s$
Valid

13. Premises: $\begin{cases} \sim p \leftrightarrow q \\ q \end{cases}$

Conclusion: p
Invalid

15.

17.

19.

21. Valid   **23.** Invalid   **25.** Valid

C H A P T E R 5

Section 5.1

1.

	a.	b.	c.	d.
Length	5	2	3	4
Third component	6	None	R. Hokenson	$150

3. a. Yes **b.** No **c.** First two are equal.

5. a. 1.4 **b.** 84 **c.** 14 **d.** −1

7. a. 141 **b.** 87 **c.** 102

9. (−5, 2, 8) (7, 2, −2) Not poss. Not poss.
(3, 6, 9) (1, −1.75, 0, −.75) (15, −6, −24) (−27, −6, 11)
  9 Not poss 45 45

11. a. 11 **b.** .28

15.

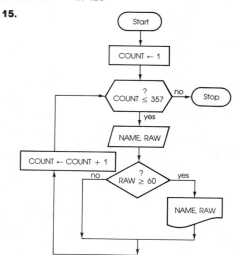

17.

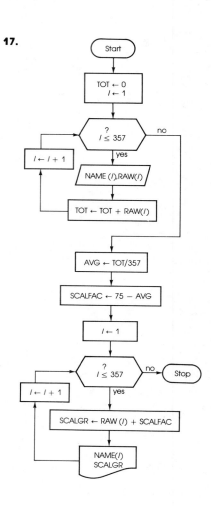

19.

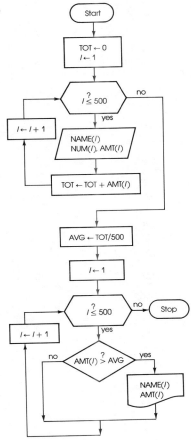

21.

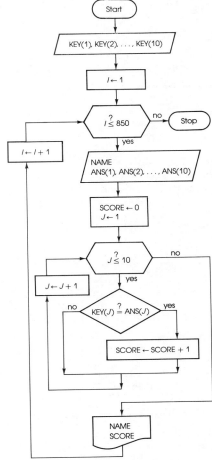

Section 5.2

1. The result is (1, 2, 4, 6, 7, 9).

3.

k	1		2		3		4		5
j	2 3 4 5 6		3 4 5 6		4 5 6		5 6		6
$a(j)$	1 9 2 7 4		9 2 7 4		6 7 4		7 9		9
i	2 2 2 2 2		3 3 3 3		4 4 4		5 5		6
$a(i)$	1 1 1 1 ①		6 2 2 ②		6 6 ④		6 ⑥		⑦

The circled elements are the least elements after each of the five passes.

5.

	Pass 1		Pass 2		Pass 3		Pass 4		Pass 5	
	Ⓧ	Ⓨ	Ⓧ	Ⓨ	Ⓧ	Ⓨ	Ⓧ	Ⓨ	Ⓧ	Ⓨ
k	1	1	2	2	3	3	4	4	5	5
$a(1)$	6	1	1	1	1	1	1	1	1	1
$a(2)$	1	6	6	2	2	2	2	2	2	2
$a(3)$	9	9	9	9	9	4	4	4	4	4
$a(4)$	2	2	2	6	6	6	6	6	6	6
$a(5)$	7	7	7	7	7	7	7	7	7	7
$a(6)$	4	4	4	4	4	9	9	9	9	9

7. Result is (1, 2, 4, 6, 7, 9).

9.

	Pass 1		Pass 2		Pass 3		Pass 4	
	Ⓧ	Ⓨ	Ⓧ	Ⓨ	Ⓧ	Ⓨ	Ⓧ	Ⓨ
FLAG	0	0	0	0	0	0	0	1
$a(1)$	6	1	1	1	1	1	1	1
$a(2)$	1	6	6	2	2	2	2	2
$a(3)$	9	2	2	6	6	4	4	4
$a(4)$	2	7	7	4	4	6	6	6
$a(5)$	7	4	4	7	7	7	7	7
$a(6)$	4	9	9	9	9	9	9	9

11.

	Loop 1		Loop 2		Loop 3		Loop 4	
	Ⓧ	Ⓨ	Ⓧ	Ⓨ	Ⓧ	Ⓨ	Ⓧ	Ⓨ
k	1	2	2	3	3	4	4	4
FLAG	0	0	0	0	0	0	0	1
$a(k)$	2	9	9	14	14	25	25	25

13.

	Loop 1		Loop 2		Loop 3		Loop 4		Loop 5	
	(X)	(Y)	(X)	(Y)	(X)	(Y)	(X)	(Y)	(X)	(Y)
k	1	2	2	3	3	4	4	5	5	6
FLAG	0	0	0	0	0	0	0	0	0	0
$a(k)$	2	9	9	14	14	25	25	33	33	

15. Trace same as for exercise 13.

17. In Figure 9 replace

$$k \leftarrow k + 1$$

by

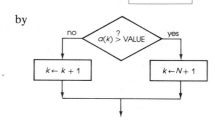

19.

	Loop 1		Loop 2		Loop 3		Loop 4	
	(X)	(Y)	(X)	(Y)	(X)	(Y)	(X)	(Y)
j	50	100	100	150	150	200	200	200
BFLAG	0	0	0	0	0	0	0	1
$a(j)$	250	500	500	750	750	1000	1000	1000

21. In Figure 11, replace

23. In Figure 11, replace every occurrence of 2000 by N and replace

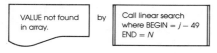

25. Alter Figure 8 by adding variable I (counter of passes) as follows:

   a. $I \leftarrow 1$ preceding entry into the first (outside) loop

   b. $I \leftarrow I + 1$ just before returning to controlling question of first (outside) loop

   c. Change controlling question of inside loop from $j < N?$ to $j \leq N - I?$

27.

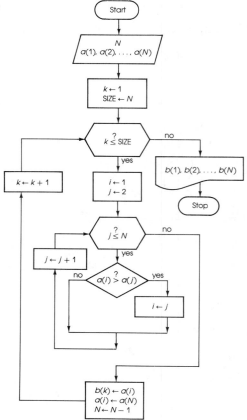

Section 5.3

1. $\begin{bmatrix} 1.5 & 2 & 1 \\ -4 & -4 & 5 \end{bmatrix}$

3. $\begin{bmatrix} -.5 & -2 & -5 \\ -4 & 6 & -7 \end{bmatrix}$

5. $\begin{bmatrix} 7 & -1 & -5 \\ 3 & 5 & -7 \\ -3 & -4 & 6 \end{bmatrix}$

7. Not possible

9. $\begin{bmatrix} 3 & 6 & 9 \\ 0 & -15 & 18 \end{bmatrix}$

11. $\begin{bmatrix} 24 & -4 & -16 \\ -8 & -12 & -28 \\ 0 & 0 & 20 \end{bmatrix}$

13. Not possible

15. $\begin{bmatrix} -6 & 1 & 9 \\ -14 & 29 & 76 \\ 10 & -15 & -45 \end{bmatrix}$

17. $\begin{bmatrix} 2 & 4 & 2 \\ -43 & -64 & 6 \end{bmatrix}$

19. Not possible

21. $\begin{bmatrix} 23 & 9 & 1 & 11 & 15 & 0 \\ 42 & 18 & -5 & -31 & -30 & 16 \end{bmatrix}$

23. Not possible   **25.** C

27. $\begin{bmatrix} -5 & 0 & 5 \\ -25 & -40 & 0 \\ 15 & 20 & -5 \end{bmatrix}$ **29.** I_4

31. $\begin{bmatrix} 8.5 & 12 & -.5 \\ -39 & -52 & 9 \end{bmatrix}$

41. a. False   **b.** True

45. a.

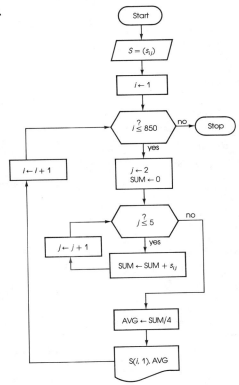

b.

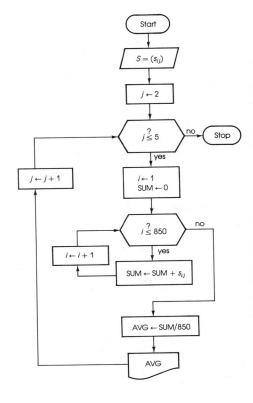

Section 5.4

1. -22 **3.** -1.5 **5.** 140   **7.** 8   **9.** -90

11. 118

13.

	Loop 1	Loop 2	Loop 3
i	1	2	3
a_{i1}	5	0	1
MIN(a_{i1})	30	8	-10
COF(a_{i1})	30	-8	-10
$|A|$	150	150	(140)

15.

	Loop 1	Loop 2	Loop 3	Loop 4
i	1	2	3	4
a_{i1}	1	2	3	4
MIN(a_{i1})	20	5	20	40
COF(a_{i1})	20	-5	20	-40
$|A|$	20	10	70	(−90)

21. In Figure 16, replace N by 5 and $N-1$ by 4.

23. 12

Section 5.5

In each of the first sixteen exercises, the matrix representation is $AX = C$, where X is the column matrix of the unknowns.

1. $A = \begin{bmatrix} 2 & -1 \\ 1 & 3 \end{bmatrix}$ $C = \begin{bmatrix} 0 \\ 7 \end{bmatrix}$

Unique solution since $|A| \neq 0$.
That solution is $x = 1$, $y = 2$.

3. $A = \begin{bmatrix} -1 & 4 \\ 3 & -12 \end{bmatrix}$ $C = \begin{bmatrix} 1 \\ 6 \end{bmatrix}$

No unique solution since $|A| = 0$.
No solution

5. $A = \begin{bmatrix} 3 & -2 \\ 5 & 1 \end{bmatrix}$ $C = \begin{bmatrix} 11 \\ 14 \end{bmatrix}$

Unique solution since $|A| \neq 0$.
That solution is $x = 3$, $y = -1$.

7. $A = \begin{bmatrix} -1 & 4 \\ 3 & -12 \end{bmatrix}$ $C = \begin{bmatrix} 1 \\ -3 \end{bmatrix}$

No unique solution since $|A| = 0$.
Infinitely many solutions

9. $A = \begin{bmatrix} 1 & -1 \\ 2 & 1 \end{bmatrix}$ $C = \begin{bmatrix} 80 \\ 610 \end{bmatrix}$

Unique solution since $|A| \neq 0$.
That solution is $x = 230$, $y = 150$.

11. $A = \begin{bmatrix} 2 & -3 & 1 \\ 1 & -1 & 4 \\ -3 & 5 & 2 \end{bmatrix}$ $C = \begin{bmatrix} 2 \\ 1 \\ 3 \end{bmatrix}$

No unique solution since $|A| = 0$.
No solution

13. $A = \begin{bmatrix} 1 & 4 & -2 \\ -3 & 0 & 1 \\ -1 & 2 & 5 \end{bmatrix}$ $C = \begin{bmatrix} -4 \\ -7 \\ -11 \end{bmatrix}$

Unique solution since $|A| \neq 0$.
That solution is $x = 2$, $y = -2$, $z = -1$.

15. $A = \begin{bmatrix} -3 & 0 & 2 \\ 1 & 4 & 5 \\ -1 & 8 & 12 \end{bmatrix}$ $C = \begin{bmatrix} 1 \\ -1 \\ -1 \end{bmatrix}$

No unique solution since $|A| = 0$.
Infinitely many solutions

17. a. Any $t \neq \frac{1}{2}$ **b.** No such t **c.** $t = \frac{1}{2}$

19. 8 dimes, 6 quarters

21. 250 pounds of peanuts, 100 pounds of cashews, 150 pounds of raisins

23. 42, 30

Section 5.6

In each of the first 14 exercises, the matrix representation is $AX = C$, where X is the column matrix of the unknowns.

1. $A = \begin{bmatrix} 1 & -1 \\ 2 & 3 \end{bmatrix}$ $C = \begin{bmatrix} 25 \\ 475 \end{bmatrix}$

Unique solution since $|A| \neq 0$.
That solution is $x = 110$, $y = 85$.

3. $A = \begin{bmatrix} 2 & -1 \\ -12 & 6 \end{bmatrix}$ $C = \begin{bmatrix} 4 \\ -24 \end{bmatrix}$

No unique solution since $|A| = 0$.
Infinitely many solutions

5. $A = \begin{bmatrix} 2 & -3 & 1 \\ 1 & -1 & 4 \\ -3 & 5 & 2 \end{bmatrix}$ $C = \begin{bmatrix} 2 \\ 1 \\ -3 \end{bmatrix}$

No unique solution since $|A| = 0$.
Infinitely many solutions

7. $A = \begin{bmatrix} -3 & 0 & 2 \\ 1 & 4 & 5 \\ -1 & 8 & 12 \end{bmatrix}$ $C = \begin{bmatrix} 1 \\ -1 \\ 1 \end{bmatrix}$

No unique solution since $|A| = 0$.
No solution

9. $A = \begin{bmatrix} 5 & -1 & 2 \\ -1 & 2 & -2 \\ 2 & 3 & 4 \end{bmatrix}$ $C = \begin{bmatrix} 1 \\ 3 \\ -6 \end{bmatrix}$

Unique solution since $|A| \neq 0$.
That unique solution is $x = 1$, $y = 0$, $z = -2$.

11. $A = \begin{bmatrix} -2 & 1 & 0 & 3 \\ 1 & 0 & 2 & -1 \\ 0 & -3 & 0 & 2 \\ -1 & -2 & 2 & 4 \end{bmatrix}$ $C = \begin{bmatrix} 1 \\ 4 \\ -2 \\ 0 \end{bmatrix}$

No unique solution since $|A| = 0$.
No solution

13. $A = \begin{bmatrix} 1 & 1 & 0 & -3 & 2 \\ 0 & -2 & 1 & 0 & 1 \\ -3 & 0 & -1 & 1 & 0 \\ 4 & 2 & 3 & 0 & -2 \\ 0 & -1 & 1 & -2 & 3 \end{bmatrix}$ $C = \begin{bmatrix} 5 \\ 1 \\ -2 \\ -3 \\ 5 \end{bmatrix}$

Unique solution since $|A| \neq 0$.
That unique solution is $x = 1$, $y = 0$, $z = -1$, $v = 0$, $w = 2$.

Section 5.7

1. a. Yes, since determinant $\neq 0$ **b.** $\begin{bmatrix} 3 & -1 \\ 2 & -1 \end{bmatrix}$

3. a. No, since determinant $= 0$

5. a. Yes, since determinant $\neq 0$ **b.** $\begin{bmatrix} -\frac{2}{3} & \frac{1}{3} \\ \frac{1}{2} & 0 \end{bmatrix}$

7. a. Yes, since determinant $\neq 0$ **b.** $\begin{bmatrix} 1 & -\frac{1}{2} \\ 0 & \frac{1}{4} \end{bmatrix}$

9. a. Yes, since determinant $\neq 0$ **b.** $\begin{bmatrix} 1 & 0 \\ 0 & 1 \end{bmatrix}$

11. a. Yes, since determinant $\neq 0$ **b.** $\begin{bmatrix} 0 & -1 \\ \frac{1}{5} & 0 \end{bmatrix}$

13. a. No, since determinant $= 0$

15. a. Yes, since determinant $\neq 0$

b. $\begin{bmatrix} -2 & 7 & -1 \\ 1 & -4 & 1 \\ 4 & -13 & 2 \end{bmatrix}$

17. a. Yes, since determinant $\neq 0$ **b.** $\begin{bmatrix} \frac{1}{5} & \frac{3}{5} & \frac{13}{25} \\ 0 & 1 & \frac{4}{5} \\ 0 & 0 & -\frac{1}{5} \end{bmatrix}$

19. a. No, since determinant $= 0$

21. a. Yes, since determinant $\neq 0$ **b.** $\begin{bmatrix} \frac{1}{2} & 0 & 0 \\ 0 & -1 & 0 \\ 0 & 0 & \frac{1}{3} \end{bmatrix}$

23. a. Yes, since determinant $\neq 0$

b. $\begin{bmatrix} 49 & -18 & -9 & 7 \\ 27 & -10 & -5 & 4 \\ -48 & 18 & 9 & -7 \\ 82 & -31 & -15 & 12 \end{bmatrix}$

25. a. $t = 10$ **b.** Any $t \neq 10$

27. a. $t = 1, -3$ **b.** Any $t \neq 1, -3$

29. a. $x = -5$, $y = 9$ **b.** $x = 0$, $y = 1$

31. a. $x = 7$, $y = -3$, $z = 7$
b. $x = 7$, $y = -4$, $z = 11$

35. No; for example, I_2 and $-I_2$ are each nonsingular, but their sum is not.

CHAPTER 6

Section 6.1

1. $70 **3.** $67.50

5. a.

Period	Principal at beginning	Interest earned	Principal at end
1	12500	218.75	12718.75
2	12718.75	222.58	12941.33
3	12941.33	226.47	13167.80
4	13167.80	230.44	13398.24

b. Deposit value = $13,398.24
Interest = $ 898.24

7.

K	1	2	3	4
I	218.75	222.58	226.47	230.44
P	12718.75	12941.33	13167.80	13398.24

9. Same as exercise 5(b)

11. Bank that pays 8% compounded quarterly

13. $3379.96 **15.** 14 years **17.** 8.86%

19. a. $13,406.35 **b.** $17,738.34

21. a. 8.16% **b.** 8.24% **c.** 8.33%

23. a. $1060 **b.** $1060.90 **c.** $1061.36
d. $1061.68 **e.** $1061.83 **f.** $1061.83
g. $1061.84

Section 6.2

1. a. Deposit value = $33,584.46; interest = $3584.46
   b. $25,789.36

3. Same as exercise 1(a)

5. Cash value = $16,555.62; interest = $7555.62

7. $101.69   **9.** Same as exercise 1(b)

11. $69,790.38   **13.** $442.81

15. Lump sum of $60,000 now

Section 6.3

1. a. 32   **b.** 33   **c.** 34.6   **d.** 5.8

3.

Class	Class limits	Frequency	Class mark
1	25–29	12	27
2	30–34	22	32
3	35–39	13	37
4	40–44	9	42
5	45–49	4	47
		60	

5.

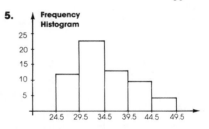

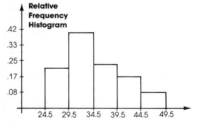

7. a. Class 2   **b.** 33.3   **c.** 34.6   **d.** 5.9

9.

Class	Class limits	Frequency	Class mark
1	24–26	2	25
2	27–29	10	28
3	30–32	17	31
4	33–35	8	34
5	36–38	7	37
6	39–41	6	40
7	42–44	6	43
8	45–47	3	46
9	48–50	1	49
		60	

11.

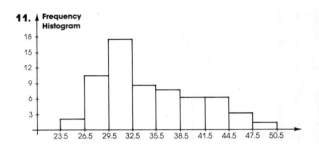

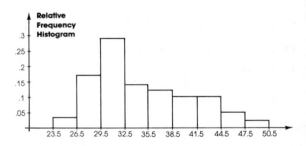

13. a. Class 3   **b.** 33.25   **c.** 34.55   **d.** 5.9

15. Yes   **17.** Yes

19.

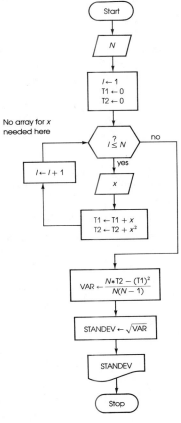

No array for x needed here

Section 6.4

1. a. .3907 **b.** .2257 **c.** .6470 **d.** .4177
e. .6736
In each case, the answer can be interpreted as the area of the region below the standard normal curve and between the given z-values.

3. Same answers as exercise 1

5. a. .3554 **b.** .4847 **c.** .1297 **d.** .4562
In each case, the answer can be interpreted as the area of the region below the normal probability distribution curve and between the given x-values.

7. Same answers as exercise 5

9. a. 16.2 **b.** 19.836 **c.** 8.172

11. a. 16.2 **b.** 20.8

13. No, since these histograms are skewed to the right

15. .2175 **17.** .0455

Section 6.5

1. a. $-36, 0, 3, 24$ **b.** $-13, 2, 3.25, 12$
c. $9, 3, 2.5, -1$ **d.** $7, 7, 7, 7$

3. a. RP: $R^+ \to R^+$ defined by $RP(M) = M - .1M$
where R^+ is the set of nonnegative numbers.
b. B: $R^+ \to R^+$ defined by $B(S) = 100 + .055$
where R^+ is the set of nonnegative numbers.

5. a. $RP(D, M) = M - \left(\dfrac{D}{100}\right)M$
b. $B(A, S) = A + .05S$
c. $NS(C, T) = C + 200 - T$
d. $D(R, T) = R * T$

7. **a.** 0, 10.9   **b.** .3, 10   **c.** 4, 6   **d.** 1, 3
e. −9, 2.01   **f.** 90, −17

9.

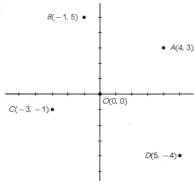

13. **a.**

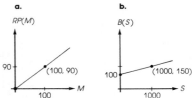

b.

15. **a.**

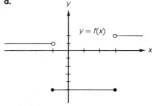

b.

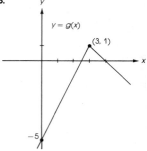

17. (a) and (c)

19. Any vertical line through the domain $[a, b]$ crosses the graph exactly once.

21. **a.** 6, 6, −3   **b.** $PST(w) = 20 * \text{ROOF}(w)$

23. In Figure 49 replace $x \leftarrow 4$ by $x \leftarrow -63$, and replace $x \leq 336$? by $x \leq 70$?

25. 25

27. Same as Figure 51, except replace $x \leftarrow x+1$ by $x \leftarrow x+\frac{1}{2}$.

C H A P T E R   7

Section 7.1

1. $N = \{B, C, D, E, F, G\}$;
$A = \{\{B, B\}, \{B, C\}, \{B, E\}, \{B, F\}, \{B, G\}, \{D, D\},$
$\{D, E\}, \{E, F\}, \{F, F\}\}$;
$\{B, B\}, \{D, D\}, \{F, F\}$ are loops; $\{B, E\}, \{E, F\},$
$\{F, B\}$ is a cycle; connected

3. No loops; $(D, B), (B, E), (E, D)$ is a cycle; not connected

5.

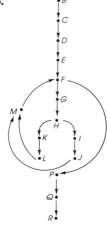

7. One possible tree structure is

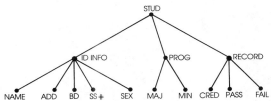

9. a. **b.** **c.**

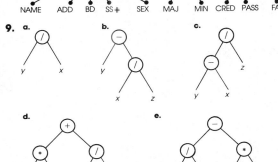

d. **e.**

11.

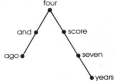

The ordered list is:

   ago, and, four, score, seven, years

13. Letting nodes B, C, D, E, F, and G, be B_1, B_2, B_3, B_4, B_5, and B_6, respectively, the binary adjacency matrix is

$$\begin{bmatrix} 1 & 1 & 0 & 1 & 1 & 1 \\ 1 & 0 & 0 & 0 & 0 & 0 \\ 0 & 0 & 1 & 1 & 0 & 0 \\ 1 & 0 & 1 & 0 & 1 & 0 \\ 1 & 0 & 0 & 1 & 1 & 0 \\ 1 & 0 & 0 & 0 & 0 & 0 \end{bmatrix}$$

15. a. $-xy$ **b.** $*+xyz$ **c.** $-*xyz$ **d.** $*x-yz$
   e. $/-3x(**)+zw5$

17. a. 5, 4, 3, 2, 1
   b. At most, m nodes must be examined to find any particular node.

Section 7.2

1. Semigroup (not monoid)   **3.** Monoid

5. Neither   **7.** Semigroup (not monoid)

9. Neither   **11.** Semigroup (not monoid)

13. Neither   **15.** Monoid

17.

$\bigcirc$	x	xx	$yxxy$
y	yx	yxx	$yyxxy$
yx	yxx	$yxxx$	$yxyxxy$
yxy	$yxyx$	$yxyxx$	$yxyyxxy$

19. a. Yes   **b.** No

21.

$+_5$	[0]	[1]	[2]	[3]	[4]
[0]	[0]	[1]	[2]	[3]	[4]
[1]	[1]	[2]	[3]	[4]	[0]
[2]	[2]	[3]	[4]	[0]	[1]
[3]	[3]	[4]	[0]	[1]	[2]
[4]	[4]	[0]	[1]	[2]	[3]

23. $\langle V, +\rangle$, $\langle M, +\rangle$, $\langle Z_5, +_5\rangle$, and $\langle Z_6, +_6\rangle$ are all commutative, whereas $\langle M, \cdot\rangle$ is not commutative.

25. No. For instance, $\{20, 30, 35\} \subset Z^2$
   yet $20 * 30 = 600 \in Z^3$
   $30 * 35 = 1050 \in Z^4$

Section 7.3

1. The output tape is:

   odd   even   odd   odd   even   odd   even

The last (leftmost) output is odd, which is the parity of the input tape.

3. The output tape is

   0   2   2   1   1   0

The last (leftmost) output is 0, which is the sum of the input bits, mod 3.

5. The output tape is

   1   0   0   1

which is the sum of 101 and 100.

7. a. 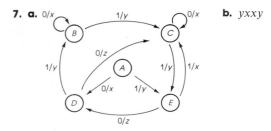   **b.** $yxxy$

15.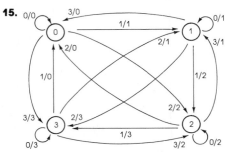

9. $S = \{K, L, P, V\}$ $I = \{0, 1, 3\}$ $O = \{Y, N\}$

f_O	K	L	P	V
0	N	N	N	N
1	Y	Y	Y	Y
3	Y	Y	N	N

f_S	K	L	P	V
0	K	L	P	V
1	L	V	K	P
3	V	P	K	L

11.

13.

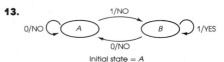

17.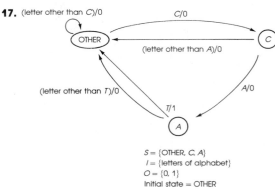

$S = \{OTHER, C, A\}$
$I = \{letters\ of\ alphabet\}$
$O = \{0, 1\}$
Initial state = OTHER

Section 7.4

1. In each case there are many possible answers; for example:
  a. $\{e, dp, pep, seeds\}$
  b. $\{eeeee, ddddd, ppppp, sssss, edede, epepe, pspsp, esese, dpdpd, dsdsd, sdsds, edpse\}$
  c. $\{d, dd, ddd, dddd, ddddd, dddddd, \ldots\}$

3. The set is the collection of all finite strings of letters, each of which is n or o. The operation is concatenation.

5.

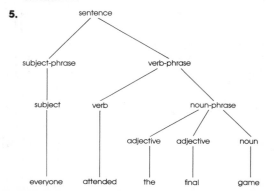

7. a. $x^n y^{n+1}$ for any positive integer n
   b. $x^{n+1} y^n$ for any nonnegative integer n
   c. $x^n y^{n+1}$ for any positive integer n

9. The grammar $\langle V, V_T, s, P \rangle$ where
   $V = \{$everyone, attended, the, final, game, sentence, subject-phrase, verb-phrase, subject, verb, noun-phrase, adjective, noun$\}$
   $V_T = \{$everyone, attended, the, final, game$\}$
   $P = \{$sentence $\rightarrow$ subject-phrase verb-phrase
       subject-phrase $\rightarrow$ subject
       verb-phrase $\rightarrow$ verb noun-phrase
       noun-phrase $\rightarrow$ adjective adjective noun
       subject $\rightarrow$ everyone   verb $\rightarrow$ attended
       adjective $\rightarrow$ the   adjective $\rightarrow$ final
       noun $\rightarrow$ game$\}$

11. All binary strings of the form $(101)^n$, where n is a positive integer.

13. All strings of the form $x^m y^n$, where m and n are positive integers.